Lecture Notes in Computer Science   11641

More information about this series at http://www.springer.com/series/7409

Jie Shao · Man Lung Yiu ·
Masashi Toyoda · Dongxiang Zhang ·
Wei Wang · Bin Cui (Eds.)

Web and Big Data

Third International Joint Conference, APWeb-WAIM 2019
Chengdu, China, August 1–3, 2019
Proceedings, Part I

 Springer

Editors
Jie Shao
University of Electronic Science
and Technology of China
Chengdu, China

Masashi Toyoda
The University of Tokyo
Tokyo, Japan

Wei Wang
National University of Singapore
Singapore, Singapore

Man Lung Yiu
Hong Kong Polytechnic University
Hong Kong, China

Dongxiang Zhang
Zhejiang University
Hangzhou, China

Bin Cui
Peking University
Beijing, China

ISSN 0302-9743   ISSN 1611-3349   (electronic)
Lecture Notes in Computer Science
ISBN 978-3-030-26071-2   ISBN 978-3-030-26072-9   (eBook)
https://doi.org/10.1007/978-3-030-26072-9

LNCS Sublibrary: SL3 – Information Systems and Applications, incl. Internet/Web, and HCI

This Springer imprint is published by the registered company Springer Nature Switzerland AG
The registered company address is: Gewerbestrasse 11, 6330 Cham, Switzerland

Preface

This volume (LNCS 11641) and its companion volume (LNCS 11642) contain the proceedings of the Third Asia-Pacific Web (APWeb) and Web-Age Information Management (WAIM) Joint Conference on Web and Big Data, called APWeb-WAIM. This joint conference aims at attracting professionals of different communities related to Web and big data who have common interests in interdisciplinary research to share and exchange ideas, experiences, and the underlying techniques and applications, including Web technologies, database systems, information management, software engineering, and big data.

The Third APWeb-WAIM conference was held in Chengdu, China, during August 1–3, 2019. APWeb and WAIM are two separate leading international conferences on research, development, and applications of Web technologies and database systems. Previous APWeb conferences were held in Beijing (1998), Hong Kong (1999), Xi'an (2000), Changsha (2001), Xi'an (2003), Hangzhou (2004), Shanghai (2005), Harbin (2006), Huangshan (2007), Shenyang (2008), Suzhou (2009), Busan (2010), Beijing (2011), Kunming (2012), Sydney (2013), Changsha (2014), Guangzhou (2015), and Suzhou (2016). Previous WAIM conferences were held in Shanghai (2000), Xi'an (2001), Beijing (2002), Chengdu (2003), Dalian (2004), Hangzhou (2005), Hong Kong (2006), Huangshan (2007), Zhangjiajie (2008), Suzhou (2009), Jiuzhaigou (2010), Wuhan (2011), Harbin (2012), Beidaihe (2013), Macau (2014), Qingdao (2015), and Nanchang (2016). Starting in 2017, the two conference committees agreed to launch a joint conference. The First APWeb-WAIM conference was held in Bejing (2017) and the Second APWeb-WAIM conference was held in Macau (2018). With the increased focus on big data, the new joint conference is expected to attract more professionals from different industrial and academic communities, not only from the Asia Pacific countries but also from other continents.

The high-quality program documented in these proceedings would not have been possible without the authors who chose APWeb-WAIM for disseminating their findings. After the double-blind review process (each paper received at least three review reports), out of 180 submissions, the conference accepted 42 regular (23.33%), 17 short research papers, and six demonstrations. The contributed papers address a wide range of topics, such as big data analytics, data and information quality, data mining and application, graph data and social networks, information extraction and retrieval, knowledge graph, machine learning, recommender systems, storage, indexing and physical database design, text analysis and mining. We are deeply thankful to the Program Committee members for lending their time and expertise to the conference. The technical program also included keynotes by Dr. Divesh Srivastava (AT&T Labs-Research, USA), Dr. Xindong Wu (Mininglamp Technology, China), Prof. Christian S. Jensen (Aalborg University, Denmark), and Prof. Guoliang Li (Tsinghua University, China). We are grateful to these distinguished scientists for their invaluable contributions to the conference program.

We thank the general co-chairs (Heng Tao Shen, Kotagiri Ramamohanarao, and Jiliu Zhou) for their guidance and support. Thanks also go to the workshop co-chairs (Jingkuan Song and Xiaofeng Zhu), tutorial co-chairs (Shaojie Qiao and Jiajun Liu), demo co-chairs (Wei Lu and Jizhou Luo), industry co-chairs (Jianjun Chen and Jia Zhu), and publicity co-chairs (Lei Duan, Yoshiharu Ishikawa, Jianxin Li, and Weining Qian).

We hope you enjoy the exciting program of APWeb-WAIM 2019 as documented in these proceedings.

June 2019

Jie Shao
Man Lung Yiu
Masashi Toyoda
Dongxiang Zhang
Wei Wang
Bin Cui

Organization

General Chairs

Heng Tao Shen	University of Electronic Science and Technology of China, China
Kotagiri Ramamohanarao	University of Melbourne, Australia
Jiliu Zhou	Chengdu University of Information Technology, China

Program Committee Chairs

Jie Shao	University of Electronic Science and Technology of China, China
Man Lung Yiu	Hong Kong Polytechnic University, Hong Kong SAR, China
Masashi Toyoda	The University of Tokyo, Japan

Workshop Chairs

Jingkuan Song	University of Electronic Science and Technology of China, China
Xiaofeng Zhu	Massey University, New Zealand

Tutorial Chairs

Shaojie Qiao	Chengdu University of Information Technology, China
Jiajun Liu	Renmin University of China, China

Demo Chairs

Wei Lu	Renmin University of China, China
Jizhou Luo	Harbin Institue of Technology, China

Industry Chairs

Jianjun Chen	Huawei America Research, USA
Jia Zhu	South China Normal University, China

Publication Chairs

Dongxiang Zhang	Zhejiang University, China
Wei Wang	National University of Singapore, Singapore
Bin Cui	Peking University, China

Publicity Chairs

Lei Duan	Sichuan University, China
Yoshiharu Ishikawa	Nagoya University, Japan
Jianxin Li	University of Western Australia, Australia
Weining Qian	East China Normal University, China

Local Arrangements Chairs

Yang Yang	University of Electronic Science and Technology of China, China
Hong Xiao	University of Electronic Science and Technology of China, China

Webmaster

Xiaochen Wang	University of Electronic Science and Technology of China, China

Senior Program Committee

Toshiyuki Amagasa	University of Tsukuba, Japan
Wolf-Tilo Balke	TU Braunschweig, Germany
Xin Luna Dong	Amazon, USA
Mizuho Iwaihara	Waseda University, Japan
Peer Kroger	Ludwig Maximilian University of Munich, Germany
Byung Suk Lee	University of Vermont, USA
Sebastian Link	University of Auckland, New Zealand
Wookey Lee	Inha University, South Korea
Yang-Sae Moon	Kangwon National University, South Korea
Xiaokui Xiao	National University of Singapore, Singapore
Rui Zhang	University of Melbourne, Australia
Shuigeng Zhou	Fudan University, China
Xiangliang Zhang	King Abdullah University of Science and Technology, Saudi Arabia
Xingquan Zhu	Florida Atlantic University, USA

Program Committee

Zhifeng Bao	RMIT University, Australia
Ilaria Bartolini	University of Bologna, Italy
Ladjel Bellatreche	ISAE-ENSMA, France
Zouhaier Brahmia	University of Sfax, Tunisia
Yi Cai	South China University of Technology, China
Tru Cao	Ho Chi Minh City University of Technology, Vietnam

Lisi Chen	Inception Institute of Artificial Intelligence, United Arab Emirates
Tanzima Hashem	Bangladesh University of Engineering and Technology, Bangladesh
Reynold Cheng	University of Hong Kong, Hong Kong SAR, China
Lizhen Cui	Shandong University, China
Jiangtao Cui	Xidian University, China
Alex Delis	University of Athens, Greece
Lei Duan	Sichuan University, China
Amr Ebaid	Purdue University, USA
Ju Fan	Renmin University of China, China
Yaokai Feng	Kyushu University, Japan
Yunjun Gao	Zhejiang University, China
Tingjian Ge	University of Massachusetts, Lowell, USA
Zhiguo Gong	University of Macau, Macau SAR, China
Chenjuan Guo	Aalborg University, Denmark
Jialong Han	Tencent, China
Haibo Hu	Hong Kong Polytechnic University, Hong Kong SAR, China
Jianbin Huang	Xidian University, China
Chih-Chieh Hung	Tamkang University, Taiwan
Dawei Jiang	Zhejiang University, China
Cheqing Jin	East China Normal University, China
Peiquan Jin	University of Science and Technology of China, China
Feifei Li	University of Utah, USA
Tianrui Li	Southwest Jiaotong University, China
Hui Li	Xiamen University, China
Zheng Li	Amazon, USA
Yu Li	Hangzhou Dianzi University, China
Xiang Lian	Kent State University, USA
An Liu	Soochow University, China
Hailong Liu	Northwestern Polytechnical University, China
Guanfeng Liu	Macquarie University, Australia
Hua Lu	Aalborg University, Denmark
Mihai Lupu	Vienna University of Technology, Austria
Zakaria Maamar	Zayed University, United Arab Emirates
Mirco Nanni	ISTI-CNR Pisa, Italy
Sanjay Madria	Missouri University of Science and Technology, USA
P. Krishna Reddy	International Institute of Information Technology, Hyderabad, India
Wee Siong Ng	Institute for Infocomm Research, Singapore
Baoning Niu	Taiyuan University of Technology, China
Hiroaki Ohshima	University of Hyogo, Japan
Yuwei Peng	Wuhan University, China
Jianzhong Qi	University of Melbourne, Australia
Yanghui Rao	Sun Yat-sen University, China

Feida Zhu	Singapore Management University, Singapore
Zhaonian Zou	Harbin Institute of Technology, China
Lei Zou	Peking University, China
Bolong Zheng	Huazhong University of Science and Technology, China
Kai Zheng	University of Electronic Science and Technology of China, China

Contents – Part I

Information Extraction and Retrieval

Knowledge Graph

Contents – Part II

Machine Learning

Recommender Systems

Text Analysis and Mining

Demos

Big Data Analytic

A Framework for Image Dark Data Assessment

Yu Liu[1], Yangtao Wang[1], Ke Zhou[1(✉)], Yujuan Yang[1], Yifei Liu[1], Jingkuan Song[2], and Zhili Xiao[3]

[1] Huazhong University of Science and Technology, Wuhan, China
{liu_yu,ytwbruce,k.zhou,gracee,yifeiliu}@hust.edu.cn
[2] University of Electronic Science and Technology of China, Chengdu, China
jingkuan.song@gmail.com
[3] Tencent Inc., Shenzhen, China
tomxiao@tencent.com

Abstract. Blindly applying data mining techniques on image dark data whose content and value are not clear, is highly likely to bring undesired result. Therefore, we propose an assessment framework which includes offline and online stages for image dark data. In offline stage, we first transform images into hash codes by Deep Self-taught Hashing (DSTH) algorithm, then construct a semantic graph, and finally use our designed Semantic Hash Ranking (SHR) algorithm to calculate the importance score. During online stage, we first translate the user's query into hash codes, then match the suitable data contained in the dark data, and finally return the weighted average value of these matched data to help the user cognize the dark data. The results on real-world dataset show our framework can apply to large-scale datasets, help the user conduct subsequent data mining work.

Keywords: Image dark data · Deep self-taught hashing ·
Semantic hash ranking · Assessment

1 Introduction

Dark data is defined as the information assets that can be easily collected and stored, but generally fail to use for data analytics and mining[1]. Most of these data are unstructured data represented by images. Many social platforms store image data (i.e., albums and chat images) as an independent resource separated from other businesses. These massive image data quickly turn into dark data, which contain lots of historical records and thus are not allowed to be removed. However, they consistently occupy the storage space but can not produce great value.

---

[1] https://www.gartner.com/it-glossary/dark-data/.

The original version of this chapter was revised: The abstract section and the keywords of this chapter have been exchanged. This have been now corrected. The correction to this chapter is available at https://doi.org/10.1007/978-3-030-26072-9_31

© Springer Nature Switzerland AG 2019
J. Shao et al. (Eds.): APWeb-WAIM 2019, LNCS 11641, pp. 3–18, 2019.
https://doi.org/10.1007/978-3-030-26072-9_1

Therefore, developers are eager to mine image dark data in order to improve the cost performance of storage. However, owing that the image dark data lack labels and associations, owners have no idea how to apply these data. For a given target, blindly conducting data mining techniques on the dark data is highly to get little feedback. Faced with image dark data whose content and value are not clear, the primary issue is to judge whether this dataset are worth mining or not. Therefore, it is of great significance to evaluate the value of image dark data according to the user's requirement. Given this, faced with the user's requirement, which way shall be taken to make the user aware of the dark data? There exist following challenges when executing association analysis on dark data.

(1) **How to extract semantic information with generalization ability?** Reasonable semantic extraction method is the key to correctly understand the user's requirement and analyze semantic distribution of dark dataset. An excellent deep model will obtain desired classification effects. However, deep model suffers from a poor generalization ability when extracting semantic information for unknown samples. Our goal is to express semantic distance for different images including unknown samples. For example, when training the semantic model, we make the semantic feature of a cat similar to that of a dog but different from that of an airplane.

(2) **How to evaluate relevance?** There are two-level evaluation for the relevance: (1) the amount of data that meets the user's requirements; (2) the matching degree of these relevant data. Traditional clustering methods need many iterations and will take a long time to complete the evaluation. In addition, graph-based computing [4] is another way to achieve global evaluation. The most well-known one is PageRank algorithm [14], which determines the importance of web pages according to the links. However, PageRank can only express directional attributes on a directed graph, so it fails to measure the mutual extent of relevance between objects. Events detecting [2] can find the hot events though connections on an undirected graph, but the representative data are very limited. Once the query data can not match any hot event, no assessment result will be returned, so it does not apply to our task.

(3) **How to reduce the query cost?** Even if the above problems have been solved, we still need to find the corresponding related data in the whole dataset to measure the feedback of the query. Online computing millions of high-dimensional floating-point vectors means a huge resource consumption. Besides, the assessment task will receive frequent query requests for different mining tasks. Thus, the assessment work is supposed to be built on more efficient distance measurement for practical feasibility.

In this paper, we propose an assessment framework for image dark data. The framework consists of four parts. First, we use deep self-taught hashing (DSTH) algorithm to transform unlabeled images into deep semantic hash codes with generalization ability. Second, we build the semantic undirected graph using restricted Hamming distance. According to what DCH [3] describes, we cut off a lot of unreasonable connections and improve the efficiency of construction and subsequent calculation on graph. Third, on the built graph, we design semantic hash ranking (SHR) algorithm to calculate the importance score for each node by random walk and obtain the rank for each image. It is worth mentioning that we improve the PageRank algorithm and extend it to undirected weighted graph, which takes both the number of connected links and the weight on edges into consideration. At last, according to the user's requirements, we match the corresponding data contained in the dataset which are restricted within a given Hamming distance range, calculate the weighted semantic importance score of these data, and return the ranking. The user can decide whether conducting data mining on this dark dataset based on the returned result. The major contributions of this paper are summarized as follows:

- We design a deep self-taught hashing (DSTH) algorithm, which can extract semantic features without labels and solve the out-of-sample problem.
- Based on the built semantic graph, we propose a semantic hash ranking (SHR) algorithm to calculate the overall importance score for each node (image) according to random walk, which takes both the number of connected links and the weight on edges into consideration.
- We propose an analysis-query-assessment framework including offline calculation and online assessment, which applies to assessing large-scale datasets.
- Our framework can help the user to detect the potential value of the dark data, avoid unnecessary mining cost. To the best of our knowledge, this is the first attempt that assesses image dark data.

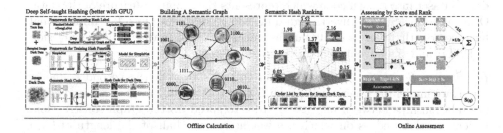

Fig. 1. The framework for image dark data assessment.

2 Design Overview

For a large-scale image dark dataset, in order to make our assessment framework effective for real-time analysis, we perform offline analysis on the dataset to get the score (rank) of each image. Then given an online matched request, we evaluate whether the dataset can be used for the query on-the-fly. As shown in Fig. 1, the framework consists of four steps. The first three steps give an offline evaluation on the dark dataset, which calculates the importance score (rank) for each image. The last step provides an online matching and weighted computing for the user, which returns the ranking score of the user's requirement based on the computed score.

Offline Evaluation. We design three steps to effectively calculate the semantic importance score and provide each image with a rank. Formally, we first train a Deep Self-taught Hashing (DSTH) model and transform all images into hash codes, then build a semantic undirected graph with restricted Hamming distance, and finally calculate the overall importance score (rank) for each image by our designed Semantic Hash Ranking (SHR) algorithm.

(1) Step 1: Deep Self-taught Hashing. As shown in the first frame of Fig. 1, we adopt the DSTH algorithm to encode each image of the dark dataset. DSTH contains three stages: hash label acquiring stage, hash function training stage and hash code generating stage. First, it is important to acquire hash labels, because the premise of feature extraction using deep learning is based on semantic labels. We choose ImageNet and the same amount of sampled image dark data as the training data and GoogLeNet trained on ImageNet to extract semantic features of these data. Next, we use the features to construct a graph via K-NN ($K = 12$), then map data to predefined l-dimensional space by means of Laplacian Eigenvalue decomposition (LE), and finally binarize all data to generate hash labels. We conduct clustering on extracted semantic features, which not only preserves original semantic classification information but also makes these semantics automatically closer or estranged according to the similarity between themselves (challenge (1) in Sect. 1). Those labels have the semantics with generalization ability, which directly affects the next hash function learning. Note that the hash function is specially trained on above sampled dark data. At last, according to the obtained deep hash functions, we transform each image of the dark dataset into a hash code which represents the semantic feature of the data. Our method (DSTH) converts high-dimensional dark data into low-dimensional hash vectors that can be easily but fast measured. The mathematical expression of DSTH and the advantages are described in Sect. 3.

(2) Step 2: Semantic Graph Construction. As shown in the second frame of Fig. 1, we model the images as a graph G where each node is an image and edges are relationships between images. In order to speed up the graph construction, we cut off those edges on which the weight exceeds half of the length of hash code, according to the conclusion of Long [3]. Let N_* denote the $*$-th node of G, $H(N_*)$ denote hash code of N_* and l denote length of hash codes. We define XOR operation as $\oplus$. Therefore, the Hamming distance weight on the undirected link between N_i and N_j can be defined as

$$d_{ij} = \begin{cases} H(N_i) \oplus H(N_j) & i \neq j, H(N_i) \oplus H(N_j) \leq \Omega, \\ NULL & otherwise. \end{cases} \quad (1)$$

where $\Omega = \lceil l/s \rceil$ and $s \in [1, l]$. In practice, the determination of Ω is based on efficiency of building a graph with tolerable loss. Formally, we define the precision of i-th node as C_i/L_i, where L_i represents the number of all nodes connected to i-th node and there exist C_i nodes of the L_i nodes that have the same label as the i-th node. Therefore, the precision of graph $P(G|\Omega)$ is defined as

$$P(G|\Omega) = \frac{1}{N} \sum_{i=1}^{N} \frac{C_i}{L_i} \tag{2}$$

(3) Step 3: Semantic Hash Ranking. As shown in the third frame of Fig. 1, after building the graph with restricted Hamming distance in *Step 2*, we calculate the importance score for each node by random walk in order to obtain the overall objective evaluation value. We extend the PageRank algorithm and propose the SHR algorithm which takes both the number of connected links and the weight on edges into consideration (challenge (2) in Sect. 1). Note that we specially design how to reasonably calculate the extent of relevance between nodes, aiming at making full use of the Hamming distance of similarity hash with generalization ability. On the built semantic graph, we use SHR to calculate the importance score for each node. At the same time, according to the physical meaning of Hamming distance, we redesign the iteration matrix elements for obtaining reasonable importance scores. SHR makes the dominant semantics more prominent, thus reinforcing the user's cognition of the dark dataset. We introduce the detailed calculation process of SHR in Sect. 4.

Online Query Assessment. As shown in the last frame of Fig. 1, for the image dark data consisting of N images, the query will be mapped to hash codes by hash function calculated in Sect. 3 and match images contained in the dark data (challenge (3) in Sect. 1). The matching range is defined as hd and we set $hd = 1$ to conduct matching shown in the last frame of Fig. 1. Mathematically, let q denote a query with n images, img_i denote the i-th image where $i \in [1, n]$, m_i denote the number of matched images for the i-th image of the query q. Meanwhile, let $S_j(img_i)$ denote the score of the j-th image where $j \in [1, m_i]$. Therefore, the score of q is defined as follows:

$$S(q) = \sum_{i=1}^{n} \frac{1}{m_i} \sum_{j=1}^{m_i} \beta_i S_j(img_i) \quad s.t. \sum_{i=1}^{n} \beta_i = 1 \tag{3}$$

where $\beta_i \in [0, 1]$ represents the importance weight of the i-th image.

Compared with the ranked scores denoted as $\{S_1, S_2, ..., S_N\}$ of image dark data calculated by SHR, we can acquire the rank of $S(q)$ denoted as k, where $S_{k-1} > S(q) \geq S_k$. Further, $T(q) = 1 - \frac{k}{N}$ represents importance degree of image dark data for the requirement. As a result, the user can decide whether the image dark data are worth fine-mining.

3   Deep Self-taught Hashing (DSTH)

In this section, we detailedly describe DSTH algorithm including how to integrate clustering information into semantic learning under deep learning framework, how to generate hash labels, and how to conduct the training process. And then, we elaborate on the advantages of DSTH.

Fig. 2. Red line represents the deep classification result based on classification labels, while blue line represents the classification result by LE. (Color figure online)

3.1 DSTH Algorithm

The algorithm mainly contains hash label generating stage and hash function training stage.

(1) Hash Label Generating Stage. The prime task of DSTH is to acquire semantic labels, because labels determine which semantic informations should be extracted from data and directly affect the subsequent function learning. Semantic labels acquiring aims at extracting semantic information and semantic feature with generalization ability. Supervised deep learning algorithm is able to better extract semantic information owing to the accurate hand-crafted labels denoted as red line in Fig. 2. LE algorithm extracts semantics according to the similarity between data themselves, which applies to those scenes without hand-crafted labels denoted as blue line in Fig. 2. Our method combines these two advantages, which can not only obtain semantic information without labels but also reach the balance between human semantic cognition and data semantic cognition, so as to acquire our expected semantic features (labels).

We apply the deep and shallow mixed learning method, which integrates clustering information and improves the generalization ability of feature extraction. In the absence of labels, we use the trained deep model to get features. After that, we use LE method and binarization to transform the extracted deep features to hash codes which serve as hash labels for next stage. Mathematically, we use n m-dimensional vectors $\{x_i\}_{i=1}^n \in \mathbb{R}^m$ to denote the image features and use $ED(i,j) = \|x_i - x_j\|_2^2$ to denote the Euclidean distance between i-th and j-th image. For θ_t ($t \in [1, n-1]$), we denote $\{ED(i, \theta_1), ED(i, \theta_2), ..., ED(i, \theta_{n-1})\} = \{ED(i,1), ED(i,2), ..., ED(i,i-1), ED(i,i+1), ..., ED(i,n)\}$, where $ED(i, \theta_t) \leq ED(i, \theta_{t+1})$. We define that $TK_{ED}(i,j) = t$, if $ED(i,j) = ED(i, \theta_t)$. Further, we use $N_K(i,j)$ to denote neighbor relationship between i-th and j-th data, which is defined as

$$N_K(i,j) = \begin{cases} True & TK_{ED}(i,j) \leq K, \\ False & TK_{ED}(i,j) > K. \end{cases} \quad (4)$$

Next, we use x_i and y_i to represent the i-th sample and its hash codes where $y_i \in \{0,1\}^\gamma$ and γ denotes the length of hash codes. We set $y_i^\rho \in \{0,1\}$ as the ρ-th element of y_i. The hash code set for n samples can be represented as $[y_1, \ldots, y_n]^T$. Our $n \times n$ local similarity matrix W is

$$W_{ij} = \begin{cases} 0 & \text{if } N_K(i,j) \text{ is False,} \\ \dfrac{x_i^T x_j}{\|x_i\| \cdot \|x_j\|} & \text{otherwise.} \end{cases} \tag{5}$$

Furthermore, we apply diagonal matrix

$$D_{ii} = \sum_{j=1}^{n} W_{ij} \tag{6}$$

Meanwhile, we use the number of different bits for calculating Hamming distance between y_i and y_j as

$$H_{ij} = \|y_i - y_j\|^2 / 4 \tag{7}$$

We define an object function ζ to minimize the weighted average Hamming distance.

$$\zeta = \sum_{i=1}^{n} \sum_{j=1}^{n} W_{ij} H_{ij} \tag{8}$$

To calculate ζ, we transform it to $\xi = tr(Y^T L Y)/4$, where $L = D - W$ is Laplacian matrix and $tr(\cdot)$ means trace of matrix. At last, we transform ξ to LapEig problem ψ with slacking constraint $y_i \in \{0,1\}^t$, and obtain the optimal t-dimensional real-valued vector $\tilde{y}$ to represent each sample. ψ is the following:

$$\psi = arg \min_{\tilde{Y}} \ Tr(\tilde{Y}^T L \tilde{Y}) \quad s.t. \begin{cases} \tilde{Y}^T D \tilde{Y} = I \\ \tilde{Y}^T D 1 = 0 \end{cases} \tag{9}$$

where $Tr(\tilde{Y}^T L \tilde{Y})$ gives the real relaxation of the weighted average Hamming distance $Tr(T^T L Y)$. The solution of this optimization problem is given by $\tilde{Y} = [v_1, \ldots, v_t]$ whose columns are the t eigenvectors corresponding to the smallest eigenvalues of following generalized eigenvalue problem. The solution of ψ can be transformed to $Lv = \lambda D v$ where vector v is the t eigenvectors which are corresponding to the t smallest eigenvalues (nonzero).

Then, we convert the t-dimensional real-valued vectors $\tilde{y}_1, \ldots, \tilde{y}_n$ into binary codes according to the threshold. We set δ^p to present threshold and $\tilde{y}_i^p$ equivalent to p-th element of $\tilde{y}_i$. The hash label as final result value of y_i^p is

$$y_i^p = \begin{cases} 1 & \tilde{y}_i^p \geqslant \delta^p, \\ 0 & otherwise. \end{cases} \quad where \quad \delta^p = \frac{1}{n} \sum_{i=1}^{n} \tilde{y}_i^p \tag{10}$$

(2) Hash Model Training Stage. We implement an end-to-end hashing deep learning module. Firstly, we employ CNNs again to receive fine-grained features. After that, we adopt encoding module which is Divide and Encode Module [9] associated with activation function of BatchNorm [7] to approximate hash labels

generated in previous stage. The learning framework is the artificial neural network on the multi-output condition. Formally, we set a function $f : \mathbb{R}^I \rightarrow \mathbb{R}^O$, where I is the input set, O is the output set and x is the input vector. The formulation is

$$f^{(k)} = \begin{cases} \varphi(W^{(k)}x + b^{(k)}) & k = 1, \\ \varphi(W^{(k)}f^{(k-1)} + b^{(k)}) & k = 2, ..., K. \end{cases} \quad (11)$$

where b is the bias vector, W is the weight matrix of convolution and $\varphi(*)$ is ReLU and BatchNorm function. When the core of $\varphi(x)$ is BatchNorm, the function is calculated as follows:

$$\tilde{x}^{(k)} = \frac{x^{(k)} - E\left(x^{(k)}\right)}{\sqrt{Var\left(x^{(k)}\right)}} \quad (12)$$

where

$$E\left(x\right) = \frac{1}{m} \sum_{i=1}^{m} x_i \quad Var\left(x\right) = \frac{1}{m} \sum_{i=1}^{m} \left(x_i - E\left(x\right)\right)^2 \quad (13)$$

In the last layer of CNN, we split a 1024-dimensional vector into 16 groups, and each group is mapped to z elements. The output number $16 \times z$ is the hash code length. Denote the output as one $m \times d$ matrix (m is the number of samples in batch and d is the number of output in last full connection layer), x is the output vector, y is the corresponding label. We define the loss function as follows:

$$F\left(x\right) = min \sum_{i=1}^{m} \sum_{j=1}^{d} \left\| x_i^{(j)} - y_i^{(j)} \right\|_2^2 \quad (14)$$

At last, we define the threshold as the same as Eq. (10). Usually, we apply the threshold values of each bit calculated in the hash label generating stage.

4   Semantic Hash Ranking (SHR)

In this section, we introduce SHR algorithm in detail, which considers both the number of connected links and the weight on edges into consideration, designs impact factor between different nodes according to hash code, and calculate the importance score for each node by random walk. Let L_* denote number of links to N_*. Draw rank factor $R(N_*)$ for N_* and impact factor $I(N_{ij})$ for N_j to N_i, where $I(N_{ij})$ is defined as

$$I(N_{ij}) = \begin{cases} \dfrac{l - d_{ij}}{\sum\limits_{t \in T_j} l - d_{tj}} R(N_j) & \exists d_{ij}, \\ 0 & otherwise. \end{cases} \quad (15)$$

where T_j is the set including orders of all nodes associated with N_j. Specially, we design the formulation according to two principals. Firstly, the less d_{ij} is,

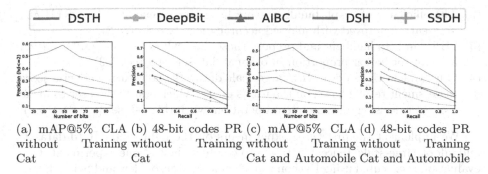

(a) mAP@5% CLA (b) 48-bit codes PR (c) mAP@5% CLA (d) 48-bit codes PR without Training without Training without Training without Training Cat   Cat   Cat and Automobile Cat and Automobile

Fig. 3. mAP@5% Code Length Analysis (CLA) and Precision-Recall (PR) curve on CIFAR-10.

the greater influence N_j contributes to N_i is. Meanwhile, the longer hash code is, the more compact the similarity presented by d_{ij} is. Secondly, PageRank considers the weights on each edge as the same, but we extend it to be applied to different weights on edges. As a result, when weights on different edges are the same, Eq. (15) should be the same as the impact factor formulation of PageRank. Consequently, $R(N_i)$ should be equal to the sum of the impact factors of all nodes linked to N_i

$$R(N_i) = \sum_{j=1,j\neq i}^{n} I(N_{ij}) \tag{16}$$

Let f_{ij} denote the coefficient of $R(N_j)$ in $I(N_{ij})$. We draw iteration formula as

$$\begin{bmatrix} R^{c+1}(N_1) \\ R^{c+1}(N_2) \\ \dots \\ R^{c+1}(N_n) \end{bmatrix} = \begin{bmatrix} 0 & f_{12} & \cdots & f_{1n} \\ f_{21} & 0 & \cdots & f_{2n} \\ \dots & \dots & \dots & \dots \\ f_{n1} & f_{n2} & \cdots & 0 \end{bmatrix} \begin{bmatrix} R^c(N_1) \\ R^c(N_2) \\ \dots \\ R^c(N_n) \end{bmatrix} \tag{17}$$

where c is the number of iteration rounds. We set the termination condition as

$$R^{c+1}(N_m) - R^c(N_m) \leq \varepsilon \tag{18}$$

where $m \in [1, n]$ and $\forall N_m$ should satisfy Eq. (18). Meanwhile, ε is constant. Let $SHR(N_*)$ denote semantic rank of N_*. The last results are

$$SHR(N_m) = R^\eta(N_m) \tag{19}$$

where η is the round on termination.

5 Evaluation

In this section, we evaluate our framework and conduct extensive experiments as follows:

(1) Using the feature extraction method with generalization ability, DSTH can solve the out-of-sample problem (see Sect. 5.1).

(2) The efficiency of graph building using hash codes generated by DSTH can be greatly improved with allowed accuracy loss (see Sect. 5.2).
(3) SHR can help highlight and prepose those data whose semantic information account for higher proportion in original dataset (see Sect. 5.3).
(4) Our framework can deal with large-scale datasets and return a concise score (assessment result) based on the user's query, which assists the user to make a correct decision on subsequent operations with this dataset (see Sect. 5.4).

We implement the first three experiments on the public CIFAR-10 dataset, and adopt large-scale Tencent datasets to conduct the last one experiment. Our evaluation is executed using Python tools including TensorFlow and Scikit-Learn library. Our experiments are run on two 10-core Intel Xeon E5-2640 machines with 128 GB of DDR4 memory. At last, we conduct the experiment on Tencent dataset using 12 machines.

5.1 Generalization Ability

In this section, we verify the effectiveness of DSTH by executing code length analysis (CLA) and precision-recall (PR) on CIFAR-10. We compare with the state-of-the-art methods on reorganized CIFAR-10 dataset, our DSTH shows a stronger generalization ability, which solves the out-of-sample problem.

In practice, we execute code length analysis (CLA) and precision-recall (PR) on CIFAR-10, compared with the state-of-the-art of single target unlabeled deep hashing algorithms including DeepBit [10], AIBC [16], DSH [11] and SSDH [20]. CIFAR-10 is a labeled data set, which consists of 60,000 32×32 color images in 10 classes, with 6000 images per class. There are 5000 training images and 1000 test images in each class. Particularly, we select the average value of the top 5% nodes in terms of precision in each class as the precision of CLA. In the experiment, we select GoogLeNet and classification model trained on ImageNet to extract deep features. Meanwhile, the CNN structure for generating hash model is similar to [12,23,24].

In order to validate the advantage in solving the out-of-samples problem mentioned in Sect. 6, we adjust the distribution of CIFAR-10 by taking the image of cat or automobile off from the training set. Besides, we stipulate that it is correct to classify a cat as a dog and an automobile as a truck. Figure 3(a) and (b) show the mAP@5% CLA and 48-bit codes PR performance without training set of cat. Figure 3(c) and (d) shows the mAP@5% CLA and 48-bit codes PR performance without training set of cat and automobile. As shown in Fig. 3, our results of mAP@5% CLA and PR also yield a significant dominance. As the same as the code length of previous, 48-bit is best for redefined CIFAR-10 data sets at precision of 58.66% and 52.56% respectively. Specially, as shown in Fig. 3(a) and (c), although the gap is reduced, performances of DSTH with 48-bit codes are higher than that of SSDH by 0.196 and 0.165 respectively in precision, illustrating the superiority of ours for solving the problem of out-of-samples.

5.2   Graph Building Efficiency

In this section, for verifying that building a graph by Hamming distance is more efficient than Cosine and Euclidean distance, we exhibit the time of graph building using three metrics with 48-bit vectors (48-bit hash codes and 48 float numbers). In order to ensure the fairness, we set $\Omega = 48$, making all nodes fully connected. As shown in Fig. 4(a), the horizontal coordinate represents the number of nodes while the ordinate represents the graph building time. With the same scale of nodes, the graph building time of Hamming distance is nearly 100 times less than that of Cosine and Euclidean, which shows that Hamming distance has overwhelming predominance over other two metrics in building a graph. Especially, with the scale of nodes increasing, the graph building time of Cosine and Euclidean grows exponentially which is unacceptable, making that Hamming distance becomes the better choice.

In order to compare precision of graph in three metrics, we choose more accurate links from top 1% to top 50% according to the weight of edges with 100,000 nodes. For example, we choose those edges on which the Hamming distance is smaller, while selecting the edges whose Cosine and Euclidean distance is larger. As shown in Fig. 4(b), Hamming distance is 0.070 lower than Euclidean at top 1% links in the worst case and 0.010 lower than Cosine at top 30% links in the best case in term of precision of graph. Averagely, Hamming distance is 0.038 lower than other two metrics in seven cases.

On the whole, there is not a marked difference of precision between three metrics, although hashing will bring certain loss to precision. However, Hamming distance has overwhelming predominance in building a graph in term of time cost. We use hashing and Hamming distance in the follow-up work with comprehensive consideration of tradeoff between efficiency and precision, since an acceptable margin of error is allowed.

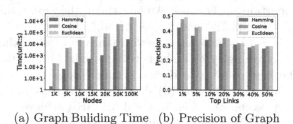

(a) Graph Buliding Time   (b) Precision of Graph

Fig. 4. Graph building time with different scale of nodes and precision of graph with 100,000 nodes using Hamming, Cosine and Euclidean distance.

5.3   Predominant Semantics

In this section, we verify SHR can highlight and prepose those data whose semantic information account for higher proportion in this section, which shows our

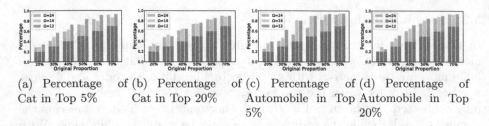

(a) Percentage of Cat in Top 5%   (b) Percentage of Cat in Top 20%   (c) Percentage of Automobile in Top 5%   (d) Percentage of Automobile in Top 20%

Fig. 5. Trend for percentage of cat and automobile in ranked result using different Ω with 48-bit codes. (Color figure online)

algorithm has practical significance for assessment tasks. In the next experiment, if the ranked results are correct, those images whose semantic distribution account for higher proportion in original data set will obtain larger scores and higher ranks. Thus, based on CIFAR-10 test set, under the premise that the amount of images of other classes remains unchanged, we choose one class as a study object to be added to the sample, making the amount of this class reach 20%, 30%, 40%, 50%, 60% and 70% respectively on the whole data set. We collect proportion of this class in the top 5% and top 20% of ranked results in six cases mentioned above. We set $\varepsilon = 1.0E-7$ and choose $\Omega = 24$, 16 and 12 to conduct the experiments.

Figure 5(a) and (b) show the percentage with different Ω in the top 5% and top 20% of ranked results respectively when choosing cat as the study object. As shown in Fig. 5(a) and (b), in all cases, SHR magnifies original proportion of cat (the part that goes beyond the blue column), indicating the efficiency of this algorithm. Detailedly, compared with $\Omega = 16$ or 24, setting $\Omega = 12$ yields better performance on the magnification of the cat percentage in the top 5% of ranked results, where the cat percentage is averagely 27.7% higher than original proportion in six cases. Among them, the best result exceeds the original proportion by 33.3% in the case of 50%. In the top 20% of ranked results, choosing $\Omega = 12$ yields better performance in most of the cases, where the cat percentage is averagely 21.2% higher than original proportion in six cases. Among them, the best result exceeds the original proportion by 26.3% in the case of 50%. Similarly, Fig. 5(c) and (d) show the results choosing automobile as the study object. As is shown in Fig. 5(c) and (d), SHR achieves the same effect. Detailedly, in the top 5% of ranked results, compared with other setting of Ω, the percentage of automobile shows the superiority in most of the cases while choosing $\Omega = 12$, which is averagely 31.7% higher than original proportion in six cases. Among them, the best result exceeds the original proportion by 40% in the case of 50%. Besides, in the top 20% of ranked results, the percentage of automobile also shows great superiority with $\Omega = 12$, which is averagely 31.0% higher than original proportion in six cases. Among them, the best result exceeds the original proportion by 34.7% in the case of 50%.

It should be explained that better precision and shorter time cost can be captured theoretically when $\Omega < 12$. However, the reduction of links causes too many isolated nodes all of whom get the same score, which may bring certain loss to the ranked results. Usually, with a larger scale of nodes, hash codes are more widely distributed, thus setting a smaller Ω will not result in too many isolated nodes. In our follow-up research, we intend to study this issue in depth.

As above experimental results show, both in the top 5% and 20% of ranked results, SHR effectively highlights and preposes the data whose semantic information account for higher proportion in original data set after ranking. Thus, our SHR is correct and effective in practical applications.

5.4 Assessment of Query

In this section, we verify that our framework can efficiently complete online assessment work according to the user's query task on large-scale real dark dataset. Also, it can guide and assist the user to conduct subsequent data mining work in order to show our framework is effective to complete the dark dataset assessment. We apply our framework to real-world data set of Tencent which express support for QQ albums, QQ chat and WeChat in a certain period. The size of data is around 5TB consisting of 1,000,000 images. Specially, according to the results shown in Sect. 5.3 that a smaller Ω is proved to be feasible at a million scale, we select $\Omega = 2$ to conduct this experiment.

We use the general method mentioned in Sect. 2 for assessment by analyzing the value of Tencent data set for three tasks which include human intimacy as task-A, lovers traveling in the outskirts as task-B, and driving on road as task-C respectively. The query images are collected from Baidu search and their respective weights are given below. Figure 6 displays the assessing process and results for above tasks. As shown in Fig. 6, the intimacy image representing the first task (query) matches three images whose ranks are high, so it is worth carrying out data mining on this data set for the task-A. For task-B which matches two images contained in Tencent data set, the images of lovers have high scores and images about landscape own medium ranks. However, the weighted score of this task is relatively high, which shows this data set can help analyze images about lovers traveling in the outskirts. Although there are three images

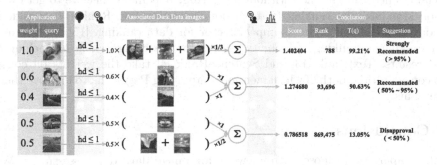

Fig. 6. The process of assessment on Tencent data set for three real applications.

that match the task-C, neither automobiles nor highways obtain high scores, so this data set are not suitable for the task-C.

For exposing semantic information of the dark data, we use deep learning to coarsely explore the content of the whole data set. According to the results, except that the amount of images including people is obviously dominant (66.13%), no more information can be captured for more specific assessment. Because of the better expressive ability, our framework can achieve better assessment results for different queries than classification algorithms.

6 Related Works

Dark Data. [5] has demonstrated the value of dark data by the long tail theory in economics and given the concept of dark data lightening which means constructing relationship according to a new task. [17] proposes using *File WinOver* System to complete the dark data judgment and risk assessment through fingerprint. [1] mentioned that the value of dark data depends on both the requirements of the task and the ability of value extraction. [6] presents the implementation of astronomical dark data management using unified databases. *GeoDeepDive* [21] and *DeepDive* [22] proposed a pragmatic scheme of dark data mining system by correcting annotations and associations of data according to feedback from users.

Content-Based Hashing for Image. Content-based Hashing is a technique that generates compact hash codes from the original data to represent the main content which preserves the data semantic relationship. With the success of Convolution Neutral Network (CNN) [8] in feature extraction, deep hashing [18, 19] becomes the mainstream for image hashing. For unlabeled images, DSTH has better ability to solve the problem of out-of-samples, because it is able to regard the instances beyond scope of cognition as the samples which have been learnt in the model as close as possible. Therefore, DSTH is a better hashing method to reduce the sensitivity of non-cognitive objects which are widely distributed in large-scale data set.

Graph-Based Ranking. Calculating the importance score of each node is a special quantization method without clustering. It is more effective to get evaluation standards by ranking for each node globally. PageRank [14] considers out-degree of related nodes as impact factor for data ranking. [15] applies random walking to ranking community images for searching, which has achieved good results. TextRank [13] and SentenceRank [4] take the weights on edges into consideration, both of which mentioned applying PageRank to improve their algorithms.

7 Conclusions

In this paper, we proposed a framework for image dark data assessment. We first transformed unlabeled images into hash codes by our developed DSTH algorithm, then constructed a semantic graph using restricted Hamming distance, and finally used our designed SHR algorithm to calculate the overall importance score for each image. During online assessment, we first translated the user's query into hash codes using DSTH model, then matched the suitable data contained in the dark data, and finally returned the weighted average value of these matched data to help the user cognize the dark data. Experimental results showed DSTH could extract semantic features with generalization ability, and SHR could correctly calculate the importance scores according to the similarity between data, and our framework could apply to large-scale datasets and had an overwhelming advantage over deep model.

Acknowledgments. This work is supported by the Innovation Group Project of the National Natural Science Foundation of China No. 61821003 and the National Key Research and Development Program of China under grant No. 2016YFB0800402 and the National Natural Science Foundation of China No. 61672254.

References

1. Cafarella, M.J., Ilyas, I.F., Kornacker, M., Kraska, T., Ré, C.: Dark data: are we solving the right problems? In: ICDE, pp. 1444–1445 (2016)
2. Cai, H., Huang, Z., Srivastava, D., Zhang, Q.: Indexing evolving events from tweet streams. In: ICDE, pp. 1538–1539 (2016)
3. Cao, Y., Long, M., Liu, B., Wang, J.: Deep cauchy hashing for hamming space retrieval. In: The IEEE Conference on Computer Vision and Pattern Recognition (CVPR), June 2018
4. Ge, S.S., Zhang, Z., He, H.: Weighted graph model based sentence clustering and ranking for document summarization. In: ICIS, pp. 90–95 (2011)
5. Heidorn, P.B.: Shedding light on the dark data in the long tail of science. Libr. Trends **57**(2), 280–299 (2008)
6. Heidorn, P.B., Stahlman, G.R., Steffen, J.: Astrolabe: curating, linking and computing astronomy's dark data. CoRR abs/1802.03629 (2018)
7. Ioffe, S., Szegedy, C.: Batch normalization: accelerating deep network training by reducing internal covariate shift. In: ICML, pp. 448–456 (2015)
8. Krizhevsky, A., Sutskever, I., Hinton, G.E.: Imagenet classification with deep convolutional neural networks. In: NIPS, pp. 1106–1114 (2012)
9. Lai, H., Pan, Y., Liu, Y., Yan, S.: Simultaneous feature learning and hash coding with deep neural networks. In: CVPR, pp. 3270–3278 (2015)
10. Lin, K., Lu, J., Chen, C., Zhou, J.: Learning compact binary descriptors with unsupervised deep neural networks. In: CVPR, pp. 1183–1192 (2016)
11. Liu, H., Wang, R., Shan, S., Chen, X.: Deep supervised hashing for fast image retrieval. In: CVPR, pp. 2064–2072 (2016)
12. Liu, Y., et al.: Deep self-taught hashing for image retrieval. IEEE Trans. Cybern. **49**(6), 2229–2241 (2019)
13. Mihalcea, R.: Graph-based ranking algorithms for sentence extraction, applied to text summarization. Unt Sch. Works **170–173**, 20 (2004)
14. Page, L., Brin, S., Motwani, R., Winograd, T.: The pagerank citation ranking: bringing order to the web. Technical report, Stanford InfoLab (1999)
15. Richter, F., Romberg, S., Hörster, E., Lienhart, R.: Multimodal ranking for image search on community databases. In: MIR, pp. 63–72 (2010)

16. Shen, F., Liu, W., Zhang, S., Yang, Y., Shen, H.T.: Learning binary codes for maximum inner product search. In: ICCV, pp. 4148–4156 (2015)
17. Shukla, M., Manjunath, S., Saxena, R., Mondal, S., Lodha, S.: POSTER: winover enterprise dark data. In: Proceedings of the 22nd ACM SIGSAC Conference on Computer and Communications Security, Denver, CO, USA, 12–16 October 2015, pp. 1674–1676 (2015)
18. Song, J., Gao, L., Liu, L., Zhu, X., Sebe, N.: Quantization-based hashing: a general framework for scalable image and video retrieval. Pattern Recogn. **75**, 175–187 (2018)
19. Song, J., He, T., Gao, L., Xu, X., Shen, H.T.: Deep region hashing for efficient large-scale instance search from images (2017)
20. Yang, H., Lin, K., Chen, C.: Supervised learning of semantics-preserving hash via deep convolutional neural networks. TPAMI **40**, 437–451 (2017)
21. Zhang, C., Govindaraju, V., Borchardt, J., Foltz, T., Ré, C., Peters, S.: Geodeepdive: statistical inference using familiar data-processing languages. In: SIGMOD, pp. 993–996 (2013)
22. Zhang, C., Shin, J., Ré, C., Cafarella, M.J., Niu, F.: Extracting databases from dark data with deepdive. In: SIGMOD, pp. 847–859 (2016)
23. Zhou, K., Liu, Y., Song, J., Yan, L., Zou, F., Shen, F.: Deep self-taught hashing for image retrieval. In: MM, pp. 1215–1218 (2015)
24. Zhou, K., Zeng, J., Liu, Y., Zou, F.: Deep sentiment hashing for text retrieval in social ciot. Future Gener. Comput. Syst. **86**, 362–371 (2018)

Medical Treatment Migration Prediction in Healthcare via Attention-Based Bidirectional GRU

Lin Cheng[1], Yongjian Ren[1], Kun Zhang[1,2], and Yuliang Shi[1,2(✉)]

[1] School of Software, Shandong University, Jinan, China
chenglin123_sdu@163.com, ryjsdu@outlook.com, kunzhangcs@126.com,
shiyuliang@sdu.edu.cn
[2] Dareway Software Co., Ltd., Jinan, China

Abstract. With the rapid expansion of the number of floating populations in China, a large number of people are gradually migrating to different hospitals to seek medical treatment. How to accurately predict the future medical treatment behaviors of patients has become an important research issue in healthcare. In this paper, an Attention-based Bidirectional Gated Recurrent Unit (AB-GRU) medical treatment migration prediction model is proposed to predict which hospital patients will go to in the future. The model groups patients who are prone to medical treatment migration, and achieves disease prediction and medical treatment migration prediction for each group. In terms of disease prediction, considering the predictive performance problem of a single prediction algorithm, a standard deviation weight recombination method is used to achieve disease prediction. When disease prediction has been completed, considering the impact of medical visit on the future medical behavior, on the basis of bidirectional gated recurrent unit (GRU) framework, we introduce an attention mechanism to determine the strength of hidden state at different moments, which can improve the predictive performance of the model. The experiment demonstrates that the predictive model proposed in this paper is more accurate than the traditional predictive models.

Keywords: Healthcare · Group patients · Disease prediction ·
Medical treatment migration prediction

1 Introduction

Medical treatment migration generally can be simply defined as the behavior of seeking medical treatment outside the insured areas by insured persons. In recent years, with the rapid expansion of the number of floating populations in China, the number of people who migrate to different hospitals to seek medical treatment has also increased. Due to the differences in medical levels between hospitals, more patients are willing to go to hospitals with higher medical level,

© Springer Nature Switzerland AG 2019
J. Shao et al. (Eds.): APWeb-WAIM 2019, LNCS 11641, pp. 19–34, 2019.
https://doi.org/10.1007/978-3-030-26072-9_2

resulting in the waste of hospital resources and unreasonable distribution of medical insurance funds. How to accurately predict the future medical treatment behavior of patients has become an important research issue in healthcare of China.

A conventional approach of medical treatment behavior prediction is the one-size-fit-all model [11]. That is, using all available training data to build a global model, and predict the medical treatment situation for each patient with this model. The benefit of applying a one-size-fit-all model is that it captures the overall information of the entire training population. However, patients may have different features, different medical conditions, etc. Using a global model may miss some specific information that is important for individual people. Thus, building predictive models for different groups is urgent and important for the study of medical treatment migration behavior. Recent studies [8] show that constructing a local prediction models at each group can improve predictive performance over global models. Then, how to group people based on different application scenarios is also a research question. In recent years, many methods for grouping people have emerged [6,10]. But, these methods did not take into account the impact of time factor.

Currently, in order to construct predictive models using medical visit sequences, recurrent neural networks (RNNs) are widely used [1,3,14]. However, RNNs cannot effectively address long-term dependencies. When the visit sequence is too large, the predictive performance of the RNNs will decrease.

To tackle the above questions, in this paper, a medical treatment migration prediction model is proposed. The model firstly groups patients according to the analyzed influencing factors, and then respectively achieves disease prediction and medical treatment migration prediction for each group. In terms of disease prediction, considering the predictive performance problem of a single prediction algorithm, a standard deviation weight recombination method is proposed to achieve disease prediction. Under the condition that disease prediction has been completed, considering the impact of each medical visit on migration behavior, we introduce an attention-based bidirectional GRU prediction model(AB-GRU) to predict which hospital patients will go to in the future. In summary, our contributions are as follows:

- In order to mine people who are prone to medical treatment migration and quantify the impact of feature weight on grouping, on the basis of C4.5 algorithm, we combine the information gain ratio with feature weight to construct a novel classification tree to realize the grouping of people.
- Considering the predictive performance problem of a single prediction algorithm, a standard deviation weight reorganization method is used to construct a combined forecasting model to predict future patients diseases.
- In order to quantify the impact of each medical visit on the future hospitalization behavior, on the basis of bidirectional GRU framework, we introduce an attention mechanism. The attention score is used to determine the strength of the hidden state at different moments, which can improve the prediction performance of the model.

The rest of this paper is organized as follows: In Sect. 2, we introduce relevant work on the prediction of medical behavior. Section 3 presents the details of the medical migration prediction model, including feature selection, grouping of patients prone to migration, disease prediction, and the process of AB-GRU. The experimental results are presented in Sect. 4. Section 5 is a summary of the work in this paper.

2   Related Work

As far as the current research on medical treatment behaviors of is concerned [14], it mainly focuses on two aspects. Firstly, the research on the types and characteristics of medical treatment behaviors [9,15], mainly concentrated on method and the choice of hospitalization behavior. Secondly, the research on the influencing factors of medical treatment behavior [2,5], mainly from the aspects of personal attributes, economic factors, social factors, etc., and obtained rich research results.

Engeda et al. used a statistical survey to study the behavior of medical treatment, and predicted patients medical behavior by Multivariable Logistic Regression algorithm [4]. Lu et al. studied the influencing factors of the choice of hospitalization behaviors among agricultural transfer [12]. That is, they analyzed the influencing factors of the choice behavior of medical treatment, and predicted the choice of hospitalization behaviors after illness by regression algorithms. In addition, Zhai et al. proposed combining the gray correlation analysis method with the multi-class selection model to realize the research and prediction of the community residents' medical treatment behavior [13]. Zhang et al. used social action theory and an analytical model to examine the main influencing factors of rural residents' medical treatment, and constructed a model to study the behavior of rural residents [17]. However, these research methods ignored the time information inside the medical treatment sequence.

Currently, using medical visit sequences, recurrent neural networks are widely used. Baytas, et al. proposed a novel Time-Aware LSTM to handle irregular time intervals in longitudinal visit records [1], Che et al. developed novel deep learning models for multivariate time series with missing values [3]. But, these methods did not achieve the interpretability of models.

Compared with the prediction methods mentioned above, the model proposed in this paper not only considers the influence of historical medical visit on future medical behavior, but also realizes the interpretability by introducing attention mechanism.

3   Methodology

3.1   Basic Symbols

We denote all the unique medical visit codes from the medical insurance data as $c_1, c_2, c_{|C|} \in C$, where $|C|$ is the number of unique medical visit codes.

Assuming there are N patients, the nth patient has $T^{(n)}$ visit records in the medical insurance data. The patient can be represented by a sequence of visits $X_1, X_2, ..., X_{T^{(n)}}$. Each visit X_i contains a set of feature vectors $x \in R^{|C|}$. We also classify diseases into 21 categories according to the International Classification Standard (ICD-10).

3.2   Overall Process

The overall process of the predictive model is shown in Fig. 1. The model firstly use the Grey Relational Analysis (GRA) method to analyze the influencing factors. In order to mine the people who are prone to medical treatment migration, we build a classification tree based on the analyzed influencing factors to realize grouping of people, and then respectively establish predictive models for different groups. In the actual medical migration prediction process, the disease prediction is a prior condition for the medical treatment migration. Therefore, the model firstly uses the autoregressive and grey algorithms to predict the disease of patients according to the probability trend of various diseases in various populations. However, considering the predictive performance problem of a single prediction algorithm, a standard deviation weight recombination method is used to construct a combined prediction model to achieve disease prediction. In terms of migration prediction, we propose AB-GRU. AB-GRU employs bidirectional gated recurrent unit to remember all the information of both the past visits and the future visits, and it introduces an attention mechanism to measure the relationships of different visits for prediction.

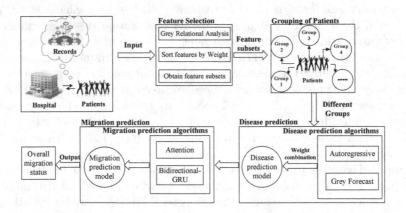

Fig. 1. Overall process of the medical migration prediction model

3.3   Feature Selection

In healthcare, there are many factors of affecting medical treatment migration as shown in Table 1. A great number of features are collected before building the predictive model but not all the variables are informative and useful. It is imperative to eliminate the redundancy of the features and select more informative variables for increasing the accuracy and efficiency of the predictive model. In our model, we use the Grey Relational Analysis (GRA) method to analyze the influencing factors. GRA aims to determine whether they can better distinguish target instances than other features by calculating the degree of association [13,16]. Then, the process of feature selection is as follows:

Table 1. Factors about medical treatment in our dataset.

Categories	Factors	Description
Patients	Age	The insured persons age
	Gender	Male = 1, Female = 2
	Income	The patients income
	Insured category	Staff = A, Resident = B
	Insured place	Insured place of patients
	Disease category	21 categories
	Distance	Distance between the insured and the hospital
	Industry category	Insured person's industry (9 categories)
Hospital	Hospital name	Name of medical hospital
	Hospital level	Hospital level : 123
	Average hospitalization days	The complexity of hospital treatment of certain diseases
	Average cost	The complexity of hospital treatment of certain diseases
	Maximum number	The maximum number of hospital in a period of time

(1) Normalize the original medical insurance data X by the min-max standardization method, and the calculation is as follows:

$$x' = \frac{x - x_{min}}{x_{max} - x_{min}} \tag{1}$$

(2) Suppose that the system reference sequence after the mean value change of each feature in the original sample is $X_0 = (x_0(1), x_0(2), \cdots, x_0(n))$, whether medical treatment is a comparative sequence $X_k = (x_k(1), x_k(2), \cdots, x_k(n))$, k = 1, 2, $\cdots$, m. Then, the correlation coefficient between X_0 and X_k is calculated by Eq. (2):

$$\delta_i(k) = \frac{min_i min_k |x_0(k) - x_i(k)| + \rho max_i max_k |x_0(k) - x_i(k)|}{|x_0(k) - x_i(k)| + \rho max_i max_k |x_0(k) - x_i(k)|} \tag{2}$$

where $\delta_i(k)$ represents the correlation coefficient of x_i to x_0 in the kth data. ρ represents the resolution coefficient ($\rho = 0.5$).

(3) The relationship θ between each factor X_0 and whether or not to migrate X_k:

$$\theta_i = \frac{1}{n}\sum_{k=1}^{n}\delta_i(k) \tag{3}$$

Finally, we can sort the relevance degree θ and select the main feature set that affects the medical migration according to the sorting size, which lays a foundation for the construction of the medical migration prediction model.

3.4 Grouping of Patients Prone to Migration

In this section, we construct a novel classification tree based on the features analyzed in Sect. 3.3, and find the people who are prone to medical treatment migration according to the classification rules.

In the process of medical migration prediction, how to quickly mine the people who are prone to migrate based on the analyzed feature sets is an important research question in this paper. In order to mine the people who are prone to medical treatment migration and quantify the impact of feature weight on grouping, on the basis of C4.5 algorithm, we combine the information gain ratio with feature weight to construct a novel classification tree to realize the grouping of people. In this paper, we define Z to represent the training data set, and Z_i denotes that Z is divided into m subsets by the feature $A, i = 1, 2, \cdots, m$, so the process of the specific population grouping is as follows:

(1) Calculate the information gain of each feature according to Eq. (4):

$$Gain(Z, A) = -\sum_{i=1}^{n} p_i log_2(p_i) - \sum_{j=1}^{m}\frac{|Z_j|}{|Z|}p_j log_2(p_j) \tag{4}$$

(2) Calculate the information gain rate:

$$SplitInformation(Z, A) = -\sum_{j=1}^{m}\frac{|Z_j|}{|Z|}log\frac{|Z_j|}{|Z|} \tag{5}$$

$$GainRatio(Z, A) = \frac{Gain(Z, A)}{SpltInformation(Z, A)} \tag{6}$$

(3) The information gain rate is combined with the influencing factor weight θ to affect the classification result:

$$GainRatio^{'}(Z, A, \theta) = \theta GainRation(Z, A) \tag{7}$$

Finally, we obtain the information gain rate $GainRatio^{'}$ of each feature by Eq. (7), and select the largest information gain rate as the classification node, construct a classification tree, realize the people grouping according to the classification rules, and then establish predictive models for different groups.

3.5   Disease Prediction of Patients

In this section, we predict the future disease of patients through autoregressive model or grey model based on the probabilistic trend of various types of diseases in various groups. Considering the predictive performance problem of a single predictive model, a standard deviation weight recombination method is used to construct a combined forecasting model to predict the future disease of patients. We define y_i as the probability of suffering from a Class i disease, where $i = 1, 2, \cdots, 21$. The disease prediction process is as follows:

Autoregressive model by Eq. (8):

$$y_i = c + \sum_{i=1}^{p} \varphi_i y_{t-i} + \varepsilon_t \tag{8}$$

where c is a constant term. φ is a memory function. ε_t is assumed to be an average number equal to 0, and the standard deviation is equal to the random error value σ. σ is assumed to be constant for any t.

Grey model by Eq. (9):

$$x^{(0)}(k) + \alpha y^{(1)}(k) = b, \quad k = 2, 3, \cdots \tag{9}$$

where α is the development coefficient and b is the grey effect.

The combined predictive formula is as follows:

$$y = \left(\frac{\vartheta - \vartheta_1}{\vartheta} \times \frac{1}{m-1} \right) y_1 + \left(\frac{\vartheta - \vartheta_2}{\vartheta} \times \frac{1}{m-1} \right) y_2 \tag{10}$$

Among them, the standard deviation of the prediction error of the autoregressive prediction model and the gray model are respectively ϑ_1, ϑ_2, and $\vartheta = \sum_{i=1}^{2} \vartheta_i \ (i = 1, 2)$, m is the number of models; y is the combined predicted value; y_1 is the predictive value of the autoregressive model; y_2 is the predictive value of grey model.

3.6   Migration Prediction for Patients

In this section, under the premise that the disease has known (completed in Sect. 3.5), we propose an Attention based Bidirectional GRU medical treatment migration prediction model (AB-GRU) to achieve the migration prediction of the patients.

The goal of the proposed model is to predict the $(t+1)^{th}$ visit's hospital. As shown in Fig. 2, given the visit information from time 1 to t, the ith visit x_i can be embedding into a vector v_i. The vector v_i is fed into the Bidirectional GRU, which outputs a hidden state H_i. Along with the set of hidden state H_i, we can compute the degree of correlation a_i between the hidden state and medical treatment behavior at each moment by attention operation. Finally, from the visit sequence v_i and hidden state H_i at all time, we can get the final prediction through the softmax function.

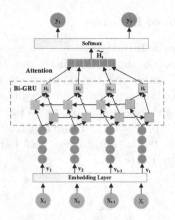

Fig. 2. The process of AB-GRU.

Embedding Layer. Given a visit sequences $X_i(i = 1, 2, ..., |C|)$. We can get its vector representation $v \in R^K$, as follows:

$$v = A^T x \tag{11}$$

where K represents the dimension of the embedding layer, $A \in R^{|C| \cdot K}$ is the weight matrix.

Bidirectional GRU. As a variant of the standard recurrent neural network (RNN), the gated recurrent unit (GRU) was originally proposed [7]. For each position t, GRU computes h_t with input x_t and previous state h_{t-1}, as:

$$r_t = \beta \left(W_r x_t + U_r h_{t-1} \right) \tag{12}$$
$$\pi_t = \beta \left(W_\pi x_t + U_\pi h_{t-1} \right) \tag{13}$$
$$\tilde{h}_t = tanh \left(W_c x_t + U \left(r_t \odot h_{t-1} \right) \right) \tag{14}$$
$$h_t = (1 - \pi_t) \odot h_{t-1} + \pi_t \odot \tilde{h}_t \tag{15}$$

where h_t, r_t and π_t are d-dimensional hidden state, reset gate, and update gate, respectively. W_r, W_π, W_c and U_r, U_π, U are the parameters of the GRU. β is the sigmoid function, and $\odot$ denotes element-wise production.

A bidirectional GRU consists of a forward and backward GRU. The forward GRU $h_{\overrightarrow{T}}$ reads the input visit sequence from x_1 to x_T and calculates a sequence of forward hidden states $(h_1^{\rightarrow}, h_2^{\rightarrow}, \cdots, h_T^{\rightarrow})$. The backward GRU $h_{\overleftarrow{T}}$ reads the visit sequence in the reverse order from x_T to x_1, resulting in a sequence of backward hidden states $(h_1^{\leftarrow}, h_2^{\leftarrow}, \cdots, h_T^{\leftarrow})$. We can obtain the final hidden state H_t, as follows:

$$H_t = [h_t^{\rightarrow}, h_t^{\leftarrow}] \tag{16}$$

Attention Mechanism. In the process of medical treatment migration prediction, the traditional neural network model ignores the impact of the length of the time interval within the visit sequences on the modeling, since the contribution of each visit to the current moment is not necessarily the same. Therefore, considering that not all features contribute to the prediction, we add the attention layer to the bidirectional GRU framework. The attention score is used to determine the strength of the hidden state during the modeling process of the medical treatment sequence, thereby significantly improving the modeling ability of the prediction model.

Implemented in the Attention mechanism as follows:

$$u_t = tanh\left(WH_t + b\right) \tag{17}$$

$$a_t = \frac{exp\left(u_t^T u\right)}{\sum_{t=1}^{T} exp\left(u_t^T u\right)} \tag{18}$$

$$\gamma = \sum_{t=1}^{T} a_t H_t \tag{19}$$

where $W \in R^{L \times |C|}$ and $b \in R^L$ are corresponding weights and bias vectors, u_t is the importance vector. a_t represents the normalized weight by Eq. (18). γ is the weighted sum of each H_t with a_t.

The output of the attention layer $\tilde{H}$ is:

$$\tilde{H} = \sum_{n=1}^{N} \gamma \tag{20}$$

Finally, $\tilde{H}$ is fed through the softmax layer to produce the $(t+1)^{th}$ choice of the hospitalization behaviors defined as:

$$y = softmax\left(W_c\tilde{H} + b_c\right) \tag{21}$$

where $W_c \in R^{2L}$ and $b_c \in R^L$ are the parameters to be learned.

Interpretation. In healthcare, we need to understand the clinical meaning of each dimension of visits, and analyze which visit are crucial to the medical treatment migration prediction.

In our proposed AB-GRU model, the attention can be used to assign weights to the hidden state of each visit. It is easy to find the importance of each medical visit by analyzing the weight of each medical visit. We sort the weight of each dimension in the hidden state in reverse order, and then select the top K weights, as shown below:

$$argsort\left(a_t\left[:, n\right]\right)\left[1:K\right] \tag{22}$$

where $a_t\left[:, n\right]$ represents the attention weight of each dimension in the tth visit. By analyzing the top K visits, we can obtain which visits have an important impact on the migration prediction. Detailed examples and analysis are given in Sect. 4.7.

4   Experiments

4.1   Data Description

In this section, we evaluate our proposed model on real data. The dataset comes from a certain area of China from 2012 to 2017. We picked out 498080 medical records, which contain 31130 patients who have migrated, 44 hospitals, and 21 diseases (classified according to ICD-10 standard). Table 2 describes the statistics about the datasets. Note that in order to protect the privacy and safety of patients and hospitals, we anonymize the corresponding experimental data. Among them, Hospital-A, Hospital-B and other forms are used to represent the name of the hospital in the medical migration prediction.

Table 2. Statistics of medical treatment datasets.

Category	Migration situation	Total	Density
Patients	31130	520954	5.98%
Hospitals	44	204	21.6%
The number of disease	1295	6497	19.9%
Medical records	498080	7718863	6.45%

4.2   Experimental Setup

To evaluate the accuracy of predicting the hospital for the next visit, we used a measure for prediction task: (1) Accuracy, which is the ratio of the predicted results equal to the actual results. (2) Weighted F1-score, which calculates F1-score for each class and reports their weighted mean.

In this paper, we implemented all the prediction methods with the Python language. We used Adadella optimizer the batch size of 500 patients. We randomly divided the dataset into training, validation and test set in a 0.75, 0.1, 0.15 ratio. and were initialized as 1 and 0.1, respectively. We set the dimension of embedding m as 100, and the dimensionality of hidden state of GRU as 100. We used 100 iterations for each method and report the best performance.

4.3   Feature Selection

The consequences of discriminative feature selection according Grey Relational Analysis method are presented below. Table 3 displays the weight value of all features calculated by GRA algorithm, and are ordered by weight from high to low. We set up the threshold (=0.5) to exclude some uninformative feature. In our experiments, the top 11 features obtained by GRA are considered as informative features. In addition, we can also conclude that the complexity of disease treatment in hospitals (Average Hospitalization Days and Average Cost) are more likely to affect the patient's medical treatment migration.

Table 3. Factors and correlation.

Factors	Correlation
Average cost	0.9498
Average hospitalization days	0.9084
Hospital level	0.8877
Income	0.8273
Age	0.7909
Treatment rate	0.6599
Disease category	0.6133
Gender	0.5945
Insured category	0.5510
Maximum number	0.5133
Distance	0.5087
Industry category	0.4766
Insured place	0.4587

4.4 Grouping of Patients Prone to Migration

In order to study the overall migration situation of patients in the region and improve the accuracy of the prediction results, we combine the information gain rate with the factor weight based on the C4.5 algorithm, construct a novel classification tree. Then, we can get groups of people who are prone to migrate according to the classification rules as shown in Table 4. From Table 4 we can conclude that more than 80% of patients have a medical treatment migration behavior in the five groups.

Table 4. Groups of patients who are prone to medical migration in 2017.

Category	Contents	Migrations	Total	Proportion
Group1	Age: 50–80, Income: 12K–50K, Gender = Female, Insured Category = Resident	8430	10012	84.2%
Group2	Age: 0–20, Insured Category = Resident	5312	6057	87.7%
Group3	Age: 60–80, Income > 20K, Insured Category = Retired Employee	4007	4857	82.5%
Group4	Age > 80, Gender = Male, Insured Category = Retired employee	1773	2064	85.9%
Group5	Age: 20–50, Insured Category = Employee	3875	4687	84.8%

4.5   Disease Prediction

In this subsection, we predict the future disease of patients through auto-regressive model or grey model based on the probabilistic trend of various types of diseases in various groups. Considering the predictive performance problem of a single predictive model, a standard deviation weight reorganization method is used to construct a combined forecasting model to achieve disease prediction.

The Fig. 3(a) and (b) respectively show the predictive situation of various diseases of Group1 and Group2 in 2017 using autoregressive model and grey model and combined prediction model. It can be seen from (a) and (b) that the predictive effect of the combined algorithm is better than that of the other two separate algorithms.

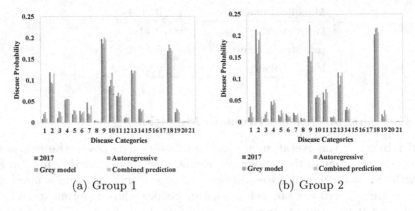

(a) Group 1   (b) Group 2

Fig. 3. Prediction of various diseases of group1 and group2 in 2017.

4.6   Migration Prediction

In this subsection, we compare it to several prediction methods in order to evaluate the predictive performance of our proposed model in migration prediction. The methods are described as follows:

MLR: This is the traditional Multiple Logistic Regression model.

Navie Bayes: This is a Classification Method Based on Bayes' Theorem and Characteristic Condition Independent Hypothesis.

SVM: This is a supervised learning algorithm used to solve classification problems.

RNN: This is the traditional unidirectional Recurrent Neural Network.

BGRU: This model uses only bidirectional GRU to predict future medical information without using any attention mechanisms.

AB-GRU (our model): An attention-based bidirectional GRU prediction model.

Table 5 shows the accuracy and weighted F1-score of methods in migration prediction. It can be concluded from Table 5 that the prediction performance of AB-GRU proposed in this paper is better than other prediction methods. It is because that the AB-GRU considers the impact of historical medical visits and time information on future medical migration.

Table 5. The accuracy and weighted F1-score of methods in migration prediction.

Methods	Accuracy	F1-score
MLR	0.6224	0.5651
Naive Bayes	0.6178	0.5640
SVM	0.6588	0.6202
RNNs	0.6837	0.6418
BGRU	0.7018	0.6594
AB-GRU	**0.7595**	**0.7173**

Figure 4 describes the ROC of several models. From Fig. 4, we can conclude that the area of ROC of AB-GRU proposed in this paper is the largest, that is, the AUC value is the largest. So AB-GRU in this paper has better predictive performance.

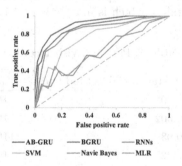

Fig. 4. The compassion of ROC.

4.7 Model Interpretation

In this subsection, we discuss interpretability of the predictive results. For each visit we are able to calculate the attention weight associated with it using Eq. (18). We first select two patients with tumor disease, and we illustrate how our model utilizes the information in the patient visits for prediction. For each patient, we show each visit, the time stamp of each visit, medical hospital during each visit, and the weight assigned by AB-GRU to each visit.

Table 6. Patient 1-visit records.

Medical visit records	Average weight of hidden state	Target hospital
Visit1 (305 days ago)	0	Hospital-A
Visit2 (195 days ago)	0.388	Hospital-B
Visit3 (111 days ago)	0.602	Hospital-B
Visit4 (41 days ago)	0.003	Hospital-A
Visit5 (14 days ago)	0.007	Hospital-A
Prediction	Hospital-B (Actual)	**Hospital-B (0.7901)**

From Table 6, we can observe that the weight assigned to most of the visits of patient 1 is close to 0, such as visit 1, visit 4 and visit 5, which means that they are ignored during prediction. However, for visit 2 and visit 3, although the two visits occurred in long days ago, they are assigned a larger weight, indicating that they have a long-term impact on the prediction of future medical institutions. Therefore, the predictive result is Hospital-B, which is mainly affected by visit 2 and visit 3.

From Table 7, we can see that for patient 2, this weight is very useful for interpretation since our model focuses on visits with nonzero weights. For visit 4–5, which occurred within the last two months, AB-GRU gives large weights to the hidden of state of last two visits, indicating that the last two visits have a larger impact on choice of the hospitalization behavior of patient 2. So, the predictive result is Hospital-B, which is mainly affected by visit 4 and visit 5.

Table 7. Patient 2-visit records.

Medical visit records	Average weight of hidden state	Target hospital
Visit1 (175 days ago)	0.002	Hospital-B
Visit2 (124 days ago)	0.002	Hospital-B
Visit3 (74 days ago)	0.03	Hospital-B
Visit4 (50 days ago)	0.227	Hospital-B
Visit5 (28 days ago)	0.739	Hospital-B
Prediction	Hospital-B (Actual)	**Hospital-B (0.9923)**

5   Conclusions

Medical treatment behavior prediction is an important research topic in healthcare. In this paper, we propose an Attention-based Bidirectional Gated Recurrent Unit (AB-GRU) medical migration prediction model to predict which hospital patients will go to in the future. The model groups patients who are prone to medical treatment migration and achieves disease prediction and medical treatment migration prediction for each group.

Acknowledgments. This work was supported by the National Key Research and Development Plan of China (No. 2018YFB1003804), the Natural Science Foundation of Shandong Province of China for Major Basic Research Projects (No. ZR2017ZB0419), the TaiShan Industrial Experts Program of Shandong Province of China (No. tscy-20150305).

References

1. Baytas, I.M., Xiao, C., Zhang, X., Wang, F., Jain, A.K., Zhou, J.: Patient subtyping via time-aware LSTM networks. In: Proceedings of the 23rd ACM SIGKDD International Conference on Knowledge Discovery and Data Mining, Halifax, NS, Canada, 13–17 August 2017, pp. 65–74 (2017)
2. Bhojani, U., et al.: No longer diseases of the wealthy: prevalence and health-seeking for self-reported chronic conditions among urban poor in Southern India. BMC Health Serv. Res. **13**(1), 306 (2013)
3. Che, Z., Purushotham, S., Cho, K., Sontag, D., Yan, L.: Recurrent neural networks for multivariate time series with missing values. Sci. Rep. **8**(1), 6085 (2018)
4. Chu Bei, Y.Y.: Health seeking behavior of elderly floating population and the influence factors. Chinese Rural Health Service Administration (2015)
5. Engeda, E.H., Dachew, B.A., Woreta, H.K., Kelkay, M.M., Ashenafie, T.D.: Health seeking behaviour and associated factors among pulmonary tuberculosis suspects in Lay Armachiho district, Northwest Ethiopia: a community-based study. Tuberc. Res. Treat. **2016**(4), 7892701 (2016)
6. Hossain, A.S.M.S.: Customer segmentation using centroid based and density based clustering algorithms (2017)
7. Kowsrihawat, K., Vateekul, P., Boonkwan, P.: Predicting judicial decisions of criminal cases from Thai supreme court using bi-directional GRU with attention mechanism. In: 2018 5th Asian Conference on Defense Technology (ACDT), pp. 50–55 (2018)
8. Lee, J., Maslove, D.M., Dubin, J.A.: Personalized mortality prediction driven by electronic medical data and a patient similarity metric. Plos One **10**(5), e0127428 (2015)
9. Li, Y.: Study on health-seeking behaviors of rural chronic patients in the view of the social determinants of health model. Med. Soc. (2015)
10. Mahurkar, K.K., Gaikwad, D.P.: Normalization using improvised k-means applied in diagnosing thyroid disease with ANN (2017)
11. Suo, Q., et al.: Personalized disease prediction using a CNN-based similarity learning method. In: 2017 IEEE International Conference on Bioinformatics and Biomedicine, BIBM 2017, Kansas City, MO, USA, 13–16 November 2017, pp. 811–816 (2017)
12. Xiao-Jun, L.U., Zhang, N.: Study on influencing factors of the choice of hospitalization behaviors among agricultural transfer population. Chin. J. Health Policy (2018)
13. Zhai, G., Duan, L., Hao, G., Duan, G., Xuan, C., Ying, Z.: The grey relational analysis of influential factors for Chinese medicine in general hospital. In: IEEE International Conference on Grey Systems and Intelligent Services (2011)

14. Zhai, Y.K., Ge, W.U.: Analysis on medical behaviors of patients based on big data mining of electronic medical records (EMR) information. J. Med. Inform. (2017)
15. Zhang, X.W., Qiu, L.J., Yang, Y.N., Zhao, J., Lian-Chen, F.U., Qian-Qian, L.I.: Internet health information seeking behaviors of medical students under a medical internet perspective. Chin. Prev. Med. (2018)
16. Zheng, C., Zhu, J.: Grey relational analysis of factors affecting IPO pricing in china a-share market. In: International Conference on Grey Systems and Intelligent Services (2017)
17. Zheng, X., Ling, X.U.: Analysis of health seeking behavior based on the planned-action theory in rural area of China. Beijing da xue xue bao. Yi xue ban = J. Peking Univ. Health Sci. **42**(3), 270 (2010)

WRL: A Combined Model for Short-Term Load Forecasting

Yuecan Liu[1], Kun Zhang[1,2], Shuai Zhen[2], Yongming Guan[2(✉)], and Yuliang Shi[1,2(✉)]

[1] School of Software, Shandong University, Jinan, China
lycan120624@163.com, kunzhangcs@126.com, shiyuliang@sdu.edu.cn
[2] Dareway Software Co., Ltd., Jinan, China
{zhenshuai,guanyongming}@dareway.com.cn

Abstract. Load forecasting plays a vital role in economic construction and national security. The accuracy of short-term load forecasting will directly affect the quality of power supply and user experience, and will indirectly affect the stability and safety of the power system operation. In this paper, we present a novel short-term load forecasting model, which combines influencing factors analysis, Wavelet Decomposition feature extraction, Radial Basis Function (RBF) neural networks and Bidirectional Long Short-Term Memory (Bi-LSTM) networks (WRL below). The model uses wavelet decomposition to extract the main features of load data, analyzes its correlation with influencing factors, and then constructs corresponding adjustment factors. The RBF neural networks are used to forecast the feature subsequence related to external factors. Other subsequences are input into Bidirectional LSTM networks to forecast future values. Finally, the forecasting results are obtained by wavelet inverse transform. Experiments show that the proposed short-term load forecasting method is effective and feasible.

Keywords: Load forecasting · Wavelet Decomposition · RBF neural networks · Bi-LSTM networks · Adjustment factors

1 Introduction

The main task of the power system is to provide users with economical, reliable, and quality-compliant electrical energy to meet the needs of all walks of society. In recent years, with the in-depth development of the power market reform, the electric quantity and load of electricity users have become more and more demanding on the quality of power delivered by the power system. Hence, load forecasting becomes an important part of the operation and development of power systems [9,10].

Load forecasting has always been a topic of social attention in the field of smart grid and data mining. This field possesses rich research history. When applying these traditional forecasting methods directly to the short-term load

© Springer Nature Switzerland AG 2019
J. Shao et al. (Eds.): APWeb-WAIM 2019, LNCS 11641, pp. 35–42, 2019.
https://doi.org/10.1007/978-3-030-26072-9_3

forecasting, there is a certain gap between the forecasting effect and the expectation. It is necessary to consider the specific problems existing in load forecasting, such as the components of load data and the influencing factors.

In response to the prior questions, the main contributions of this paper are as follows:

- Wavelet decomposition is used to process the load data to eliminate the noise influence aiming at the noise problem in the load data and extract the main features inside the load data, and the main feature subsequences are obtained.
- In view of the influence of external factors on the load data, the corresponding adjustment factors are constructed which indicate the factor's influencing degree to the load.
- In order to improve the accuracy of the model, for the different parts after wavelet decomposition and reconstruction, RBF neural networks and Bidirectional LSTM networks are respectively used to forecast.

The other parts of this paper are organized as follows: Sect. 2 reviews related work. Section 3 describes the framework of the overall load forecasting model. Section 4 introduces the feature extraction and correlation analysis among the data. Section 5 states our main forecasting model. Section 6 processes experiments based on real load data and other data and analyses the results. Section 7 draws concluding remarks.

2   Related Work

In terms of improved traditional methods, Dudek et al. constructed a short-term load forecasting univariate model based on linear regression [2], and introduced time series correlation factors to improve the accuracy of short-term load forecasting. Jiang et al. combined support vector regression and two-step hybrid parameter optimization method to make the forecasting result more accurate [6]. Hu et al. used the generalized regression neural networks method to forecast and prove the accuracy of the results through experiments [4].

With the deepening of research in the field of load forecasting, researchers are increasingly inclined to use time series and combined forecasting models to perform forecasting. For example, Ghofrani et al. considered the load data as a time series. The Bayesian neural network was used for forecasting, and compares the accuracy and processing speed by experiments [3]. In [11], Xiao et al. based on the combined model of multi-objective optimization algorithm, the weight of the single model was optimized. The forecasting result simultaneously has high precision and high stability.

However, these methods do not consider the user's classification and industry factor. Hence it is not possible to forecast separately according to the characteristic of users in different industries. In view of the above problems, this paper aims to solve the problems of imperfect related factors processing and other issues, our proposed model names WRL is introduced.

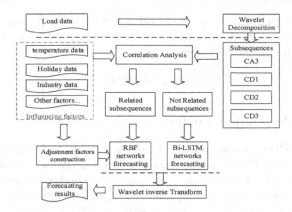

Fig. 1. Overall framework of the model.

3   The Framework of Load Forecasting Model

The framework of the overall model is shown in Fig. 1. There may be a strong correlation between the load data and temperature, holidays, industry information or other factors. Considering the influence of these factors can make the forecasting results more accurate. Hence, at first, we denoise and extract features of the load data based on wavelet 3-layer decomposition and reconstruction. The obtained four subsequences and external factors carry out correlation analysis respectively, thus related subsequences and unrelated subsequences are distinguished. Then as for each influencing factor, corresponding adjustment factors are built. The related subsequences use the RBF neural networks to forecast load values while the unrelated subsequences use the Bidirectional LSTM networks to forecast. Finally, the forecasting results are obtained by wavelet inverse transform.

4   Feature Extraction and Correlation Analysis

The wavelet decomposition is used to process the load data. While denoising, the detail component can be separated from the profile component, and the internal features of the data are extracted [5]. Then the correlation analysis with external factors can obtain the correlation between each influencing factor and each feature subsequence.

4.1   Feature Extraction Based on Wavelet Decomposition

Wavelet 3-layer decomposition is used in this paper cause it has a good denoising effect and the information loss can be ignored. The load data is decomposed and reconstructed based on Daubechies wavelet using Mallat algorithm into different frequency band components. Thus a more in-depth analysis of signal characteristics is performed. The one third-layer low-frequency sequence and the three high-frequency sequences of each layer obtained by decomposing and reconstructing the load signal are:

$$[CA3, CD1, CD2, CD3] \tag{1}$$

4.2   Factors Correlation Analysis

Holiday Data Preprocessing. Load data and temperature data acquired are all divided by hours. Hence, as for the data like holiday factor, we define a 0–1 quantification method in hours to distinguish the holidays from the workdays. Due to the different holiday arrangements of different industries, we determine the quantified value according to the comparison of hourly electricity load in the holiday with it before and after the holiday, if it is much smaller than the normal load, the quantified value is 0, otherwise, the quantified value is 1. Other similar factors can be quantified by similar method.

Temperature and Holiday Data Correlation Analysis. Taking the calculation of temperature correlation as an example, the Pearson formula is:

$$\gamma_{xt} = \frac{\sum (x_k - \overline{x})(t - \overline{t})}{\sqrt{\sum (x_k - \overline{x})^2 \sum (t - \overline{t})^2}} \tag{2}$$

Where x_k is the value of the reconstructed subsequence, t is the temperature sequence, $\overline{x}$ and $\overline{t}$ are the average values of the reconstructed subsequence and the temperature sequence respectively, and the threshold φ is set. When $\gamma_{xt} > \varphi$, the subsequence is related to the temperature.

Industry Data Correlation Analysis. SOM clustering can imitate the relevant attributes of human brain neurons, and automatically cluster the input patterns through self-training [1]. Then, for the reconstructed subsequences of all users under the same industry label, SOM clustering is performed according to the four reconstructed subsequences respectively. By means of visual display of the clustering result, if the largest class in the clustering result according to a certain sequence has maximum users, then the industry is related to the sequence, resulting in a collection of related pairs of subsequences and industries.

5   Forecasting Model Construction

In this paper, the RBF neural networks are used to forecast the factors-related reconstruction subsequences after wavelet decomposition. Bidirectional LSTM networks are used to forecast the factors-unrelated subsequences. Then the finally results are obtained by wavelet inverse transform.

5.1   Adjustment Factors Construction

The construction methods of the three adjustment factors are as follows:

(1) Industry adjustment factor construction
For each industry, the working mechanism has strong similarities, and the development of the industry is synchronized by the influence of the economic development. Therefore, the typical feature curve can reflect the impact of industry development on the load. Hence, we define the industry typical feature curve as the industry adjustment factor, ie:

$$\delta_W = [w_1, w_2, w_3, \cdots, w_n] \tag{3}$$

$$w_i = Avg(L_i) \tag{4}$$

Where w_i is the ith value on industry typical curve, L_i is the total load of all users of the industry at the ith moment.
(2) Temperature adjustment factor construction
For the temperature factor, load data is affected by the temperature of previous several days. Considering the variance and average value of previous several temperatures are influencing factors that affecting load changes. Therefore, we define the temperature adjustment factor as a three-dimensional feature set, ie:

$$\delta_T = [(t_1, V_{y,t_1}, Mean_{y,t_1}), (t_2, V_{y,t_2}, Mean_{y,t_2}), \cdots, (t_n, V_{y,t_n}, Mean_{y,t_n})] \tag{5}$$

Where V_{y,t_i} is the variance of y temperatures before t_i. $Mean_{y,t_i}$ is the average value of y temperatures before t_i.
(3) Holiday adjustment factor construction
The holiday data is quantized into 0–1 values. So the holiday adjustment factor is directly used as:

$$\delta_H = [h_1, h_2, h_3, \cdots, h_n] \tag{6}$$

Where h_i is the ith time period quantized value of holiday.
After processing, the temperature, holiday and industry adjustment factor sequences acquired are δ_T, δ_H, δ_W.

5.2   Forecasting Part

In this paper, the Radial Basis Function (RBF) neural networks are used to forecast the factors-related subsequences while the Bidirectional LSTM networks are used to forecast the unrelated subsequences. The whole flow of the forecasting part is shown in Fig. 2.

RBF Neural Networks. For the partial area of the input space of RBF network, only a few connection weights affecting the output of the area [8], and the change of load is often related to the values of the influencing factors of the adjacent time period. Therefore, the RBF neural networks are suitable to forecast subsequences related to external factors.

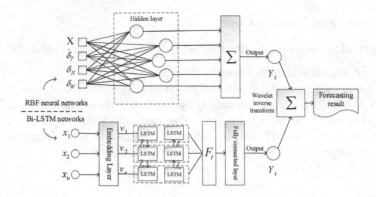

Fig. 2. Forecasting part of WRL

The factors-related subsequences and the relevant adjustment factor sequences are simultaneously input to the RBF neural networks for regression. The hidden layer neurons adopt the Gaussian kernel function and the output is a single sequence.

Bidirectional LSTM Networks. The Long-short term memory (LSTM) networks can be used to forecast time series [7]. Many studies have used LSTM networks for time series forecasting and proved to be effective. In this paper, the factor-unrelated subsequences after wavelet decomposition and reconstruction represents the internal characteristics of the load sequence. The subsequences show a certain periodicity. The Bidirectional LSTM (Bi-LSTM) networks can be trained by all available input information in the past and future. We use Bi-LSTM networks to forecast and have achieved better performance.

6   Experiments Analysis

We just considered temperature, holidays, and industry factors in this paper. The experimental load is grabbed from Electricity information collection System of a provincial grid company in China with their industry information from April 1, 2016 to March 31, 2018. Temperature data of corresponding cities is grabbed from China Meteorological Data Service Center website, including hourly temperature data from surface meteorological stations. Holidays information is collected from calendars.

In our experiments, users of the province are divided into two parts: the low-voltage station users and the high-voltage station users. It can be seen from Fig. 3 that the model proposed in this paper has a good effect in terms of both stations' load forecasting.

Six methods for load forecasting are compared with the proposed method WRL, including support vector regression (SVR), pure LSTM, wavelet decomposition+LSTM, ARIMA, Holt-Winters, and expert-prediction algorithm. These

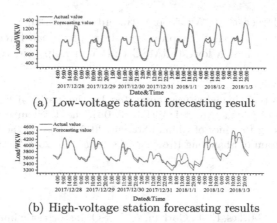

(a) Low-voltage station forecasting result

(b) High-voltage station forecasting results

Fig. 3. Forecasting results of different stations

methods are taken separately for forecasting five days (24 h per day) in each of the four seasons of the city's general electric users. Then the average of the root mean square error (RMSE) and error rate of each method are obtained, as shown in Table 1. Table 1 indicates that the proposed method WRL has the minimum RMSE and error rate among one-year forecasting.

Table 1. Comparison of different methods in forecasting results.

Method	RMSE	Error rate (%)
ARIMA	381.9311	24.421
SVR	313.3408	21.257
Holt-Winters	255.7437	21.892
Expert prediction	255.9021	18.376
LSTM	184.5701	14.682
WD+LSTM	149.5589	13.479
WRL	**109.4477**	**9.658**

Experiments show that our proposed model (WRL) that combines factors analysis, Wavelet Decomposition and deep learning algorithms can effectively meet the needs of power companies to make decisions by forecasting load of future several days.

7 Conclusion

In this paper, a novel model names WRL combines influencing factors analysis, Wavelet Decomposition and deep learning algorithms is proposed for the short-term load forecasting. We do experiments upon real data set. And the experiments results show that our proposed short-term load forecasting method has a high accuracy and a lowest RMSE that can help grid companies make decisions.

Acknowledgements. This work was supported by the National Key Research and Development Plan of China (No. 2018YFB1003804), the TaiShan Industrial Experts Program of Shandong Province of China (No. tscy20150305) and the Key Research & Development Program of Shandong Province of China (No. 2016ZDJS01A09).

References

1. Akinduko, A.A., Mirkes, E.M., Gorban, A.N.: SOM: stochastic initialization versus principal components. Inf. Sci. **364–365**, 213–221 (2016)
2. Dudek, G.: Pattern-based local linear regression models for short-term load forecasting. Electr. Power Syst. Res. **130**, 139–147 (2016)
3. Ghofrani, M., Carson, D., Ghayekhloo, M.: Hybrid clustering-time series-bayesian neural network short-term load forecasting method. In: North American Power Symposium (2016)
4. Hu, R., Wen, S., Zeng, Z., Huang, T.: A short-term power load forecasting model based on the generalized regression neural network with decreasing step fruit fly optimization algorithm. Neurocomputing **221**, 24–31 (2017)
5. Huang, L., Wang, J.: Forecasting energy fluctuation model by wavelet decomposition and stochastic recurrent wavelet neural network. Neurocomputing **309**, 70–82 (2018)
6. Jiang, H., Zhang, Y., Muljadi, E., Zhang, J.J., Gao, D.W.: A short-term and high-resolution distribution system load forecasting approach using support vector regression with hybrid parameters optimization. IEEE Trans. Smart Grid **9**(4), 3341–3350 (2018)
7. Kong, W., Dong, Z.Y., Jia, Y., Hill, D.J., Xu, Y., Zhang, Y.: Short-term residential load forecasting based on lstm recurrent neural network. IEEE Trans. Smart Grid **PP**(99), 1 (2017)
8. Mohammadi, R., Ghomi, S.M.T.F., Zeinali, F.: A new hybrid evolutionary based RBF networks method for forecasting time series: a case study of forecasting emergency supply demand time series. Eng. Appl. Artif. Intell. **36**(36), 204–214 (2014)
9. Srivastava, A.K., Pandey, A.S., Singh, D.: Short-term load forecasting methods: a review. In: International Conference on Emerging Trends in Electrical Electronics and Sustainable Energy Systems (2016)
10. Strasser, T., et al.: A review of architectures and concepts for intelligence in future electric energy systems. IEEE Trans. Ind. Electron. **62**(4), 2424–2438 (2015)
11. Xiao, L., Wei, S., Yu, M., Jing, M., Jin, C.: Research and application of a combined model based on multi-objective optimization for electrical load forecasting. Energy **119**, 1057–1074 (2017)

Data and Information Quality

DeepAM: Deep Semantic Address Representation for Address Matching

Shuangli Shan[1,2], Zhixu Li[1,3](✉), Yang Qiang[4], An Liu[1], Jiajie Xu[1], and Zhigang Chen[5]

[1] Institute of Artificial Intelligence, School of Computer Science and Technology, Soochow University, Suzhou, China
shanshuangli1@gmail.com, {zhixuli,anliu,xujj}@suda.edu.cn
[2] Neusoft Corporation, Shenyang, China
[3] IFLYTEK Research, Suzhou, China
[4] King Abdullah University of Science and Technology, Jeddah, Saudi Arabia
qiang.yang@kaust.edu.sa
[5] State Key Laboratory of Cognitive Intelligence, iFLYTEK, Hefei, People's Republic of China
zgchen@iflytek.com

Abstract. Address matching is a crucial task in various location-based businesses like take-out services and express delivery, which aims at identifying addresses referring to the same location in address databases. It is a challenging one due to various possible ways to express the address of a location, especially in Chinese. Traditional address matching approaches relying on string similarities and learning matching rules to identify addresses referring to the same location, could hardly solve the cases with redundant, incomplete or unusual expression of addresses. In this paper, we propose to map every address into a fixed-size vector in the same vector space using state-of-the-art deep sentence representation techniques and then measure the semantic similarity between addresses in this vector space. The attention mechanism is also applied to the model to highlight important features of addresses in their semantic representations. Last but not least, we novelly propose to get rich contexts for addresses from the web through web search engines, which could strongly enrich the semantic meaning of addresses that could be learned. Our empirical study conducted on two real-world address datasets demonstrates that our approach greatly improves both precision (up to 5%) and recall (up to 8%) of the state-of-the-art existing methods.

Keywords: Address matching · LSTM · Attention mechanism

1 Introduction

Nowadays, address localization becomes a core function in various location-based businesses like take-out services and express delivery. However, due to various address expression standards as well as typing errors in human inputs, address

© Springer Nature Switzerland AG 2019
J. Shao et al. (Eds.): APWeb-WAIM 2019, LNCS 11641, pp. 45–60, 2019.
https://doi.org/10.1007/978-3-030-26072-9_4

Table 1. An example of five different address expressions of the same location

ID	Address
add1	RT-MART, Danling road No.18, Haidian district, Beijing (北京市海淀区丹棱路18号乐天玛特)
add2	RT-MART, Chuangfu Building, Danling road No.18, Beijing (北京市丹棱路18号创富大厦乐天玛特)
add3	RT-MART, Seismological Bureau of Beijing south 50 meters, Haidian district, Beijing (北京市海淀区地震局向南50m乐天玛特)
add4	Chuangfu Building 1106, Beijing (北京市创富大厦1106)
add5	RT-MART, crossroad of Caihefang Road and Danling Road, Haidian district, Beijing (北京市海淀区彩和坊路和丹棱路交叉口乐天玛特)

parsing and understanding becomes a big obstacle in address localization. The problem becomes more serious for addresses in Chinese, given more complex and diverse expression such as those listed in Table 1.

For better address parsing and understanding, **address matching** task has been studied [3], aiming at identifying addresses referring to the same location across different address databases. Traditional methods for address matching or standardization rely on approximate string matching metrics such as edit distance, which are not robust to deal with various kinds of expressions. A more widely-used way for address matching is to build a decision tree consisting of learned matching rules, where each rule corresponds to a path from the root node to a leaf node on the tree. Based on this so-called *address matching tree*, the similarities between two addresses could be computed. To deal with the fuzzy expressions of the addresses, some work proposes to use the forward maximum matching algorithm to segment the address to match entries in a standard address dictionary, referring to the learned matching rules [1]. Some other work also combines edit distance with space vector model to calculate the similarity between strings [9]. They measure the dissimilarity caused by the differences in the characters of address strings using edit distance, and calculate the dissimilarity caused by differences in the address terms using TF-IDF term weighting. The final result is obtained by weighting the two dissimilarities.

However, the existing methods relying on string similarities and rule matching can be easily influenced by the diverse expressions including redundant, incomplete or unusual expression of addresses. For example in Table 1, where all the five addresses refer to the same location. Assume we have a matching rule that CITY+STREET+STREET_NO.+POI_NAME → ADDRESS, which means, addresses having the same city name, street name, street No., and POI name should refer to the same POI location. Based on this rule, we may identify that *addr*1 and *addr*2 refer to the same location, but we could hardly judge that *addr*3, *addr*4 and *addr*5 also refer to the same location with *addr*1, given that *addr*3 is represented by the orientation information of the adjacent address,

$addr4$ lacks the information of poi, and $addr5$ uses the road intersection to describe the address.

To address the weaknesses of the existing approaches, instead of using the syntactic features of addresses only, we turn to measure the semantic similarity between addresses in this paper. Recently, deep learning techniques have achieved great success on sentence representation, such as Seq2Seq model for machine translation [16] and the skip-thought model for distributed sentence representation [7]. Inspired by the state-of-the-art sentence representation models, we propose a novel address matching approach based on deep semantic address representation. The challenge lies on how to learn proper semantic representations for all the addresses.

Particularly, we propose to use an encode-decoder architecture with two long short term memory (LSTM) networks [5], where one is regarded as encoder and the other as decoder, to learn the semantic vector representation for an address string. In addition, due to the problems of address redundancy and addresses incomplete, we up-sample and sub-sample the address during the training process, then take the sampled address as the input of the encoder and use the original address as the output of the decoder to improve the robustness of the model. We also apply an attention layer between encoder and decoder to enable the semantic vectors to represent richer semantic information by assigning higher weights to more important features. Last but not the least, the semantic features contained in an address itself is still very limited. To fully get the semantic meaning for an address in the address domain, we would like to get extra contexts for the address from the web with the help of Web Search Engines, such that richer semantic meaning of the address could be learned from much richer contexts.

We summarize our contributions in this paper as follows:

- We novelly propose to measure the semantic similarity between addresses for address matching based on deep semantic address representation.
- We propose to use an encode-decoder architecture with two LSTM networks to learn the semantic vector representation for an address string. We up-sample and sub-sample the address in the encoder-decoder model to improve the robustness. The attention mechanism is also applied to the model to highlight important features of addresses in their semantic representations.
- We novelly propose to get rich contexts for addresses from the web through web search engines, which could strongly enrich the semantic meaning of addresses that could be learned.

We perform experiments on two real-world datasets, and the empirical results demonstrate that our proposed model works much better than the state-of-the-art methods on both precision (up to 5%) and recall (up to 8%).

Roadmap. The rest of the paper is organized as follows: we give a formal definition to address matching problem in Sect. 2. After presenting a compound framework for address matching in Sect. 3, we then present our deep semantic matching approach in Sect. 4. After reporting our experiments in Sect. 5, we cover the related work in Sect. 6. We conclude in Sect. 7.

2   Problem Definition

Address localization is a core task in various location-based businesses. Depending on whether there is a standard address database, address localization can be divided into address standardization task and address matching task. In this paper, we study on the case without a standard address database, i.e., address matching task, which is formally defined below:

Definition 1 (Address Matching). *Given a set of addresses $D = \{add_1, add_2, ..., add_n\}$, the goal of address matching is to find every address pair (add_i, add_j) satisfying $add_i \doteq add_j$, where $add_i \in D, add_j \in D, i \neq j$ and $\doteq$ is a comparison operator having its two operands referring to the same real-world object.*

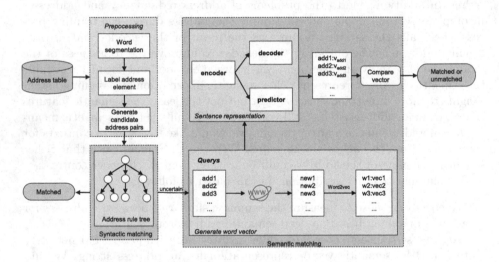

Fig. 1. A compound framework with both syntactic and semantic address matching

3   A Compound Framework for Address Matching

Instead of working independently, we prefer to let our deep semantic matching approach work under a compound framework, where the address matching rule tree is also adopted in the *Syntactic matching step* before going into the *semantic matching step* using deep semantic representations of addresses.

We illustrate this compound framework in Fig. 1: Given a set of addresses, we first generate a number of candidate address pairs for matching based on some simple heuristic rules as introduced in [18]. For instance, only addresses in the same `city` and `district` (if any), sharing at least one word (after removing stop words) in the left part of their address strings need to be compared. For every candidate address pair, we use a basic address matching tree following [1] to

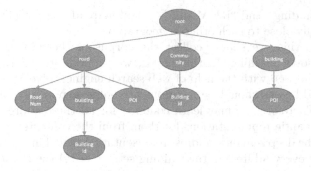

Fig. 2. A basic address matching tree for syntactic address matching

decide whether the two addresses could be syntactically matched in the *Syntactic matching step*. If yes, the address pair will be put into the final matching results. Otherwise, the address pair goes into the *semantic matching step*, where vectors of the two addresses would be obtained from the learned *deep semantic address representation model*. We then compute the similarity between candidate pairs and decide the final matching results according to the predefined threshold. We give more details below:

– **Syntactic Matching.** For syntactic matching, we adopt an address matching tree containing a number of expert-defined address matching rules as the one shown in Fig. 2, where every path from the root node to a leaf node corresponds to a matching rule. For every candidate address pair, we would let them traverse the whole tree in a deep-first traverse way to find if it could strictly match with a matching rule. Once successfully matched with a matching rule at its leaf node, the traverse process will be terminated.
– **Semantic Matching.** All candidate address pairs that cannot be matched in the syntactic matching step will go to the semantic matching step. For every candidate pair, we use the learned *deep semantic address representation model* to get the vectors of the two addresses, and then compute their similarities to see if they could be matched or not according to a predefined threshold. The core module of semantic matching is how we build the *deep semantic address representation model*, which will be introduced in the following section.

4   Deep Semantic Address Representation Model

It is a nontrivial task to learn a deep semantic representation model for addresses. The general sentence representation models are built for general purposes, but address representation is a task of a specific domain. Take the addresses in Table 1 for example, the general sentence representation models would not support that all the five addresses are close in semantic. But as a domain representation model, our deep semantic representation model should learn that

"Chuangfu Building" and "RT-MART", "Danling road" and "Caihefang Road" are semantically close to each other in geography.

To achieve this, we should collect the corpus of the address localization domain for model learning. In this paper, we propose to collect the address corpus from the web with the help of web search engines, given the assumption that all the addresses should be mentioned, wholly or partially, at somewhere on the web. Therefore, we may learn contexts for all these addresses and then learn deep semantic representations for them from the contexts.

We learn the deep semantic address representation model in two stages. In the first stage, for every address in the training set, we crawl the latest news about it on the web through web search engines, and treat these news as our corpus. Next, we use the corpus to enrich the semantic information of the addresses and employ **Word2Vec** [12] to train our word vector. In the second training stage, we propose to use a encoder-decoder model with attention mechanism to represent sentence vectors. Different from the traditional model, we here replace one hot representation with the pre-trained word vectors as the initial vectors, which helps to capture more semantics.

In the rest of this section, we first introduce how we get extra contexts for a given address from the web and perform domain word2vec training in the first stage in Sect. 4.1, and then present the encoder-decoder model with attention mechanism for deep semantic address representation in Sect. 4.2.

4.1   Stage 1: Word Embedding for the Address Domain

There are plenty of contexts on the Web where rich geographic semantic information can be captured. Let $A = \{e_1, e_2, e_3, ..., e_n\}$ denote a collection of elements of an address which can be gotten with word segment tool [12]. For every address element e_i, we use Web Search Engines to obtain a collection of web page, denote as $W_i = \{p_{i1}, p_{i2}, p_{i3}, ..., p_{im}\}$. Therefore, for an address, there are a total of $n \times m$ pages. In order to remove irrelevant web page, we use a vector space model (VSM) for representing a collection of web pages to calculate a score for each page. Specifically, each web page is represented by a document vector consisting of a set of weighted keywords. Keywords in each web page are address elements. The weights of the address element in each web page vector are determined according to the term frequency inverse document frequency (TF-IDF) model. The weight of e_i in web page $p_j (1 < j < n \times m)$, denoted as $w(e_i, p_j)$, which can be calculated by the following formula:

$$w(e_i, p_j) = Num(e_i, p_j) \times \log \frac{m \times n + 1}{k} \tag{1}$$

where $Num(e_i, p_j)$ denotes the number of occurrences of e_i in web page p_j and k is the number of web pages which contain element e_i. After that, we can calculate a score for every web page by the following equation:

$$score(p_j) = \sum_{i=1}^{n} w(e_i, p_j) \tag{2}$$

Next, we select the web pages with high scores, e.g. larger than a predefined threshold, and then extract contexts from the web pages. These contexts are used as the training corpora to train the word vectors related to address elements by using the Word2vec technique. A general corpus, Chinese Wikipedia [13], is also utilized to pre-train word vectors before using the web pages for address domain. After that, for every address element $e_i \in A$, we can get its vector x_i. Finally, we can obtain a collection of word vectors $W = \{x_1, x_2, ..., x_n\}$ for every sentence.

4.2   Stage2: Deep Address Representation Learning

In the second training state, we focus on learning the address representation with encoder-decoder model. Specifically, we apply LSTM on the encoder model and decoder model respectively. In addition, we utilize predictor model to predict the next sentence. In order to solve the problems of address redundancy and addresses incomplete, we upsample and subsample the addresses from the original training set to construct a new training set. That is, we use address matching rules to determine which address elements are unessential. After that, we remove the unessential address elements so that the addresses can be represented in different forms and we can also artificially add duplicate elements to the address. Next, we take the sampled address as the inputs of the encoder and use the original address as the output of the decoder to satisfy that the semantics of addresses with incomplete or redundancy are close. Furthermore, we apply the attention mechanism on encoder-decoder model and encoder-predictor model such at the important features can be emphasized by assigning higher weights.

(1) Encoder-Decoder Model with Attention. Firstly, the encoder reads the input sentence one-by-one which is an address or a sentence from web corpus. Note that the address or the sentence here has been initialized by the word embedding shown in Sect. 4.1, i.e. $W = \{x_1, x_2, ..., x_n\}$. Next, in a certain time step t of LSTM, there are three gates in a LSTM time step, input gate, forget gate and output gate respectively, denoted as i_t, f_t and o_t. They are composed of a sigmoid neural net layer and a pointwise multiplication operation and their values range from 0 to 1. Let $H = \{h_1, h_2, ..., h_t\}$ be the set of hidden state for each sentence from the inputs. We use the following equations to get the hidden state h_t for the time step t. We first need to forget the old subject when we meet a new subject. And we decide which information we are going to throw away from the previous hidden state h_{t-1} with the Eq. (3):

$$f_t = \sigma(w_f * h_{t-1} + w_f * x_t) \tag{3}$$

The next step is to decide what new information we are going to store in the current state. And input gate decides which values will be updated as follows.

$$i_t = \sigma(w_i * h_{t-1} + w_i * x_t) \tag{4}$$

Next, a tanh layer creates a vector of new candidate values $\tilde{c}_t$:

$$\tilde{c}_t = \tanh(w_c * h_{t-1} + w_c * x_t) \tag{5}$$

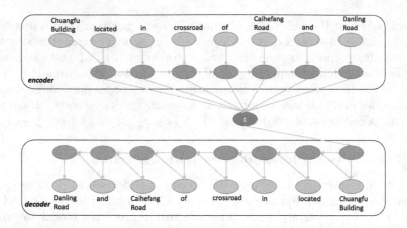

Fig. 3. The illustration of encoder-decoder model

After the forget gate and input gate have been computed, it is time to update the old memory state c_{t-1} into the new state c_t.

$$c_t = f_t * c_{t-1} + i_t * \tilde{c}_t \tag{6}$$

where c_t denotes the memory state in time step t. Finally, we need to decide what we are going to output. Specifically, we run a sigmoid layer to decide which parts of the cell state are going to output.

$$o_t = \sigma(w_o * h_{t-1} + w_o * x_t) \tag{7}$$

Then, we put the memory state through tanh to push the values between -1 and 1 and multiply it by the output of the output gate. Therefore, we can output the parts.

$$h_t = o_t * \tanh(c_t) \tag{8}$$

where w_f, w_i, w_c, and w_o denote weights of each part. In this way, the entire input sentence are mapped as a fixed-length vector which is then provided as an input to the decoder model. Then, the decoder outputs a sentence by decoding the hidden stage into the input sentence in the same manner. This process is shown in Fig. 3.

Due to the hierarchical relationship of address elements, we want to determine which part of the input is most responsible for the current decoder output. Therefore, we adjust our model by adding attention mechanism. Suppose that e_t denotes the input word embedding at current step t and h_{t-1} denotes the hidden state decoder at previous time step $t-1$. Attention model would first mix encoder output vector h_{t-1} with current input word embedding e_t.

$$v_t = f(h_{t-1}, e_t) \tag{9}$$

where v_t is the weight of each word vector. Generally, f is a simple feed-forward network with 1 or 2 dense layers. After that, we normalize the v_t with the following equation:

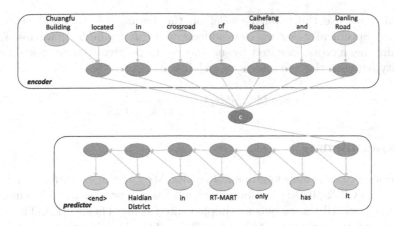

Fig. 4. The illustration of encoder-predictor model.

$$w_t = \frac{exp(v_t)}{\sum_{t=1}^{N} exp(v_t)} \tag{10}$$

where N is the number of words contained in a input sentence. Then, the final fixed length vector $\boldsymbol{v}$ can be calculated as follows:

$$\boldsymbol{v} = \sum_{t=1}^{N} w_t \boldsymbol{h}_t \tag{11}$$

In this way, the important words are weighted with the higher scores and then we can use information from these words to construct the fixed length vector. In addition, it can take variable length inputs instead of the fixed-length inputs.

(2) Encoder-Predictor Model with Attention. This model is used to predict the next sentence and obtain the geographic semantic information. Different from the encoder in the encoder-decoder model, the input of encoder in the encoder-predictor model is just a sentence in the web corpus. And the predictor is designed to predict the following words based on the previous state. And the attention mechanism is also used in the same way. Here, we employ the conditional probability to output the sentence with the highest probability:

$$p(\boldsymbol{y}|\boldsymbol{v}) = \sum_{t=1}^{N} p(\boldsymbol{y}_i|\boldsymbol{y}_1, ..., \boldsymbol{y}_{i-1}, \boldsymbol{v}) \tag{12}$$

where $\boldsymbol{y}_i$ is defined as the output of the predictor at time step i. All the words in the sentence are sequentially generated using the LSTM until the end of the sentence symbol is generated. The training objective is to maximize the conditional probability over the corpus using stochastic gradient descent. Especially, $\boldsymbol{y}_0$ is defined as a zero vector. After that, we store the optimal parameters of our model for the training set. The process of the encoder-decoder model is shown in Fig. 4.

Finally, we use the trained model on the testing set to build vectors of addresses. Specifically, for a given address pair $(addr_i, addr_j)$, we use Eq. 11 to obtain the vectors, denoted by v_i and v_j respectively, then calculate the similarity between the two addresses as follows:

$$sim(addr_i, addr_j) = \frac{v_i \times v_j}{\parallel v_i \parallel \cdot \parallel v_j \parallel} \tag{13}$$

5   Experiments

This section presents our experimental results. We run the experiments on the a server with a GTX-1080ti gpu, four-core Intel Core i7 processor, 16 GB memory, running centos. All the models are implemented using Theano [17]. The size of word embedding dimensionality is set as 300 and the learning rate for all models is 0.0002.

5.1   Datasets and Metrics

For evaluation, we compare our proposed method with several existing state-of-the-art methods on two real-world datasets below:

– **POI.** We collect addresses from a Chinese POI website[1] where every POI has one corresponding address. This dataset contains 200k pieces of POI addresses. The expressions of addresses in this dataset are diverse and they also have some redundant or incomplete information.
– **Company.** We also crawl addresses on two food review websites[2,3] and one company information query website[4]. This database contains 10k company addresses. The problems of the redundant, incomplete information and diverse expressions for addresses in this dataset becomes even worse.

Metrics. We basically use three metrics to evaluate the effectiveness of the methods: **Precision:** the percentage of correctly matched pairs among all address pairs. **Recall:** the percentage of correctly matched pairs among all address pairs that should be matched. **F1 Score:** a combination of precision and recall, which is calculated by $F1 = \frac{2*precision*recall}{precision+recall}$.

5.2   Methods for Comparison

In this section, we introduce our proposed method *DeepAM* with the existing methods including *String-Based* method, *Dictionary-Based* method, and *Address Matching Tree* method. We also illustrate the effectiveness of our **Stage1** training process with the *DeepAM without Stage1* method.

---

[1] www.poi86.com.
[2] www.dianping.com.
[3] www.meituan.com.
[4] www.qichacha.com.

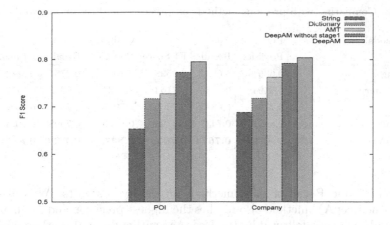

Fig. 5. Comparing with previous methods on F1 score

- The *String-Based* method combines the vector space model and edit distance to calculate the address similarity.
- The *Dictionary-Based* method uses a place-name dictionary to parse the address elements such that it can reduce the effect of place-name ambiguity on address matching.
- The *Address Matching Tree* (AMT for short) method builds up a rule-based matching tree and applies it to do address matching by transforming address matching tree into a set the matching rules.
- The *DeepAM* method utilizes Web Search Engines to get extra contexts for the address as the corpus and uses an encoder-decoder architecture with two LSTM networks to get the address representation. The vectors of address pairs are built based on the model to compute the similarities.
- The *DeepAM without Stage1* method deletes the Stage1 training process and uses the traditional one-hot representation to initialize the encoder-decoder model in the Stage2 process.

5.3 Comparisons with Previous Methods

In this section, we compare the effectiveness of our proposed address matching method with the mentioned methods above.

As is shown in Fig. 5, the String-Based method has the lowest F1 scores, because it just considers the address similarity based on string which cannot tackle the problem of diversity well. The accuracy of the Dictionary-based method is lower than the AMT method since it greatly relies on the quality of place-name dictionary. Our proposed DeepAM method performs best all of them in that it extracts the geographic information from the web and use sentence representation model to learn the deep semantic information. We can also see that the F1 score of the DeepAM without Stage1 method is worse than the DeepAM model because of the lack of rich contexts on the web. In addition, we compare

Table 2. Comparing with previous methods on precision and recall on two datasets

Methods	POI			Company		
	Precision	Recall	F1 Score	Precision	Recall	F1 Score
String	0.6854	0.6232	0.6528	0.7015	0.6753	0.6881
Dictionary	0.7504	0.6854	0.7164	0.7432	0.6939	0.7177
AMT	0.7752	0.6843	0.7269	0.7945	0.7332	0.7626
DeepAM without stage1	0.8032	0.7437	0.7723	0.8159	0.7698	0.7921
DeepAM	**0.8249**	**0.7674**	**0.7954**	**0.8372**	**0.7736**	**0.8041**

the Precision and Recall of these methods on the two datasets. As is listed in
Table 2, the DeepAM method also reaches the highest precision and recall among
all methods, which is followed by the DeepAM without stage1 method with the
second highest values. We can see that the Stage1 training process in our model
works well and improves the precision and recall both by about 2%. The precision and recall of AMT method is better than Dictionary-based method and
String-based method is the worst one.

5.4 Evaluation of Quality of Web Contexts

In this section, we evaluate the influence on the matching precision for the quality of web contexts. As shown in Fig. 6(a), the precision of DeepAM method
first rises when the percentage of web contexts increases, and it reaches the
highest points at 0.4. After that, it has a little drop. This indicates that the
model learns richer semantic information when it has lager web contexts in the
address domain. But when this figure continues to go up, it shows a decreasing trend since the irrelevant web content has a negative effect. We can also see
from Fig. 6(b), for the Company dataset, the experimental results show a similar
variation tendency for the precision.

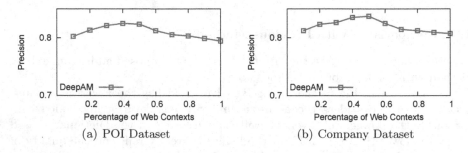

(a) POI Dataset   (b) Company Dataset

Fig. 6. Effect of the quality of web contexts

Table 3. An example of web contexts

Address element	Context
Chuangfu Building	Chuangfu Building is located at No. 18 Danling Street, Haidian District. The total construction area of Chuangfu Building is 30,408 square meters, including 17 floors above ground and 3 underground floors. Chuangfu Building is adjacent to Sihuan, Zhongguancun Street and Caihefang Road, enjoying the convenience of the city. The supporting facilities around Chuangfu Building are complete: close to the Municipal Industry and Commerce Bureau, the Earthquake Administration, major ministries and political organizations. (创富大厦由北京海淀科技园建设股份有限公司开发的,创富大厦项目总建筑面积30408平方米,其中地上17层,地下3层。创富大厦毗邻四环、中关村大街和彩和坊路,尽享都市便捷。创富大厦周边配套设施齐备:紧邻市工商局,地震局和各大部委及政要机构。)
	Chuangfu Building is located in the central location of modern science and technology development. It is located in the southwestern corner of Zhongguancun West District. It is located at the crossroad of Danling Road and Caihefang Road, west of Suzhou Street and south of Haidian South Road. (创富大厦位于现代科技发展中心性区位,中关村西区内的西南金角位置,雄踞于丹梭路与彩和坊路交叉路口,西邻苏州街,南依海淀南路。) ......
Wangfujing Street	Wangfujing Commercial Street has a large number of large shopping malls from south to north. There are Dongfang Xintiandi in the southeast of Wangfujing Commercial Street, Wangfujing Department Store in the southwest, and New Dong'an Market in the northeast, forming the Wangfujing Commercial Street. (王府井商业街有南到北的大型商场很多,有位于王府井商业街东南的东方新天地,西南的王府井百货商场,还有在东北的新东安市场,组成了王府井商业街。)
	Wangfujing street, located in the north side of east Changan street in the center of Beijing, is a famous commercial district with a long history for hundreds of years. (王府井步行街,位于北京市中心的东长安街北侧,是具有数百年悠久历史的著名商业区) ......

Lastly, we list some examples of web contexts we get for some example addresses in Table 3. As can be observed, theses web contexts contain rich semantic information about addresses, which is beneficial to get more accurate sentence representation, such as the location, adjacent objects and so forth.

6   Related Work

Given the complexity and diversity of address expressions, it is difficult to form a unified standard address model and geocoding standard, which results in the hardness of address data sharing among different government sectors and industries. To tackle the problem, a wide variety of methods have been proposed for address matching and address standardization.

A widely-used way for address matching is to build a decision tree consisting of learned matching rules, where each rule corresponds to a path from the root node to a leaf node on the tree. Based on this so-called *address matching tree*, the similarities between two addresses could be computed. Kang et al. propose an address matching tree model based on the analysis of the spatial constraint relationship of address elements [11]. This method has a higher address matching rate, but it needs to establish a variety of complex address models and needs to determine the spatial constraint between the address elements. An improved hash structure-based maximum inverse matching algorithm is proposed [2]. This method can make full use of the hash function to improve the retrieval efficiency, however, it also has the disadvantage of being sensitive to the address hierarchy.

Some recent efforts propose to do address matching based on the semantic analysis to the addresses. They get the semantic vectors of address expressions and then apply them to compute the similarity to decide the matching results. Song et al. apply the Chinese word segmentation and semantic inference (HMM model) in natural language processing to deal with unstructured Chinese addresses [15]. The disadvantage of this approach has a strong dependence on the HMM model training set, and it requires a large number of addresses to train the model.

The process of translating manually written addresses into a certain digital format is known as address standardization. There are also some researches on Address Standardization [4,6,8]. A method based on trie-tree and finite state machine is proposed in [10] which focuses on the problem of inaccurate word segmentation. They use trie-tree to realize Chinese word segmentation and apply finite state machine to match each layer of address. However, address hierarchy is too complex resulting in low matching efficiency. In addition, Yong et al. propose a standardization method based on the Euclidean distance between the address to be processed and the address in the standard library, but it only works well on some specific data sets [19]. Furthermore, Sharma et al. use a fully connected neural network with a single hidden layer to tackle the syntactic and semantic challenges in a large number of addresses [14]. However, it needs a large of labeled training samples.

7   Conclusions and Future Work

Address matching is a crucial task in various location-based businesses like take-out services and express delivery. In this paper, we propose to use an encoder-decoder architecture with two LSTM networks to learn the semantic vector representation for an address string. We also propose to get rich contexts for addresses from the web which could strongly enrich the semantic meaning of addresses that could be learned. Our experiments conducted on two real-world datasets demonstrate that our proposed model works much better than the state-of-the-art methods on both precision (up to 5%) and recall (up to 8%). In our future study, we consider to involve more geographical information into the deep semantic address embedding.

Acknowledgments. This research is partially supported by National Natural Science Foundation of China (Grant No. 61632016, 61572336, 61572335, 61772356), the Natural Science Research Project of Jiangsu Higher Education Institution (No. 17KJA520003, 18KJA520010), and the Open Program of Neusoft Corporation (No. SKLSAOP1801).

References

1. Cheng, C., Yu, B.: A rule-based segmenting and matching method for fuzzy Chinese addresses. Geogr. Geo-Inf. Sci. **3**, 007 (2011)
2. Ding, Z., Zhang, Z., Li, J.: Improvement on reverse directional maximum matching method based on hash structure for Chinese word segmentation. Comput. Eng. Des. **29**(12), 3208–3211 (2008)
3. Drummond, W.J.: Address matching: GIS technology for mapping human activity patterns. J. Am. Plan. Assoc. **61**(2), 240–251 (1995)
4. Guo, H., Zhu, H., Guo, Z., Zhang, X., Su, Z.: Address standardization with latent semantic association. In: Proceedings of the 15th ACM SIGKDD International Conference on Knowledge Discovery and Data Mining, pp. 1155–1164. ACM (2009)
5. Hochreiter, S., Schmidhuber, J.: LSTM can solve hard long time lag problems. In: Advances in Neural Information Processing Systems, pp. 473–479 (1997)
6. Kaleem, A., Ghori, K.M., Khanzada, Z., Malik, M.N.: Address standardization using supervised machine learning. Interpretation **1**(2), 10 (2011)
7. Kiros, R., et al.: Skip-thought vectors. In: Advances in Neural Information Processing Systems, pp. 3294–3302 (2015)
8. Kothari, G., Faruquie, T.A., Subramaniam, L.V., Prasad, K.H., Mohania, M.K.: Transfer of supervision for improved address standardization. In: 2010 20th International Conference on Pattern Recognition (ICPR), pp. 2178–2181. IEEE (2010)
9. Li, D., Wang, S., Mei, Z.: Approximate address matching. In: 2010 International Conference on P2P, Parallel, Grid, Cloud and Internet Computing, pp. 264–269. IEEE (2010)
10. Luo, M., Huang, H.: New method of Chinese address standardization based on finite state machine theory. Appl. Res. Comput. **33**, 3691–3695 (2016)
11. Mengjun, K., Qingyun, D., Mingjun, W.: A new method of Chinese address extraction based on address tree model. Acta Geodaetica et Cartographica Sinica **44**(1), 99–107 (2015)
12. Mikolov, T., Sutskever, I., Chen, K., Corrado, G.S., Dean, J.: Distributed representations of words and phrases and their compositionality. In: Advances in Neural Information Processing Systems, pp. 3111–3119 (2013)
13. Qiu, Y., Li, H., Li, S., Jiang, Y., Hu, R., Yang, L.: Revisiting correlations between intrinsic and extrinsic evaluations of word embeddings. In: Sun, M., Liu, T., Wang, X., Liu, Z., Liu, Y. (eds.) CCL/NLP-NABD -2018. LNCS (LNAI), vol. 11221, pp. 209–221. Springer, Cham (2018). https://doi.org/10.1007/978-3-030-01716-3_18

14. Sharma, S., Ratti, R., Arora, I., Solanki, A., Bhatt, G.: Automated parsing of geographical addresses: a multilayer feedforward neural network based approach. In: 2018 IEEE 12th International Conference on Semantic Computing (ICSC), pp. 123–130. IEEE (2018)
15. Song, Z.: Address matching algorithm based on Chinese natural language understanding. J. Remote Sens. **17**(4), 788–801 (2013)
16. Sutskever, I., Vinyals, O., Le, Q.V.: Sequence to sequence learning with neural networks. In: Advances in Neural Information Processing Systems, pp. 3104–3112 (2014)
17. The Theano Development Team, et al.: Theano: a python framework for fast computation of mathematical expressions (2016)
18. Tian, Q., Ren, F., Hu, T., Liu, J., Li, R., Du, Q.: Using an optimized Chinese address matching method to develop a geocoding service: a case study of Shenzhen, China. ISPRS Int. J. Geo-Inf. **5**(5), 65 (2016)
19. Yong, W., Jiping, L., Qingsheng, G., An, L.: The standardization method of address information for POIs from internet based on positional relation. Acta Geodaetica et Cartographica Sinica **45**(5), 623–630 (2016)

Drawing CoCo Core-Sets
from Incomplete Relational Data

Yongnan Liu[1,2]([⊠]) [iD] and Jianzhong Li[1]

[1] School of Computer Science and Technology,
Harbin Institute of Technology, Harbin, China
cellsi2011@gmail.com, lijzh@hit.edu.cn
[2] School of Data Science and Technology,
Heilongjiang University, Harbin, China

Abstract. Incompleteness is a pervasive issue and brings challenges to answer queries with high-quality tuples. Since not all missing values can be repaired by complete values, it is crucial to provide completeness of a query answer for further decisions. To estimate such completeness results fast and objectively, CoCo core-sets are proposed in this paper. A CoCo core-set is a subset of an incomplete relational dataset, which contains tuples providing enough complete values on attributes of interest and whose ratio of complete values is close to that of the entire dataset. Based on CoCo core-sets reliable mechanisms can be designed to estimate query completeness on incomplete datasets. This paper investigates the problem of drawing CoCo core-sets on incomplete relational data. To the best of our knowledge, there is no such a proposal in the past. (1) We formalize the problem of drawing CoCo core-sets, and prove that the problem is NP-Complete. (2) An efficient approximate algorithm to draw an approximate CoCo core-set is proposed, where uniform sampling technique is employed to efficiently select tuples for coverage and completeness. (3) Analysis of the proposed approximate algorithm shows both coverage of attributes of interest and the relative error of ratio of complete attribute values between drawn tuples and the entire data can be within a given relative error bound. (4) Experiments on both real-world and synthetic datasets demonstrate that the algorithm can effectively and efficiently draw tuples preserving properties of entire datasets for query completeness estimation, and have a well scalability.

Keywords: Data quality · Data completeness · Query completeness ·
Incomplete data · CoCo core-sets

1 Introduction

Incomplete data are pervasive and severely reduce data quality [12,22], which brings challenges to data-driven applications [7,16] on big data. To repair missing values, data imputation [23] or data repairing methods [21] are studied from

© Springer Nature Switzerland AG 2019
J. Shao et al. (Eds.): APWeb-WAIM 2019, LNCS 11641, pp. 61–76, 2019.
https://doi.org/10.1007/978-3-030-26072-9_5

kinds of perspectives. However, such methods usually employ continuous time-consuming components [6,14]. Sometimes, not all missing values can be repaired, such as missing ratings for movies from users [17]. Biased query answers on incomplete data can mislead decisions [11]. To avoid such biases, the query completeness problem [19,20] is investigated to determine whether a query can get a complete answer on incomplete data under different conditions. But such methods usually suffer from high time complexity beyond NP-Completeness [13,20].

In order to determine query completeness fast, a small set of tuples with statistical properties guaranteed can be used to estimate query completeness. In such tuples from incomplete datasets, *coverage* plays an important role in providing enough information for further decisions. Therefore, to estimate query completeness fast and objectively, data used in estimation should satisfy two conditions of *coverage* and *completeness* (formally defined in Sect. 2): (1) they should provide enough kinds of information of interest; (2) the ratio of data containing different extent of information can be specified, and the ratio of complete values of chosen data is the same as that of the entire data. For instance, a movie in a recommendation system can be assigned a score by a sum of weighted ratings from different users, which can be done by a query on ratings. Since a user only watches a few movies [17], there are many missing ratings for a user. To compute the score objectively and fast, each user chosen for estimation should rate more than a specified number of movies, and such users should exceed a specified ratio. Besides, to approximately reflect the statistical properties of the entire ratings dataset, the average number of movies rated by chosen users should be almost the same as that by entire users. Therefore, to fast and objectively obtain a query completeness result, a novel data selection mechanism is crucial and should be carefully designed.

In this paper, *CoCo* core-sets with a small size satisfying the two conditions: *Coverage* and *Completeness* respectively are proposed to estimate query completeness on incomplete data fast and objectively. There may be many CoCo core-sets in the entire dataset, and any of them can provide effective and enough information. Besides, on low completeness data, CoCo core-sets can also be used as a completeness controllable subset to improve data quality for further applications, such as package queries [3] or recommendation systems [9].

Besides of the size constraint, there are other challenges to draw CoCo core-sets. (1) To satisfy the coverage condition, interesting attributes with complete values should be contained as many as possible. (2) To satisfy the completeness condition, tuples with different number of complete values should be considered. And such tuples should be carefully selected to preserve the statistical properties on the entire data. (3) The methods of drawing CoCo core-sets should be efficient and scalable.

Core-sets are designed for answering queries approximately [1]. But many methods are only tailored to numeric attribute values [1,15], which cannot be directly used in many systems containing categoric attribute values as well [10]. Furthermore, these methods cannot be used to estimate query completeness on incomplete data, since they does not consider information loss in the core-sets brought by missing values. Methods based on sampling [2,4] can select a subset of the original dataset with statistical properties approximately guaranteed. But such subset does not necessarily satisfy the coverage condition and completeness condition. Moreover, such sampling methods may choose many samples with too many missing values on incomplete data, which make little contribution to estimation.

To overcome the drawbacks of existing methods, we propose CoCo core-sets defined formally in Sect. 2 to provide the coverage and completeness information of tuples of interest drawn from the entire dataset. The main contributions of this paper are summarized as follows:

1. CoCo core-sets used to estimate query completeness are firstly investigated, where both coverage and completeness conditions are satisfied. The problem of drawing a CoCo core-set is proved to be NP-Complete. An effective approximate algorithm to draw a high-quality CoCo core-set is proposed with linear time complexity in the size of the dataset.
2. For the completeness part, a strategy to assign error bounds is designed and proved. An algorithm based on uniform sampling using this strategy is proposed. By theoretical analysis, the tuples chosen can be proved to approximately satisfy the completeness condition at any given error bound with high probability, and the time complexity is linear in the size of the dataset.
3. For the coverage part, a strategy to guess the optimal size of tuples for coverage is proposed, followed by an efficient algorithm based on uniform sampling. By theoretical analysis, we show that our algorithm can achieve at least a $(1 - 1/e - \epsilon)$ approximation guarantee in expectation to the optimum coverage, and the time complexity is linear in the size of the dataset.
4. Experiments on both real-world and synthetic datasets demonstrate that the algorithm can effectively and efficiently draw tuples preserving properties of the entire dataset for query completeness estimation, and have a well scalability.

2   Problem Definition and Computational Complexity

In this section, we first introduce basic definitions of *coverage* and *completeness*, and then formalize the problem of drawing a CoCo core-set, followed by a theorem providing time complexity of the problem.

Definition 1 (Coverage). *Given a relational dataset D, the attributes $A = \{A_j\}$, $1 \leq j \leq m$, an attribute A_j is covered by a tuple t, if the attribute value $t[A_j]$ is complete, i.e., not missing, otherwise, A_j is not covered by t. The coverage of a tuples set is the size of union of attributes covered by the tuples in the set.*

Definition 2 (Completeness). *Given a relational dataset D, with tuples size n, the attributes $A = \{A_j\}$, $1 \leq j \leq m$, the tuple completeness, denoted as $Cpl(t)$, is defined as $Cpl(t) = \dfrac{\sum_{a \in A} Cpl(t[a])}{m}$, where $Cpl(t[a]) = 1$, if attribute value*

$t[a]$ *is complete, otherwise* $Cpl(t[a])$ *is 0. The* dataset completeness, *denoted as* $Cpl(D)$, *is defined as follows:* $Cpl(D) = \dfrac{\sum\limits_{t \in D} Cpl(t)}{n} = \dfrac{\sum\limits_{t \in D} \sum\limits_{a \in A} Cpl(t[a])}{mn}$, *where* $Cpl(t[a])$ *is defined as above.*

Based on definitions of coverage and completeness, a CoCo core-set is defined below.

Definition 3 *(CoCo Core-set).* *Given a relational dataset* D, *the attributes* $A = \{A_j\}$, $1 \le j \le m$, *real numbers* $\epsilon, \theta, \alpha \in [0,1]$, *and a positive integer* y, *a CoCo core-set* T_{CC} *is a subset of the entire dataset with minimum size satisfying:*

1. *(Property A) there are at least* y *attributes covered by at least one tuple;*
2. *(Property B) the completeness of the CoCo core-set is approximate to that of the entire dataset, i.e.,* $\left| \dfrac{Cpl(D) - Cpl(T_{CC})}{Cpl(D)} \right| \le \epsilon$;
3. *(Property C) the ratio of tuples with tuple completeness above* θ *is no less than* α;

In the definition above, property A corresponding to the coverage condition requires a CoCo core-set to contain complete values on several different attributes to provide enough information for further operations. The property B and property C together correspond to the completeness condition. The property B is the relative error bound property, which contributes to estimation of completeness of query answers on incomplete data, and property C makes a CoCo core-set can contain different tuples with different tuple completeness under a given ratio. In this paper, to make use of an incomplete dataset, the conflict in property C made by extreme θ is not considered.

The hardness of drawing a CoCo core-set is given below. Due to space limit, proofs are omitted. The problem can be reduced from the *Set Covering* problem with tuples constructed from given set.

Theorem 1. *Drawing a CoCo Core-set of size* N *or less is NP-Complete.*

3   Drawing an Approximate CoCo Core-Set

Since the problem of drawing a CoCo core-set is NP-Complete, we shall propose an efficient approximate algorithm *DCC* to draw an approximate CoCo core-set. With high probability, the coverage and completeness condition will be satisfied with error bound guaranteed. We first introduce DCC in Subsect. 3.1 and then show algorithms DCP and DCV used in DCC for completeness and coverage in Subsects. 3.2 and 3.3 respectively.

3.1   An Approximate Algorithm

Intuitively, since a CoCo core-set consists of tuples for two goals: completeness and coverage, DCC will draw tuples by two steps. Firstly, tuples for completeness are drawn to make sure the dataset completeness of such drawn tuples is close to that of the entire tuples. Secondly, tuples for coverage are drawn to cover as many uncovered attributes as possible. Such uncovered attributes are not covered by the tuples for completeness. Since DCC usually cannot know the optimal size of tuples covering all specified attributes, DCC guesses the optimal size and then selects tuples for coverage by a greedy mechanism. After these two steps, tuples for completeness and coverage are combined together as an approximate CoCo core-set. To make the size of an approximate CoCo core-set as small as possible, the numbers of tuples for completeness and coverage are carefully designed.

Algorithm 1. DCC

    Input   : a relational dataset D with n tuples and attributes of size m,
                real numbers $\epsilon_{CV}, \epsilon_{CP}, \delta, \theta, \alpha > 0$, and a positive integer y
    Output: An approximate CoCo core-set T_{CC}
1: $T_{CC} \leftarrow \phi$;
   // Drawing tuples for completeness
2: $T_{CP} \leftarrow \mathsf{DCP}(D, \epsilon_{CP}, \delta, \theta, \alpha)$;
   // Drawing tuples for coverage
3: $T_{CV} \leftarrow \mathsf{DCV}(D \backslash T_{CP}, T_{CP}, \epsilon_{CV}, y)$;
   // Combining tuples
4: $T_{CC} \leftarrow T_{CV} \cup T_{CP}$;
5: **return** T_{CC};

    The algorithm DCC is shown in Algorithm 1. As the intuitive idea, an approximate CoCo core-set consists of the tuples for completeness T_{CP} drawn by DCP and the tuples for coverage T_{CV} drawn by DCV. Given a dataset with bounded number of attributes, since the time complexity of DCP and DCV is $O(n)$ and $O(n)$ respectively shown next, the time complexity of DCC is $O(n)$, where n is the number of the tuples. Note that, the size of the approximate CoCo core-set drawn by DCC is determined by the given error bounds for completeness and coverage.

3.2   Drawing Tuples for Completeness

To select tuples satisfying Property B in Definition 3 efficiently, methods such as [5] based on uniform sampling are proposed. Tailored to complete dataset, such a method cannot provide samples satisfying Property C in Definition 3.

    To draw tuples for completeness efficiently, DCP performs uniform sampling to draw tuples. Given a relative error bound ϵ and failure probability δ, the minimum sample size (MSS) is given in Eq. (1) [5], where $\phi_{\delta/2}$ is the $\delta/2$ fractile of the standard normal distribution, and $\inf(C_t)$ and $\sup(C_t)$ are the minimum and maximum number of complete values in a tuple respectively. Therefore, the minimum sample size is determined by error bound ϵ and failure probability δ.

$$MSS(\epsilon, \delta) = \left\lceil \frac{\phi_{\delta/2}^2}{\epsilon^2} \left(\frac{\sup{(C_t)}}{\inf{(C_t)}} - 1 \right) \right\rceil \qquad (1)$$

To satisfy Property C, methods to determine sizes of tuples with different tuple completeness must be carefully designed. When tuples from different parts are combined, the dataset completeness of combined samples should be close to that of the entire original dataset. To achieve this goal, the given error bound and failure probability must be divided so that sizes of tuples with different tuple completeness can be determined according to corresponding error bounds and failure probabilities.

Next, we first introduce a strategy to assign error bound and failure probability to tuples with different tuples completeness, and then DCP is proposed based on this strategy.

Assigning Errors and Failure Probabilities. When a real number θ is given, tuples are divided into two parts by its tuple completeness. One is tuples with tuples completeness at least θ, denoted as T_H, and the other is tuples with tuples completeness less than θ, denoted as T_L. Clearly, the dataset completeness of the entire data is given as below:

$$Cpl(D) = \rho_H \times \frac{|T_H|}{|D|} + \rho_L \times \frac{|T_L|}{|D|}$$

where $\rho_H = Cpl(T_H)$ and $\rho_L = Cpl(T_L)$. If ρ_H and ρ_L can be approximated by sampling, then $Cpl(D)$ can be approximated by sampling. Therefore, with different errors and failure probabilities, sample sizes of T_H and T_L can be determined.

To describe how well a random variable is approximated, (ϵ, δ) - approximation is defined in Definition 4 [5].

Definition 4. $((\epsilon, \delta)$ - approximation$)$ $\hat{Y}$ is called as an (ϵ, δ) - approximation of Y if $P\left(\left|\frac{\hat{Y}-Y}{Y}\right| \geq \epsilon\right) \leq \delta$ for any $\epsilon \geq 0$ and $0 \leq \delta \leq 1$, where $P(X)$ is the probability of a random variable X.

Now we introduce a strategy to assign error bounds and failure probabilities between samples from T_H and T_L.

Theorem 2. $\widehat{Cpl(D)}$ is an (ϵ, δ) - approximation of $Cpl(D)$, if $Cpl(S_H)$ is an (ϵ, Δ_H) - approximation of $Cpl(T_H)$, and $Cpl(S_L)$ is an (ϵ, Δ_L) - approximation of $Cpl(T_L)$, satisfying $\Delta_H + \Delta_L \leq \delta$, and sizes of S_H and S_L are given by Eq. (1).

By Theorem 2, tuples with different tuple completeness can be sampled differently, according to different goals. For T_H, the error bound is ϵ_H, and the failure probability Δ_H. And ϵ_L and Δ_L are the counterparts for T_L. From Theorem 2, we can set $\epsilon_H = \epsilon_L = \epsilon$, and $\Delta_H + \Delta_L = \delta$. By Eq. (1), since given a fixed

---

Algorithm 2. DCP

Input : a relational dataset D with n tuples, real numbers $\epsilon > 0, \delta > 0, \theta > 0$,
$\quad\quad\quad \alpha > 0$
Output: a tuples set T_{CP} for completeness
1: $T_{CP} \leftarrow \phi$;
   // compute error probabilities of each part
2: $\Delta \leftarrow$ AssignPros$(\epsilon, \delta, \alpha)$; // Δ_H and Δ_L are both in Δ
   // sampling on tuples with tuple completeness at least θ
3: $s_H \leftarrow$ MSS(ϵ, Δ_H);
4: $S_H \leftarrow$ UniformSampling(D, s_H);
5: $T_{CP} \leftarrow T_{CP} \cup S_H$;
   // sampling on tuples with tuple completeness under θ
6: $s_L \leftarrow$ MSS(ϵ, Δ_L);
7: $S_L \leftarrow$ UniformSampling(D, s_L);
8: $T_{CP} \leftarrow T_{CP} \cup S_L$;
9: **return** T_{CP};

ϵ, a sample size is determined monotonely with δ, proper sample sizes satisfying Property C can be obtained by binary search on δ.

The Algorithm DCP. The algorithm DCP based on uniform sampling is shown in Algorithm 2.

To describe DCP consistently, symbols defined above are continuously used. As is shown above, given error bound ϵ, failure probability δ, and ratio requirement α, the failure probability Δ_H and Δ_L can be determined. With ϵ and Δ_H, sample size s_H for uniform sampling on T_H can be computed by Eq. (1), where $\sup(C_t)$ can be set to m, and $\inf(C_t)$ can be set to 1. Then by uniform sampling implemented in a reservoir [24], a tuples set S_H is sampled and added into T_{CP}. And then uniform sampling on T_L is similar. Therefore, T_{CP} contains tuples satisfying Property B and Property C in Definition 3, which makes completeness condition is approximately satisfied. Since there are 2 scans on T_H and T_L to sample tuples, and either scan takes $O(n)$ by using reservoir sampling, the time complexity of DCP is $O(n)$, where n is the number of the tuples.

3.3   Drawing Tuples for Coverage

Since the coverage function defined on a tuple set is submodular, to efficiently draw tuples for coverage, sampling is a feasible methodology to obtain tuples of interest with controllable error bound and high probability. Like the sampling method stochastic-greedy [18] or SG for short, DCV uses a similar idea of greedily covering attributes by choosing a best sampled tuple covering most uncovered attributes in each round. For SG, the number of tuples k for coverage is fixed by a user. Therefore, SG tries to use k tuples to cover as many attributes as possible. But for DCV, the optimal size of tuples for covering uncovered attributes is usually unknown. What's worse, a user cannot provide such a size either. To get a proper trade-off between coverage and size, DCV shown below tries to

---

Algorithm 3. DCV

Input : a relational dataset D with n tuples, a tuples set T_{CP}, a real number
$\epsilon > 0$, and a positive integer y

Output: a tuples set T_{CV} for coverage

1: $T_{CV} \leftarrow \phi$;
2: $l \leftarrow \max(\lfloor \ln \frac{1}{\epsilon} \rfloor, 1)$;
   // Obtain the number of not covered attributes
3: $h \leftarrow$ Uncovered(T_{CP}, y);
   // Guessing the number of optimal size in each round
4: $k_G \leftarrow \frac{h}{l}$;
   // cover attributes by l rounds
5: **for** $i \leftarrow 1$ **to** l **do**
      // Sample s tuples by uniform sampling
6:     $s \leftarrow \frac{n}{k_G} \ln \frac{1}{\epsilon}$;
7:     **for** $j \leftarrow 1$ **to** k_G **do**
8:        $Samples \leftarrow$ UniformSampling(D, s);
          // Find the tuple covering most uncovered attributes
9:        $t_j \leftarrow$ FindBest$(Samples)$;
10:       $T_{CV} \leftarrow T_{CV} \cup \{t_j\}$;

11: **return** T_{CV};

---

guess a proper size, and then DCV uses SG multiple times to cover attributes. With theoretical analysis below, we can show that DCV covers more uncovered attributes, with at most a constant factor more samples.

The Algorithm DCV. The algorithm DCV based on uniform sampling is shown in Algorithm 3. Given a user specified number y of attributes to cover, DCV needs to know which attributes are not covered by the tuples in T_{CP}. And then DCV uses l rounds to cover these attributes. To take advantage of SG, DCV guesses a proper size k_G, which is size of tuples chosen in each round. With k_G, sample size for uniform sampling in each round can be computed. Among the samples, a tuple covering most uncovered attributes is selected and added into the result set. DCV can be easily modified to stop earlier than k_G rounds as soon as all attributes are covered.

There is an alternative method, denoted as DSG. Since each tuple can at least cover one uncovered attributes, we need at most h tuples. Therefore, after knowing that which attributes are not covered, DSG can simply invoke SG to cover such attributes with $k = h$. With fewer samples in each round, DSG is faster than DCV. But as the theoretical results shown below and experimental results shown in Sect. 4, DSG usually selects almost the same size of best tuples as DCV, but obtains low coverage due to a smaller samples size in each sampling.

In each uniform sampling, by using a reservoir [24], it takes $O(n)$ to obtain samples, where $n = |D|$. It takes $O(s)$ to find a best tuple to cover most remaining uncovered attributes, given a small or bounded number of attributes. The time complexity of DCV is $O(n)$, where $n = |D|$.

Since DCV usually cannot know the optimal size of tuples for covering uncovered attributes, DCV has to guess a size of tuples to cover these attributes. With such guessed size, DCV runs multiple times of SG to efficiently cover attributes as Lemma 1. The coverage and time complexity of DCV is shown in Theorem 3.

Lemma 1. *Given a size k of tuples for covering attributes, after l times run of SG, the coverage in expectation is at least $f(I^*)(1 - l\epsilon - e^{-l})$, where $f(I^*)$ is the optimum coverage with such k tuples. And if $l \leq \ln \frac{1}{\epsilon}$, the more times SG is run, the more coverage in expectation is obtained.*

From Lemma 1, though multiple runs of SG improve coverage, too many times run may not bring more benefit. Therefore, DCV guess a proper size of tuples for covering, if such guessed size is too small, DCV can take advantage of multiple runs to obtain great improvement. Otherwise, since each tuple can at least cover one attribute, the optimal size is at most h, which makes DCV achieve the same approximate ratio as SG. Based on this idea, the coverage of DCV is shown in Theorem 3.

Theorem 3. *If l is set to $l = \ln \frac{1}{\epsilon}$, and k_G is set to $k_G = \frac{h}{l}$, then with $O(n(\ln \frac{1}{\epsilon})^2)$ coverage function evaluations, DCV achieves a $(1 - \epsilon + \epsilon \ln \epsilon)$ approximation guarantee in expectation to the optimum solution if $k_G \geq k^*$, and if $k_G < k^*$, DCV achieves a $(1 - \epsilon - 1/e)$ approximation, where h is the number of uncovered attributes and k^* is the optimal size of tuples covering such attributes.*

4   Experimental Results

In this section, we will show the experimental results of the proposed algorithm DCC on different datasets with different parameters. All the experiments are implemented on a Microsoft Windows 10 machine with an Intel(R) Core i7-6700 CPU 3.4 GHz and 32 GB main memory. Programs are compiled by Microsoft Visual Studio 2017 by C++ language. Each program is run 10 times on the same dataset to show stability, and the time cost given below is the average time cost. The completeness is defined as in Sect. 2.

4.1   Datasets and Metrics

MovieLens Dataset. The relation MovieLens (ML for short) is downloaded from MovieLens website[1], which is marked as MovieLens 1M Dataset on the website. Specially, our MovieLens contains 6040 records (users ratings, varying from 1 to 5), with 3952 dimensions (movies), and the size is 89.1 MB. The missing ratio is 95.81%, namely, including 1 million valid ratings. Actually, there are 246 movies without any ratings. Though there are many attributes (movies)

---

[1] MovieLens dataset: https://grouplens.org/datasets/movielens/1m/.

compared to records, we can still view the number of attributes as a bounded number, since such data is a part of the whole dataset, and usually there are far more users than movies.

Cars Information. The relation *Cars* is crawled from a website[2] providing services on cars. Without changing any value, we built the relation Cars including information from dealers selling new cars. The relation Cars consists of 8 attributes of a dealer: name, stock, star-rate, ratings, location, distance from the main company, phone number, and email. There are 542850 tuples in this relation, and the size is 152 MB. The dataset completeness of the relation Cars is 0.843.

Synthetic Relations. To evaluate efficiency of our algorithm DCC, we generated a collection of synthetic relations with 100 attributes and tuples number from 10K to 5M, whose sizes are from 5.38 MB to 2.65 GB correspondingly. The synthetic datasets are generated in two steps. First, there are 100 tuples with only one complete attribute value in each tuple to cover all the attributes, which is used to test effectiveness of algorithms to cover attributes. Second, there are other tuples with incomplete attribute values, where attribute values are missing with different probabilities. Detailed information are shown in Subsect. 4.3.

Metrics. We shall evaluate our algorithm by four metrics: relative error, coverage, efficiency and size of the approximate CoCo core-set. *Relative Error* denoted as *RE* measures the extent to which the approximate CoCo core-set can give an estimate of the dataset completeness of the entire dataset, which is shown by the relative error between the estimated completeness by the approximate CoCo core-set and the real dataset completeness. *Coverage* measures the number of attributes of interest covered. *Efficiency* measures the time cost of an algorithm, and *Size* measures the size of the approximate CoCo core-set drawn by our algorithm DCC. To show coverage improvement, we also implemented stochastic-greedy or SG for short, from [18] in C++.

According to Algorithm 1, DCC can be further evaluated by performance of DCP and DCV respectively. The performance of DCP is evaluated by *RE*, *size* and *efficiency* under different error bounds ϵ_{CP}, while DCV is evaluated by *size*, *coverage* and *efficiency* under different error bounds ϵ_{CV}. In all the experiments below, y is set be the attributes number of the dataset. We shall first report experimental results on real datasets, and then we shall give results on synthetic datasets used to further evaluate DCC.

---

[2] Cars dataset: http://www.cars.com.

Table 1. Experimental results of DCP on real datasets

ϵ_{CP}	MovieLens			Cars		
	RE	Size	TimeCost (s)	RE	Size	TimeCost (s)
0.10	0.003	3943	11.867	0.019	1875	316.297
0.15	0.006	2803	11.443	0.018	833	317.432
0.20	0.002	1699	11.250	0.017	468	317.680
0.25	0.003	1088	11.014	0.018	299	315.806
0.30	0.021	754	11.178	0.018	207	313.929

4.2 Experiments on Real Datasets

Performance of DCP. To investigate performance of DCP, we ran a group of experiments to observe RE, sizes and efficiency under different ϵ_{CP}, and the results are shown in Table 1 and depicted in Fig. 1. We set $\epsilon_{CV} = 0.05$, the failure probability $\delta_{CP} = 0.6$, and θ is dataset completeness of such two real datasets, $\alpha = 0.4$. For T_H, we set $\Delta_H = 0.2$, and $\Delta_L = 0.4$ for T_L. Notice that we set a large failure probability δ_{CP}, because incomplete dataset can lead to a large failure probability. But since we can assign failure probabilities, we set Δ_H to be smaller, because we want more tuples with large tuple completeness to reduce failure probability on the entire dataset, which is demonstrated true as the experimental results shown below. The variable $sup(C_t)$ in Eq. (1) on the MovieLens dataset is set to be 30, since there are too many missing values, while $Sup(C_t)$ is set to be 9 on the Cars dataset, since tuples in Cars dataset are with high tuple completeness and we add an extra attribute ID. The variable $inf(C_t)$ in both MovieLens and Cars dataset are set to be 1, since there is at least one attribute without missing values.

Relative Errors of DCP. As in Table 1, real relative errors in the RE columns under both datasets are very small, far from the corresponding given relative error bounds ϵ_{CP}, which shows that the strategy to assign relative errors and failure probabilities is effective. And tuples in T_{CP} can provide good estimates of the real dataset completeness of the two real datasets. For the MovieLens dataset, there are many tuples with low tuple completeness, which leads to a low dataset completeness. Since T_{CP} can give good estimates under different relative error bounds, a user can use T_{CP} to analyze the query completeness on such low completeness dataset, which will give guides for designing queries with completeness guaranteed.

Sizes in DCP. In Fig. 1, on the MovieLens dataset, tuples sampled with high tuple completeness no less than θ is denoted as H-ML, while tuples sampled with low tuple completeness less than θ is denoted as L-ML, and the sum of the two parts are denoted as All-ML. H-Cars, L-Cars and All-Cars are defined similarly on the Cars dataset. Figure 1 depicts the different sizes of tuples in T_{CP} on the two real datasets with different ϵ_{CP}.

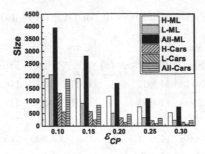

Fig. 1. Sizes of T_{CP} on real datasets

As is shown in Fig. 1, Algorithm 2 samples different sizes of tuples with different tuple completeness according to given θ and α by Theorem 2. And the completeness condition is satisfied on all the experiments. For example, when $\epsilon_{CP} = 0.2$, there are 327 tuples sampled with high tuple completeness, and 141 tuples sampled with low tuple completeness on the Cars dataset. On the Movie-Lens dataset, there are 1187 tuples with high tuple completeness and 512 tuples low tuple completeness. Since there are only 1892 tuples with tuple completeness no less than given θ, when $\epsilon_{CP} = 0.10$, all the 1892 tuples are sampled. To get a good estimate of $Cpl(ML)$, Algorithm 2 has to select 2051 tuples. On the Cars dataset, since there are sufficient tuples with different tuple completeness, Algorithm 2 always selects more tuples on high tuple completeness. Therefore, by Theorem 2, tuples with different tuple completeness can be drawn to improve data usability from low-quality datasets according to given ratio requirement for further use.

Efficiency of DCP. As is shown in Table 1, changes of ϵ_{CP} do not bring obvious changes of time cost on both datasets. Since the uniform sampling is implemented by reservoir sampling [24], which needs to scan all the dataset. Therefore, given the same dataset, the time cost stays almost the same.

Table 2. Experimental results of DSG and DCV on the MovieLens dataset

ϵ_{CV}	DSG				DCV			
	TimeCost (s)	Size	Coverage	ToCover	TimeCost (s)	Size	Coverage	ToCover
0.05	293.383	107	169	178	296.432	115	175	178
0.10	282.515	96	154	172	285.697	106	168	172
0.15	393.939	124	221	239	392.969	119	219	239
0.20	298.532	98	150	181	297.446	104	155	181
0.25	417.606	138	200	252	415.157	119	226	252
0.30	316.068	93	157	190	313.560	90	155	190

Performance of DCV. To investigate performance of DCV, we ran a group of experiments to observe coverage and efficiency with different tuple sizes under different ϵ_{CV}, and the results are shown in Table 2. In the experiments, we set $\epsilon_{CP} = 0.3$, failure probability $\delta_{CP} = 0.6$, and $\alpha = 0.4$. For T_H, we set $\Delta_H = 0.2$, and $\Delta_L = 0.4$ for T_L. The variable $sup(C_t) = 30$ on the ML dataset, and $sup(C_t)$ is set to be 9 on the Cars dataset. The variable $inf(C_t) = 1$ in both ML and Cars dataset. Since DCV uses the similar idea of SG [18], to show the improvement of DCV, we modified the Algorithm 1 into a new Algorithm DSG, which invokes SG after invoking DCP on the same dataset used in DCC. On the Cars dataset, since there are many tuples with high tuple completeness, T_{CP} always covers all the attributes, which makes Algorithm 3 skipped. Therefore, only experimental results of DCV on the ML dataset is shown below. In Table 2, experimental results of the two algorithms: DSG and DCV are shown separately. For each ϵ_{CV}, DCV was run 10 times, the worst coverage is listed in the column *Coverage*, with the time cost in seconds in the column *TimeCost(s)*, size of T_{CV} in the column *Size*, and number of uncovered attributes by T_{CP} in the column *ToCover*. And the corresponding results are shown in the counterparts under the *DSG* part.

Coverage of DCV. On the MovieLens dataset, since there are a few ratings, which brings challenges to Algorithm 3, but as shown in Table 2, both DSG and DCV covers enough attributes under different ϵ_{CV} with as small size of T_{CV} as possible. As is shown in Table 2, when $\epsilon_{CV} \leq 0.2$, DCV covers most attributes, and covers more attributes than DSG, even if some attributes are with lower frequencies. Because a samller ϵ_{CV} leads to a larger l, which brings more chances to cover attributes as in Theorem 3. When ϵ_{CV} is too large, such as $\epsilon_{CV} \geq 0.4$, l can be very small. Since l is at least 1, which makes DCV perform almost the same as DSG.

Efficiency of DCV. As shown in Table 2, with almost the same time as DSG, DCV covers most uncovered attributes. Since there are only 6040 tuples in the MovieLens, and reservoir sampling is used, multiple tries of sampling the best tuples for coverage do not bring too much time cost with different ϵ_{CV}. To evaluate DCV more deeply, synthetic relations are generated based on the MovieLens dataset, and the experimental results are shown in Subsect. 4.3.

4.3   Experiments on Synthetic Relations

To further evaluate our algorithms, we generated synthetic relations with different dataset completeness. Besides, rare attributes with fewer frequencies were generated to test coverage of DCV. The results are shown in Table 3 and Fig. 2, where the results are the average values of 10 runs under each number of tuples, and TC is short for Time Cost in seconds. In the experiments, for coverage, we set $\epsilon_{CV} = 0.1$, θ is the dataset completeness of current dataset. For completeness, we set $\epsilon_{CP} = 0.3$, $\Delta_H = 0.2$, $\Delta_L = 0.4$, and $\delta = 0.6$.

   As is shown in Table 3, though we set a loose error bound and high failure probability in DCP, DCC can still obtain small relative errors as Theorem 2

Table 3. Experimental results on synthetic relations

#Tuples	Size	TC (s)	RE	Coverage
10k	2640	14.319	0.017	92
50k	2636	69.278	0.008	94
100k	2648	161.424	0.003	96
500k	2639	686.560	0.005	95
1M	2647	1576.560	0.003	94
5M	2641	7024.790	0.006	93

shows. And the size of approximate CoCo core-set is rather small compared to the entire dataset. For example, there are 1M tuples in the entire dataset, while there are only 2647 tuples in the CoCo core-set with relative error 0.00294 and 94 attributes covered on average.

As depicted in Fig. 2, where both number of tuples and time cost are shown in logarithm with base 10, the algorithm DCP is linear as analyzed. Though DCV is a little worse than linearity, it does not bring too much time to DCC, which shows well scalability of our algorithms.

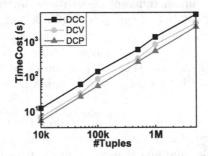

Fig. 2. Time cost on synthetic relations

5   Related Work

Core-sets for Numeric Values. The concept of numeric core-sets is firstly from [1], followed by many other methods such as [4,15]. There are two main drawbacks of methods above. One is that they are only numeric core-sets, and the other is that only aggregated values are stored, which leads to loss of detailed features described in the origin dataset.

Drawing Core-sets by Sampling. Sampling can be used to draw a subset of original entire data, such as [2,8]. In the papers above, the minimized sample sizes can be computed in that each sample on complete data can make contributions to approximately answering queries. But due to missing values on incomplete data, effective information may be lost in samples, which makes it hard to provide minimized sample sizes for different queries. Tuples containing different extent of information should be sampled differently.

Determining Query Completeness. The problems of whether a query can get a complete answer on different kinds of data are studied by many researchers [13,20]. Such problems can be called *query completeness* for short. The time complexity of such a problem is usually beyond NP-Complete, which makes it very hard to determine query completeness fast on the entire dataset. A CoCo core-set with a small size is drawn from the entire dataset, which makes it possible to design efficient methods to estimate query completeness fast based on a CoCo core-set.

6   Conclusion

This paper investigates the problem of drawing CoCo core-sets on incomplete relational data. We analyze the time complexity of the problem and prove that it is NP-Complete. To solve this problem effectively and efficiently, an algorithm is proposed to draw an approximate CoCo core-set based on uniform sampling techniques. We prove that our algorithm can draw an approximate CoCo core-set, which almost preserves coverage and completeness properties of the entire dataset. Experimental results on both real-world and synthetic datasets show that our algorithms can draw an effective CoCo core-set on low-quality dataset as theoretical analysis, and our algorithms have a well scalability on larger datasets. Based on an approximate CoCo core-set, query completeness can be estimated fast and objectively, and the algorithms of estimation are our future work.

Acknowledgment. This work was supported by the National Key R&D Program of China under Grant 2018YFB1004700, and the National Natural Science Foundation of China under grants 61832003, U1811461 and 61732003.

References

1. Agarwal, P.K., Har-Peled, S., Varadarajan, K.R.: Approximating extent measures of points. J. ACM **51**(4), 606–635 (2004)
2. Agarwal, S., Mozafari, B., Panda, A., Milner, H., Madden, S., Stoica, I.: BlinkDB: queries with bounded errors and bounded response times on very large data. In: Proceedings of the 8th ACM European Conference on Computer Systems, pp. 29–42. ACM (2013)
3. Brucato, M., Abouzied, A., Meliou, A.: Package queries: efficient and scalable computation of high-order constraints. VLDB J. **27**, 1–26 (2017)
4. Cheng, S., Cai, Z., Li, J., Fang, X.: Drawing dominant dataset from big sensory data in wireless sensor networks. In: 2015 IEEE Conference on Computer Communications (INFOCOM), pp. 531–539. IEEE (2015)
5. Cheng, S., Li, J.: Sampling based (ϵ, δ)-approximate aggregation algorithm in sensor networks. In: 29th IEEE International Conference on Distributed Computing Systems, ICDCS 2009, pp. 273–280. IEEE (2009)

6. Chu, X., et al.: Katara: a data cleaning system powered by knowledge bases and crowdsourcing. In: Proceedings of the 2015 ACM SIGMOD International Conference on Management of Data, pp. 1247–1261. ACM (2015)
7. Chung, Y., Krishnan, S., Kraska, T.: A data quality metric (DQM): how to estimate the number of undetected errors in data sets. Proc. VLDB Endow. 10(10), 1094–1105 (2017)
8. Cohen, E., Kaplan, H.: Tighter estimation using bottom k sketches. Proc. VLDB Endow. 1(1), 213–224 (2008)
9. Deng, T., Fan, W., Geerts, F.: On the complexity of package recommendation problems. In: Proceedings of the 31st ACM SIGMOD-SIGACT-SIGAI Symposium on Principles of Database Systems, PODS 2012, pp. 261–272 (2012)
10. Deng, T., Fan, W., Geerts, F.: On recommendation problems beyond points of interest. Inf. Syst. 48, 64–88 (2015)
11. Dong, X.L., et al.: Knowledge-based trust: estimating the trustworthiness of web sources. Proc. VLDB Endow. 8(9), 938–949 (2015)
12. Fan, W.: Data quality: from theory to practice. ACM SIGMOD Rec. 44(3), 7–18 (2015)
13. Fan, W., Geerts, F.: Capturing missing tuples and missing values. In: Proceedings of the Twenty-Ninth ACM SIGMOD-SIGACT-SIGART Symposium on Principles of Database Systems, pp. 169–178. ACM, June 2010
14. Hao, S., Tang, N., Li, G., Li, J.: Cleaning relations using knowledge bases. In: 2017 IEEE 33rd International Conference on Data Engineering (ICDE), pp. 933–944. IEEE (2017)
15. Indyk, P., Mahabadi, S., Mahdian, M., Mirrokni, V.S.: Composable core-sets for diversity and coverage maximization. In: Proceedings of the 33rd ACM SIGMOD-SIGACT-SIGART Symposium on Principles of Database Systems, PODS 2014, pp. 100–108. ACM (2014)
16. Khalefa, M.E., Mokbel, M.F., Levandoski, J.J.: Skyline query processing for incomplete data. In: IEEE 24th International Conference on Data Engineering, ICDE 2008, pp. 556–565. IEEE (2008)
17. Miao, X., Gao, Y., Zheng, B., Chen, G., Cui, H.: Top-k dominating queries on incomplete data. IEEE Trans. Knowl. Data Eng. 28(1), 252–266 (2016)
18. Mirzasoleiman, B., Badanidiyuru, A., Karbasi, A., Vondrak, J., Krause, A.: Lazier than lazy greedy. In: Twenty-Ninth AAAI Conference on Artificial Intelligence (2015)
19. Razniewski, S., Korn, F., Nutt, W., Srivastava, D.: Identifying the extent of completeness of query answers over partially complete databases. In: Proceedings of the 2015 ACM SIGMOD International Conference on Management of Data, pp. 561–576. ACM (2015)
20. Razniewski, S., Nutt, W.: Completeness of queries over incomplete databases. Proc. VLDB Endow. 4(11), 749–760 (2011)
21. Rekatsinas, T., Chu, X., Ilyas, I.F., Ré, C.: Holoclean: holistic data repairs with probabilistic inference. Proc. VLDB Endow. 10(11), 1190–1201 (2017)
22. Sadiq, S., et al.: Data quality: the role of empiricism. SIGMOD Rec. 46(4), 35–43 (2018)
23. Song, S., Zhang, A., Chen, L., Wang, J.: Enriching data imputation with extensive similarity neighbors. Proc. VLDB Endow. 8(11), 1286–1297 (2015)
24. Vitter, J.S.: Random sampling with a reservoir. ACM Trans. Math. Softw. (TOMS) 11(1), 37–57 (1985)

Reducing Wrong Labels for Distant Supervision Relation Extraction with Selective Capsule Network

Zihao Wang, Yong Zhang$^{(\boxtimes)}$, and Chunxiao Xing

Research Institute of Information Technology,
Beijing National Research Center for Information Science and Technology,
Department of Computer Science and Technology, Institute of Internet Industry,
Tsinghua University, Beijing 100084, China
zhangyong05@tsinghua.edu.cn

Abstract. Distant Supervision is a common technique for relation extraction from large amounts of free texts, but introduces wrong labeled sentences at the same time. Existing deep learning approaches mainly rely on CNN-based models. However, they fail to capture spatial patterns due to the inherent drawback of pooling operations and thus lead to suboptimal performance. In this paper, we propose a novel framework based on Selective Capsule Network for distant supervision relation extraction. Compared with traditional CNN-based models, the involvement of capsule layers in the sentence encoder makes it more powerful in encoding spatial patterns, which is very important in determining the relation expressed in a sentence. To address the wrong labeling problem, we introduce a high-dimensional selection mechanism over multiple instances. It is one generalization of traditional selective attention mechanism and can be seamlessly integrated with the capsule network based encoder. Experimental results on a widely used dataset (NYT) show that our model significantly outperform all the state-of-the-art methods.

1 Introduction

Relation extraction, which aims at extracting semantic relations from free texts, is a crucial research topic in the area of natural language processing. One attractive approach to solve this problem is *distant supervision* [14], which automatically generates labeled data by heuristically aligning relational facts in a knowledge base, such as Freebase [1] with an unlabeled corpus. The assumption here is that if a sentence in the corpus contains two named entities of a relational fact in the knowledge base, the sentence will correspondingly express the relation. However, this assumption might fail in many situations. For instance, a pair of entities could either express more than one kind of relations or no relation at all. But the assumption of distant supervision ignores this possibility, which leads

---

This work was supported by NSFC (91646202), National Key R&D Program of China (SQ2018YFB140235).

J. Shao et al. (Eds.): APWeb-WAIM 2019, LNCS 11641, pp. 77–92, 2019.
https://doi.org/10.1007/978-3-030-26072-9_6

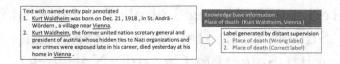

Fig. 1. An example of distant supervision with wrong label in NYT dataset

to the wrong labeling problem. Next let's look at an example shown in Fig. 1. In the first sentence, it is wrong labeled even the sentence contains exactly two entities "Kurl Waldheim" and "Vienna" in knowledge base.

Recently neural network models have shown remarkable successes in different NLP tasks [13,21,30]. A serious of studies also adopted neural networks in the problem of relation extraction [10,15,27,29]. Unlike previous feature-based methods, they can automatically encode the sentence into low-dimensional semantic embedding space $\mathbb{R}^n$ and predict the relation labels according to similarity measurement between vector representations. In this way, they can avoid the inherited data sparsity problem of feature-based methods.

To address the wrong labeling problem brought by distant supervision, previous studies [7,19,29] adopt multi-instance learning (MIL) to address the problem, by treating each produced sentence differently during training. The basic idea is to organize all sentences for the same relation mention into one bag and perform training in the unit of bag. In order to generate the representation of a bag, many techniques have been proposed, such as using one representative sentence [29], average over all sentences [10], and selectivity attention [10]. Our work falls into this category. However, these approaches just use a scalar obtained from limited linear function to evaluate the contribution of one sentence. As a result, they cannot provide enough information to determine the contribution of each sentence and provide good bag representation.

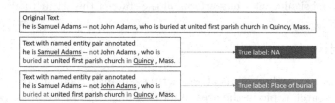

Fig. 2. An example in NYT dataset for the necessity of spatial pattern in relation extraction. The colored words are necessary to make good predictions (Color figure online)

In this paper, we present an end-to-end framework to enhance the sentence encoding as well as to reduce the noises generated in the process of distant supervision. One limitation of previous studies [9,10,12,29] is that their sentence encoders are all based on the Convolutional Neural Network (CNN). However, the pooling operations of CNN fail to capture the spatial relationship between features in a particular layer. Nevertheless, in the task of relation extraction, it is also necessary to well represent spatial patterns within a sentence. Figure 2 shows a good example. For the same sentence, we want to recognize the relations between entity "Quincy" and two different entities "Samuel Adam" and "John Adam". As their spatial patterns with colored words are different, they definitely express different relations. If we just use pooling operation to capture the most significant local feature without considering the spatial pattern (piecewise pooling will not help since the concerned spatial patterns locate between two entities), it will result in wrong predictions. As a result, previous CNN-based approaches would definitely suffer from a significant loss of information about the internal properties of present entities in a sentence and relationships between them.

The recently proposed Capsule Network [6,17] has shown promising ability to overcome the limitations of CNN models in image processing tasks. Specifically, it organizes multiple neurons into a *capsule* to represent internal properties of an instance [5]. The output of each capsule is a vector associated with activation probability that the corresponding instance exists in the image. To overcome the shortcoming of pooling operation in CNN, Sabour *et al.* [17] also describes a dynamic routing scheme that could smartly decide the interaction between capsules in different layers. To this end, we adopt capsule layers as well as the dynamic routing mechanism in our sentence encoder. We replace the scalar-output feature detectors of CNNs with capsule-output layers and devise effective routing-by-agreement strategy in the similar spirit of the original work [6,17]. In this way, our encoder can better preserve the intrinsic features of the sentence, such as texture patterns like local order of words or semantics of phrases as well as spatial patterns across the sentences, and generate a high quality representation of sentence.

Similar to previous studies, we also adopt multi-instance learning technique to alleviate the wrong labeling problem. Instead of just using one scalar to decide the weight of one sentence [10], we propose a novel mechanism *capsule alignment* which utilizes an latent vector to decide the contribution of each sentence in the bag. We formulate the problem of assigning weights to different sentences as solving a linear equation system, which makes it more explainable. Then we propose a reverse-routing algorithm to address this problem and generate the bag representation under the same formulation. Experimental results on a popular dataset demonstrate our superiority over state-of-the-art methods.

The main contributions of this paper are summarized as following:

- We employ the capsule network to construct a relation extractor which can effectively embed the semantics of entity context. With routing-by-agreement mechanism, it can enrich the representation of sentence by capturing the spatial patterns among words.
- We integrate the multi-instance learning techniques into our capsule-based framework to alleviate the wrong labeling problem. Specifically, we propose the reverse-routing mechanism to make full use of the supervision information from knowledge base.

– We conduct extensive experiments on a widely used dataset. The results show
that our proposed model clearly outperform state-of-the-art methods.

2   Related Work

Distant Supervision. Traditional supervised methods [2,28] utilized kernel
based techniques to extract relation from plain texts. However, these methods
need a large number of annotated data, which is difficult to obtain when faced
with web scale data. Some unsupervised frameworks [25,26] focused on extracting infoboxes from Wikipedia web pages.

Distant Supervision was first proposed by Mintz et al. [14], which aligns
relational facts from Freebase with plain texts to generate labeled corpus for
relation extraction. Many state-of-the-art methods [7,16,19] adopted feature-
based models to improve the effectiveness. PCNN model [29] is the widely used
neural network model which is able to generate the representation of sentences
with relation mentions automatically.

Reducing Wrong Labels. Many approaches have been proposed to address
the wrong labeling problem in distant supervision relation extraction. Takamatsu
et al. [20] tried to model the process of distant supervision and thus reduced the
number of wrong labels. Gormley et al. [3] improved the performance of relation
extraction by enriching the embedding of features. Lin et al. [10] improved the
multi-instance learning based methods by adopting selectivity attention mechanism.

Some approaches tried to recognize the true positive samples from distant
supervision dataset before relation extraction. Luo et al. [12] reduced the influence of noisy information by learning noise transition patterns. Ye et al. [27]
solved this problem by ranking ties in relations. Recently Wang et al. [22,23]
utilized the Generative Adversarial Network (GAN) and deep reinforcement
learning techniques to alleviate the wrong labeling problem by re-sampling the
dataset respectively.

Some methods adopted external resources such as handcrafted feature engi-
neerings to recognize true positive samples. Gormley et al. [3] introduced
enriched embedding of features. Liu et al. [11] utilized heterogeneous information
source. Su et al. [18] enhanced the textual embedding of relations with global
statistics and proposed an ensemble-based method. These studies are orthogonal
to our approach and can be seamlessly integrated into our framework to further
improve the performance.

Capsule Network. Recently Sabour et al. [17] proposed a novel model *Capsule
Network* to overcome some inherited drawbacks of CNN. The basic idea is to
divide each layer of a neural network into many small groups of neurons called
"capsules" [5]. Hinton et al. [6] further proposed an EM routing algorithm to
capture richer information from capsules in the routing process. Jaiswal et al. [8]
improved the GAN with the help of capsules. Recently the capsule network
has also been adopted in some NLP tasks, such as text classification [31] and

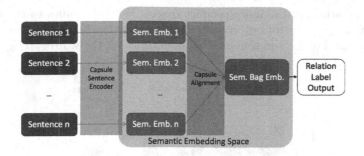

Fig. 3. Model overview

sentiment analysis [24]. To the best of our knowledge, our work is the first one to adopt the capsule network in the task of relation extraction.

3 Methodology

In this section, we introduce our framework for relation extraction. The overall architecture is shown in Fig. 3. It consists of four components: embedding layer, capsule sentence encoder, capsule alignment layer and output layer. Capsule sentence encoder and capsule alignment layer are the core components for selective capsule network. Next we will introduce the details of them.

3.1 Embedding Layer

The first component is the embedding layer that encodes the original sentence into matrix representation. To detect the relation mention in both semantic and structural way, we follow previous studies and adopt position embeddings [10,29] along with word embedding for each token. d_w and d_p is the dimension of word and position embedding, respectively. The embedding vector for i-th token is $s_i \in \mathbb{R}^{d_0}$, where $d_0 = d_w + d_p$. Next we construct the sentence matrix representation $s \in \mathbb{R}^{L_0 \times d_0}$ by concatenating all the word embeddings:

$$s = s_1 \oplus s_2 \oplus ... \oplus s_{L_0} \tag{1}$$

where L_0 is the maximum length of all sentences. We will use zero padding for shorter sentences.

3.2 Capsule Sentence Encoder (CSE)

In this work, we adopt the capsule network as our capsule sentence encoder (CSE). Figure 4 is the structure of our capsule sentence encoder. Unlike CNN based models, the capsule network handles capsule (vector) rather than single neuron (scalar) as the base unit in the network. And it could preserve richer information than traditional CNN-based encoders due to the versatile representation and rich operations in the inner-product spaces. When it comes to the relation extraction task, the vector output capsule represents the instantiate parameters such as the local order and semantic representation of the words [31].

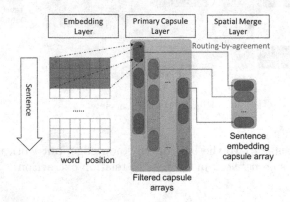

Fig. 4. The structure of the capsule sentence encoder

Our capsule sentence encoder consists of two layers: primary capsule layer (PCL) and spatial merge layer (SML). In PCL, we use the capsule convolutional kernels to transform the input sentence matrix representation $\mathbf{s}$ into filtered capsule arrays $\mathbf{U} = [\bar{\mathbf{u}}_1, ..., \bar{\mathbf{u}}_k, ..., \bar{\mathbf{u}}_m]$, where the filtered capsule array $\bar{\mathbf{u}}_k \in \mathbb{R}^{L_1 \times d_1}$ is the output of k-th capsule convolutional filter (m filters in total), L_1 is the length of the capsule array and d_1 is the capsule length in each array. Then, SML merges each filtered capsule array $\bar{\mathbf{u}}_k = [\mathbf{u}_{1,k}, ..., \mathbf{u}_{j,k}, ..., \mathbf{u}_{L_1,k}]$, $\mathbf{u}_{j,k} \in \mathbb{R}^{d_1}$ into single capsule $\mathbf{v}_k \in \mathbb{R}^{d_2}$ by routing-by-agreement mechanism. The merging operation is spatial-aware since we use different transformations at different positions. In short, the capsule sentence encoder is one mapping from $\mathbb{R}^{L_0 \times d_0}$ to $\mathbb{R}^{m \times d_2}$ which captures texture pattern by PCL and spatial pattern by SML.

Primary Capsule Layer (PCL). In PCL, each capsule convolutional filter is defined by window size $2w + 1$ and stride step. Its input is some sub-matrix under the filter while its output is a vector. For the k-th filter of weight $W_k \in \mathbb{R}^{(2w+1) \times d_0 \times d_1}$ and bias $b_k \in \mathbb{R}^{d_1}$, the sub-matrix under the j-th filter window is converted into capsule $\mathbf{u}_{j,k} \in \mathbb{R}^{d_1}$ by linear transformation. Then the output of each filter is calculated in Eq. (2).

$$\mathbf{u}_{j,k} = \sum_{\hat{j}=-w}^{w} \sum_{\hat{k}=1}^{d_0} W_k^{\hat{j},\hat{k},\cdot} \mathbf{s}_{j+\hat{j},\hat{k}} + b_k \tag{2}$$

where capsule $\mathbf{u}_{j,k}$ can be regarded as a "semantic chunk" which preserves particular semantic and spatial information of the sentence.

Spatial Merge Layer (SML). In the spatial merge layer, we devise routing-by-agreement mechanism to merge each array of capsules into single capsule. As discussed previously, spatial pattern plays a significant role in determining the relation expressed in a sentence. To this end, we merge all capsules generated by one filter in PCL into one capsule in SML. In this way, we can adaptively aggregate the information from child semantic chunks into parent as well as bound the complexity of the model.

For each capsule $u_{j,k}$, we first apply spatial specific linear transformation $T_j \in \mathbb{R}^{d_1 \times d_2}$ to get the spatial specific representation in Eq. (3).

$$u_{k|j} = T_j u_{j,k} \tag{3}$$

After the linear transformation, we combine the spatial specific representation $u_{k|j}$ into v_k by the weighted average in Eq. (4).

$$z_k = \sum_{j=1}^{L_1} \alpha_{jk} u_{k|j} \tag{4}$$

where α_{jk} is the weight for corresponding capsule.

Algorithm 1. Routing-by-agreement for the spatial specific representation of one capsule array

---

Require: routing iteration r and spatial-aware capsule array $\{u_{k|j}\}_{j=1}^{L_1}$ for k-th filter
Ensure: merged capsule v_k for k-th filter
    Initialize all $c_{jk} \leftarrow 0$
    for r iterations **do**
        Update the value of α_{jk} with Equation (6)
        $z_k \leftarrow \sum_{j_1=1}^{L_1} \alpha_{jk} u_{k|j}$
        $v_k \leftarrow \frac{\|z_k\|^2}{1+\|z_k\|^2} \frac{z_k}{\|z_k\|}$
        $c_{jk} \leftarrow c_{jk} + \langle v_k, u_{k|j} \rangle$
    return $\{v_k\}_{k=1}^m$

When z_k is obtained, we apply squashing [17] as the non-linear function to get the output capsule v_k in SML (Eq. (5)).

$$v_k = \frac{\|z_k\|^2}{1 + \|z_k\|^2} \frac{z_k}{\|z_k\|} \tag{5}$$

To obtain α_{jk}, we devise a routing-by-agreement mechanism as is shown in Algorithm 1. The basic idea is to update the values of α_{jk} iteratively to guarantee that the each parent capsule in SML gathers the information from child capsules in PCL by reasonable contribution. In this way, SML preserves the spatial information about all the child semantic chunks and automatically learns spatial patterns related to the relation.

$$\alpha_{jk} = \frac{\exp(c_{jk})}{\sum_j \exp(c_{jk})} \tag{6}$$

So far, for each input sentence $\mathbf{s}$, we use the capsule sentence encoder to generate its representation as a capsule array $\bar{\mathbf{v}} = [\mathbf{v}_1, \cdots, \mathbf{v}_m] \in \mathbb{R}^{m \times d_2}$. The capsules in the array aggregate both local texture pattern and sentence-wise spatial pattern.

3.3   Capsule Alignment

We adopt multi-instance learning techniques to alleviate the wrong labeling problem. Following the previous study [10], it is safe to make the following assumption: a relation expressed by two entities in one sentence can be inferred from evidences in all sentences that contain the same two entities collectively. Therefore, for a triplet (e_1, e_2, r), we construct a bag using all sentences which mention both e_1 and e_2 and use the relation $r \in \mathcal{R}$ as the label of the bag. Suppose there is a bag with n sentences, we denote its representations generated by some sentence encoder as $\mathcal{B} = \{\mathbf{v}^{(1)}, ..., \mathbf{v}^{(n)}\}$. Then by exploiting the information of all sentences, we can reduce the effect of wrong labeled sentences by leveraging those with correct labels within the same bag.

Given some relation label r, state-of-the-art method [10] utilized selective attention to generate the bag embedding from sentences within the bag. Specifically, it estimated the contribution weight $w_{i|r}$ for each instance $\mathbf{v}^{(i)}$, which satisfies $w_{i|r} > 0$ and $\sum_i w_{i|r} = 1$. Geometrically, the selective attention selects one point $\mathbf{b}_r$ from the convex hull $\mathcal{C}$ supported by the $\{\mathbf{v}^{(1)}, ..., \mathbf{v}^{(n)}\}$. Under this point of view, selective attention is the adaptive extension of earlier methods such as finding the one representative sentence (ONE) [29] or average over all sentences in the bag (AVE) [10].

In this section, we design *capsule alignment* mechanism to generate a bag embedding by the contribution weighted average following similar idea. However, the sentence embedding from capsule sentence encoder is *capsule array* [1] rather than scalar array. Meanwhile, the capsule is the representation whose dimension expresses particular semantic information. As we will introduce, it is questionable to directly apply selective attention to the sentence embeddings in shape of capsule array. To address such issues, we first provide an equivalent formulation for selective attention. Then we introduce the capsule alignment naturally following this route.

Target-Contribution **Formulation.** In the original derivation of selective attention, the latent scalar is firstly calculated by $a_{i|r} = \langle \mathbf{v}^{(i)}, \mathbf{y}_r \rangle$ for each relation r, where $\mathbf{y}_r$ is the vector representation for relation label. The contribution weights are directly calculated from latent scalar by softmax operation $w_{i|r} = \texttt{softmax}(a_{i|r})$. This process could also be viewed under *Target-Contribution* formulation, in which we inference the *Target* in latent space and calculate the *Contribution* by solving linear equation system (Fig. 5).

We have the observation that $\mathbf{y}_r$ is a linear function that maps the embedding instances $\{\mathbf{v}^{(1)}, \dots, \mathbf{v}^{(n)}\}$ into $\mathbb{R}^1$ as latent scalars $\{a_{1|r}, \dots, a_{n|r}\}$. In latent space

---

[1] Each capsule array here is a matrix rather than vector.

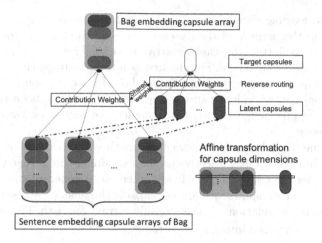

Fig. 5. The capsule alignment mechanism

$\mathbb{R}^1$, selective attention defines the target scalar t_r by

$$t_r = \sum_{i=1}^{n} \texttt{softmax}(a_{i|r}) \cdot a_{i|r} \tag{7}$$

Equivalently, we can calculate the embedding weight $w_{i|r}$ by solving the linear equation system in Eq. (8).

$$\sum_{i=1}^{n} w_{i|r} a_{i|r} = t_r, \quad \sum_{i=1}^{n} w_{i|r} = 1, \quad w_{i|r} > 0 \tag{8}$$

The solution of Eq. (8) finds the linear representation of target scalar t_r by latent scalar $\{a_{1|r}, \ldots, a_{n|r}\}$ in latent space. It looks trivial since we already have $w_{i|r} = \texttt{softmax}(a_{i|r})$ as one possible solution (and the unique solution when $n = 1, 2$). And then we can finally calculate the bag embedding using Eq. (9).

$$\mathbf{b}_r = \sum_{i=1}^{n} w_{i|r} \mathbf{v}^{(i)} \tag{9}$$

Capsule Alignment from *Target-Contribution* Formulation. Next we will discuss how to derive capsule alignment mechanism under *Target-Contribution* formulation Unlike the scalar array in the selective attention mechanism, the sentence embeddings $\bar{\mathbf{v}}^{(i)} = [\mathbf{v}_1^{(i)}, \ldots, \mathbf{v}_m^{(i)}]$ by capsule sentence encoders is an array of *capsules*. The bag embedding $\bar{\mathbf{b}}_r = [\mathbf{b}_{1|r}, \ldots, \mathbf{b}_{m|r}]$ is a capsule array generated by contribution weighted average of the input sentence representations. So the major difficulty of applying selective attention is to handle the bag of capsule arrays $\mathcal{B} = \{\bar{\mathbf{v}}^{(1)}, \ldots, \bar{\mathbf{v}}^{(n)}\}$ instead of vectors. One straightforward way to apply the selective attention to our case is to first convert all capsules into vectors.

There are two feasible approaches. The first solution is to take the l^2-norm of each capsule in the array and then vector representation is obtained. The second solution is to flatten the capsule array into single vector while sacrificing the inherited capsule structure. For the first solution, it drops the capsule information by compressing a vector to a scalar. For the second solution, it ignores the dimensional correspondence among capsules, which have been proved to be important prior knowledge [17,31] to construct the model in low-dimensional parametric space.

To overcome the deficiency of above naive methods, we propose capsule alignment, a high-dimensional bag selective mechanism under the *Target-Contribution* formulation. The basic idea of capsule alignment is to generate the weights $w_{i|r}$ by preserving the correspondence among capsule dimension by 4 steps.

Firstly, given a relation r, the capsule array for i-th instance $\bar{\mathbf{v}}^{(i)} = [\mathbf{v}_1^{(i)}, ..., \mathbf{v}_m^{(i)}]$ is converted into latent vector $\mathbf{x}_{i|r} \in \mathbb{R}^{d_2}$

$$x_{i|r,l} = \sum_{k=1}^{m} \mathbf{v}_{k,l}^{(i)} \mathbf{w}_k^{r,l} + b^{r,l} \qquad (10)$$

where $\mathbf{w}^{r,l}$ and $b^{r,l}$ are the weights and bias for relation r and the l-th dimension of capsules. Those affine transformations combine the information from capsule array by spatial merge layer while respecting the dimensional correspondence among capsules. The representation of each instance in the latent space is capsule $\mathbf{x}_{i|r} \in \mathbb{R}^{d_2}$ rather than scalar in the selective attention case. Our proposed method here treats the dimensions in capsules heterogeneously. One simplified homogeneous version will apply single affine transformation to all dimensions. We will discuss the differences in the evaluation section.

Algorithm 2. Reverse-routing to estimate the contribution weights

---

Require: target vector $\mathbf{t}_r$ and bag of latent vectors $\{\mathbf{x}_{i|r}\}_{i=1}^n$
Ensure: contribution weights $\{w_{i|r}\}_{i=1}^n$
   For i-th latent vector: $\gamma_{i|r} \leftarrow \langle \mathbf{t}_r, \mathbf{x}_{i|r}/\|\mathbf{x}_{i|r}\| \rangle$
   For i-th latent vector: $w_{i|r} \leftarrow \texttt{softmax}(\gamma_{i|r})$
   return $\{w_{i|r}\}_{i=1}^n$

Secondly, we inference the target capsule $\mathbf{t}_r$ in the latent vector space. This target is used in the following step to calculate the contribution weights for each instance.

$$t_{r,l} = \sum_{i=1}^{n} \texttt{softmax}\left(\sigma\left(x_{i|r,l}\right)\right) x_{i|r,l} \qquad (11)$$

To agree with the dimensional correspondence among capsules, $t_{r,l}$ is inferred by the corresponding dimension and then composed into the latent target vector $\mathbf{t}_r$. $\sigma(x) = 1/(1 + \exp(-x))$ is the sigmoid function. The softmax function in Eq. (11) works on index i.

Thirdly, $w_{i|r}$ are supposed to be derived by solving the linear equation system like the selective attention case.

$$\sum_{i=1}^{n} w_{i|r}\mathbf{x}_{i|r} = \mathbf{t}_r, \quad \sum_{i=1}^{n} w_{i|r} = 1, \quad w_{i|r} > 0 \tag{12}$$

However, Eq. (12) might have no solution if the latent target vector $\mathbf{t}_r$ is out of the convex hull $\mathcal{C} \subset \mathbb{R}^{d_2}$ supported by latent capsules $\{\mathbf{x}_{1|r}, \ldots, \mathbf{x}_{n|r}\}$. So we relax the task of solving Eq. (12) by considering approximate solution. Inspired by previous routing-by-agreement mechanism, we introduce the reverse-routing mechanism to estimate the contribution $w_{i|r}$.

Like routing-by-agreement, this procedure measures the inner-product similarity between the input and output capsules. It is a reversed version since the output of routing is inferred as target $\mathbf{t}_r$ and considered as one input of Algorithm 2.

Finally, we generate the bag embedding by Eq. (13) using $w_{i|r}$:

$$\bar{\mathbf{b}}_r = \sum_{i=1}^{n} w_{i|r}\bar{\mathbf{v}}^{(i)} \tag{13}$$

which is also the capsule array.

Follow the original capsule network [17], the capsules can be compressed into a vector $\mathbf{b}_r = [\|\mathbf{b}_{1|r}\|, \ldots, \|\mathbf{b}_{m|r}\|]$ by l^2-norm. Finally in the output layer, we use a softmax operation for all $r \in \mathcal{R}$ over this l^2-normed vector to get the conditional probabilities:

$$\Pr(r|s) = \texttt{softmax}(W\mathbf{b}_r + o) \tag{14}$$

In Eq. (14), each conditional probability can be regarded as the confidence score of the corresponding relation. And the final output of the network is a vector of scores whose length is the number of relations.

3.4 Training Process

As we adopt multi-instance learning technique, the objective is to discriminate bags rather than instances. Given the pairs of bag embedding $\mathbf{b}_i$ and relation label $(\mathbf{b}_i, r_i) \in \mathcal{B}$, we define the objective function using cross-entropy loss at bag level as following:

$$J(\theta) = \sum_{(\mathbf{b}_i, r_i) \in \mathcal{B}} \log p(r_i \mid \mathbf{b}_i; \theta) \tag{15}$$

where θ is the set of parameters to be learned. We solve the optimization problem with Stochastic Gradient Descent over shuffled mini-batches. All the weights are initialized using Xavier initializer [4].

4   Evaluation

In this section, we conduct an extensive set of experiments to demonstrate the effectiveness of our proposed methods. Our selective capsule network framework consisting of capsule sentence encoder and heterogeneous capsule alignment layer is denoted by CSE+Heter.

4.1   Experiment Setup

Dataset. We evaluate our proposed method on a widely used dataset: New York Times (NYT), which is first presented in [16]. It is generated by aligning entity pairs from Freebase with New York Times corpus (NYT). The sentences from the years 2005–2006 are used as the training set and sentences from 2007 are used as the testing set. We use entity pairs from Freebase [1] for distant supervision. There are 52 actual relations and a special relation NA which indicates there is no relation between the two entities.

Following previous studies, we evaluate our model with held-out evaluation, which compares the extracted facts with those in Freebase. It provides an approximate measure of performance without requiring expensive human evaluation. We use precision-recall curve (PR curve) as the main metric for evaluation.

Hyper-parameter Settings. The dimension of word embeddings is 50 and that of positional embeddings is 5. And in the primary capsule convolutional layer, the window size is 3 and stride is 1. The number of filter and length of capsule are set to 32 and 8 respectively. For the spatial capsule merge layer, we keep the length of merged capsule to be 8. With above settings, the complexity of our model is comparable with the previous study [10] except the spatial-aware transformations in SML. For the training process, the learning rate is fixed to 0.01 and the batch size is 160. The embedding data is the same as Lin.

4.2   Compare with State-of-the-art Methods

In this part, we compare our model with several state-of-the-art methods. Same as our approach, these methods only use Freebase for distant supervision and do not rely on any other external resource or feature engineering.

- CNN + ATT [10] is the method with CNN for sentence encoding and selectivity attention to reduce wrong labels.
- PCNN + ATT [10] is the method with PCNN [29] model for sentence encoding and selectivity attention to reduce wrong labels.
- PCNN+ATT+TM [12] is the method which can learn noise transition patterns from data.
- PCNN+ATT+DSGAN [22] is the method using GAN to recognize positive samples from the results of distant supervision.
- PCNN+ATT+RL [23] is the method that adopts deep reinforcement learning to recognize positive samples from the results of distant supervision.

We obtain the source code from the original authors and use the default settings. Although there are some other popular methods [7,14,19,27], they have been proved to be beaten by above selected baselines in previous studies. So here we do not compare with them.

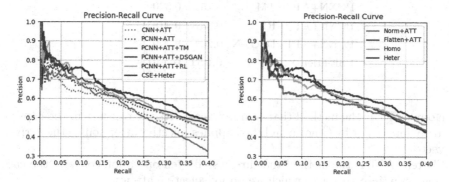

Fig. 6. Left: comparison with state-of-the-art methods; Right: effect of capsule alignment mechanisms

The results of comparison with state-of-the-art methods are shown in Fig. 6. First let's look at the results comparing with CNN-based encoders, we can see that our proposed model outperforms CNN+ATT and PCNN+ATT models with a large margin. This is because our capsule network based encoder can generate high quality sentence representation for spatial patterns. Even compared with the comprehensive models with additional noise reduction modules, our method still has advantages. With the help of capsule alignment, our methods can make full use of sentences within a bag and overcome the drawbacks of encoder-selector based models. Meanwhile, the noise modeling and data argumentation methods mentioned in such models could be combined with our model seamlessly to further improve the performance.

Table 1 compares the AUC score and best F_1 score for the models in Fig. 6. The AUC scores for these PR curves reflect the area size under them. The best F_1 score is collected among the recall level between 0 and 0.4, which is a reasonable range in most previous studies. This result demonstrates the effectiveness of our proposed methods from another aspect.

4.3 Detailed Analysis of Capsule Alignments

Next we make detailed analysis on the effect of our proposed *capsule alignment* mechanism. As we have discussed, the bag embedding can be obtained by either capsule alignment or selective attention with degenerating the capsule array into vector. The baseline methods are constructed based on this idea. We implemented the following methods:

Table 1. AUC and best F_1 score in recall range $[0, 0.4]$

Model	AUC	Best F_1 score
CNN+ATT	0.219	0.380
PCNN+ATT	0.252	0.423
PCNN+ATT+TM	0.257	0.370
PCNN+ATT+RL	0.261	0.418
PCNN+ATT+DSGAN	0.264	0.429
CSE+Heter	**0.281**	**0.440**

- Norm+ATT is the method that compress converts the capsules in the capsule array into scalar by l^2-norm and then applies selective attention on the derived vector.
- Flatten+ATT is the method that first use flatten operation converts the capsule array into vector to which we apply selective attention.
- Homo is the method that treats the dimensions in capsules with the same affine transformation and generates the weights by reverse-routing algorithm.
- Heter is the method that treats the dimensions in capsule with the different affine transformations and generates the weights by reverse-routing algorithm.

The PR curves of these approaches are shown in Fig. 6. It is observed that Heter obtains the best performance. Norm+ATT and Homo do not perform better as they suffer from the information loss. Meanwhile, Homo with $m+1$ parameters outperforms the Norm+ATT with m parameter in a large margin, which shows that the dimensional correspondence among capsules really helps to parametrize our model in low-dimensional space. On the other hand, Heter is comparable with Flatten+ATT but performs better at low recall level. It shows that dimensional correspondence among capsules could be used to improve the performance.

5   Conclusion

In this paper, we have proposed a novel framework based on capsule network for distant supervision relation extraction. We employed the capsule network to enhance the representation of sentences and well capture the texture patterns and spatial patterns. We further devised a capsule alignment mechanism on the basis of multi-instance learning to make full use of the information from sentence representations and alleviate the wrong labeling problem. Experimental results on a real world dataset show that our framework significantly outperforms state-of-the-art methods.

References

1. Bollacker, K.D., Evans, C., Paritosh, P., Sturge, T., Taylor, J.: Freebase: a collaboratively created graph database for structuring human knowledge. In: SIGMOD, pp. 1247–1250 (2008)
2. Bunescu, R.C., Mooney, R.J.: Subsequence kernels for relation extraction. In: NIPS, pp. 171–178 (2005)
3. Gormley, M.R., Yu, M., Dredze, M.: Improved relation extraction with feature-rich compositional embedding models. In: EMNLP, pp. 1774–1784 (2015)
4. Gregor, K., LeCun, Y.: Emergence of complex-like cells in a temporal product network with local receptive fields. CoRR abs/1006.0448 (2010). http://arxiv.org/abs/1006.0448
5. Hinton, G.E., Krizhevsky, A., Wang, S.D.: Transforming auto-encoders. In: Honkela, T., Duch, W., Girolami, M., Kaski, S. (eds.) ICANN 2011. LNCS, vol. 6791, pp. 44–51. Springer, Heidelberg (2011). https://doi.org/10.1007/978-3-642-21735-7_6
6. Hinton, G.E., Sabour, S., Frosst, N.: Matrix capsules with EM routing. In: ICLR (2018)
7. Hoffmann, R., Zhang, C., Ling, X., Zettlemoyer, L.S., Weld, D.S.: Knowledge-based weak supervision for information extraction of overlapping relations. In: ACL, pp. 541–550 (2011)
8. Jaiswal, A., AbdAlmageed, W., Natarajan, P.: Capsulegan: generative adversarial capsule network. CoRR abs/1802.06167 (2018)
9. Jiang, X., Wang, Q., Li, P., Wang, B.: Relation extraction with multi-instance multi-label convolutional neural networks. In: COLING, pp. 1471–1480 (2016)
10. Lin, Y., Shen, S., Liu, Z., Luan, H., Sun, M.: Neural relation extraction with selective attention over instances. In: ACL (2016)
11. Liu, L., et al.: Heterogeneous supervision for relation extraction: a representation learning approach. In: EMNLP, pp. 46–56 (2017)
12. Luo, B., et al.: Learning with noise: enhance distantly supervised relation extraction with dynamic transition matrix. In: ACL, pp. 430–439 (2017)
13. Luo, L., et al.: Beyond polarity: interpretable financial sentiment analysis with hierarchical query-driven attention. In: IJCAI, pp. 4244–4250 (2018)
14. Mintz, M., Bills, S., Snow, R., Jurafsky, D.: Distant supervision for relation extraction without labeled data. In: ACL, pp. 1003–1011 (2009)
15. Miwa, M., Bansal, M.: End-to-end relation extraction using LSTMs on sequences and tree structures. In: ACL (2016)
16. Riedel, S., Yao, L., McCallum, A.: Modeling relations and their mentions without labeled text. In: Balcázar, J.L., Bonchi, F., Gionis, A., Sebag, M. (eds.) ECML PKDD 2010. LNCS (LNAI), vol. 6323, pp. 148–163. Springer, Heidelberg (2010). https://doi.org/10.1007/978-3-642-15939-8_10
17. Sabour, S., Frosst, N., Hinton, G.E.: Dynamic routing between capsules. In: NIPS, pp. 3859–3869 (2017)
18. Su, Y., Liu, H., Yavuz, S., Gur, I., Sun, H., Yan, X.: Global relation embedding for relation extraction. In: NAACL-HLT, pp. 820–830 (2018)
19. Surdeanu, M., Tibshirani, J., Nallapati, R., Manning, C.D.: Multi-instance multi-label learning for relation extraction. In: ACL, pp. 455–465 (2012)
20. Takamatsu, S., Sato, I., Nakagawa, H.: Reducing wrong labels in distant supervision for relation extraction. In: ACL, pp. 721–729 (2012)

21. Wang, J., Wang, Z., Zhang, D., Yan, J.: Combining knowledge with deep convolutional neural networks for short text classification. In: IJCAI, pp. 2915–2921 (2017)
22. Wang, W.Y., Xu, W., Qin, P.: DSGAN: generative adversarial training for distant supervision relation extraction. In: ACL, pp. 496–505 (2018)
23. Wang, W.Y., Xu, W., Qin, P.: Robust distant supervision relation extraction via deep reinforcement learning. In: ACL, pp. 2137–2147 (2018)
24. Wang, Y., Sun, A., Han, J., Liu, Y., Zhu, X.: Sentiment analysis by capsules. In: WWW, pp. 1165–1174 (2018)
25. Wu, F., Weld, D.S.: Open information extraction using Wikipedia. In: ACL, pp. 118–127 (2010)
26. Yan, Y., Okazaki, N., Matsuo, Y., Yang, Z., Ishizuka, M.: Unsupervised relation extraction by mining Wikipedia texts using information from the web. In: ACL, pp. 1021–1029 (2009)
27. Ye, H., Chao, W., Luo, Z., Li, Z.: Jointly extracting relations with class ties via effective deep ranking. In: ACL, pp. 1810–1820 (2017)
28. Zelenko, D., Aone, C., Richardella, A.: Kernel methods for relation extraction. J. Mach. Learn. Res. **3**, 1083–1106 (2003)
29. Zeng, D., Liu, K., Chen, Y., Zhao, J.: Distant supervision for relation extraction via piecewise convolutional neural networks. In: EMNLP, pp. 1753–1762 (2015)
30. Zhao, K., et al.: Modeling patient visit using electronic medical records for cost profile estimation. In: Pei, J., Manolopoulos, Y., Sadiq, S., Li, J. (eds.) DASFAA 2018. LNCS, vol. 10828, pp. 20–36. Springer, Cham (2018). https://doi.org/10.1007/978-3-319-91458-9_2
31. Zhao, W., Ye, J., Yang, M., Lei, Z., Zhang, S., Zhao, Z.: Investigating capsule networks with dynamic routing for text classification. CoRR abs/1804.00538 (2018)

Data Mining and Application

Coupled Semi-supervised Clustering: Exploring Attribute Correlations in Heterogeneous Information Networks

Jianan Zhao, Ding Xiao, Linmei Hu, and Chuan Shi[✉]

Beijing University of Posts and Telecommunications, Beijing, China
{zhaojianan,dxiao,hulinmei,shichuan}@bupt.edu.cn

Abstract. Heterogeneous Information Network (HIN) has been widely adopted in various tasks due to its excellence in modeling complex network data. To handle the additional attributes of nodes in HIN, the Attributed Heterogeneous Information Network (AHIN) was brought forward. Recently, clustering on HIN becomes a hot topic, since it is useful in many applications. Although existing semi-supervised clustering methods in HIN have achieved performance improvements to some extent, these models seldom consider the correlations among attributes which typically exist in real applications. To tackle this issue, we propose a novel model SCAN for semi-supervised clustering in AHIN. Our model captures the coupling relations between mixed types of node attributes and therefore obtains better attribute similarity. Moreover, we propose a flexible constraint method to leverage supervised information and network information for flexible adaption of different datasets and clustering objectives. Extensive experiments have shown that our model outperforms state-of-the-art algorithms.

Keywords: Attributed Heterogeneous Information Network ·
Semi-supervised clustering · Coupled Attributes

1 Introduction

Heterogeneous Information Network (HIN) [15], as a new network modeling method, has drawn much attention due to its ability to model complex objects and their rich relations. Moreover, in many real HINs, objects are often associated with various attributes. For example, in Yelp dataset, where businesses and users can be regarded as nodes. Business objects are associated with attributes like locations, ratings and business types; besides, users own attributes including age and gender. Researchers have brought forward the concept of Attributed Heterogeneous Information Network [9] (AHIN) to address HINs with node attributes.

Clustering is a fundamental task in data analysis. Given a set of objects, the goal is to partition them into clusters such that objects in the same clusters are similar to each other, while objects in different clusters are dissimilar.

© Springer Nature Switzerland AG 2019
J. Shao et al. (Eds.): APWeb-WAIM 2019, LNCS 11641, pp. 95–109, 2019.
https://doi.org/10.1007/978-3-030-26072-9_7

Semi-supervised clustering incorporates supervision about clusters into the algorithm in order to improve the clustering results. Till now, many semi-supervised clustering algorithms for information networks have been proposed. Some semi-clustering methods [1,6,8] are proposed for semi-supervised clustering in homogeneous information networks, where links are assumed to be of the same type. As for heterogeneous information networks, GNetMine [4], PathSelClus [16] and SemiRPClus [10] are proposed. All of the algorithms above do not concern the attributes of nodes. Recently, SCHAIN [9] studies the problem of semi-supervised clustering in AHIN and achieves state-of-the-art clustering performance.

Although these methods have achieved satisfactory performances to some extent, they have some obvious shortcomings. First of all, they do not consider the coupling relationships among object attributes. In the real world, attributes are associated with each other and have complex relationships addressed as coupling relationships [10,16]. To illustrate, in a movie dataset, the "budget" and "gross" attribute of a movie is largely dependent on the "country" attribute. Ignoring the dependency between attributes will inevitably lead to inferior clustering results. In addition, previous works cannot flexibly consider the importance of supervision with datasets while many real applications may need to take into account the different importance of supervision for different tasks.

To handle these issues, we put forward an innovative model **S**emi-supervised **C**lustering with **C**oupled **A**ttributes in **A**ttributed Heterogeneous Information **N**etworks (SCAN). Inspired by the newly emerging non-IID learning [3], we take one step further to mine the coupled similarity between node attributes, thus capturing the inter-dependent relationships between attributes. Moreover, we propose a novel constraint method to flexibly leverage the supervision information through which users can adjust the importance of supervised information for different clustering objectives.

The main contributions of this paper can be summarized as follows:

- To our best knowledge, we are the first to mine the coupling relationships between node attributes in AHIN. We propose a coupled node similarity measure to better analyze the inter-dependent relationships between mixed data types of node attributes.
- We propose a novel approach to use supervision information, which is able to leverage node similarity and supervision constraint flexibly.
- We conduct extensive experiments on two real-world datasets. The proposed method is proved to be effective over the state-of-the-art methods.

2   Definitions and Model Overview

2.1   Problem Definition

In this paper, we consider the problem of semi-supervised clustering in AHINs. The terms are defined as follows.

Definition 1. Attributed   Heterogeneous   Information   Network (AHIN) [9]. Let $\mathcal{T} = \{T_1, \ldots, T_m\}$ be a set of m object types. For each type

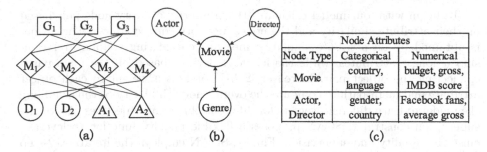

Fig. 1. An AHIN (a), its network schema (b), and its node attributes (c)

T_i, let $\mathcal{X}_i$ be the set of objects of type T_i and A_i be the set of attributes defined for objects of type T_i. An object x_j of type T_i is associated with an attribute vector $\boldsymbol{f}_j = (f_{j1}, f_{j2}, \ldots, f_{j|A_i|})$. An AHIN is a graph $G = (V, E, \mathcal{A})$, where $V = \bigcup_{i=1}^{m} \mathcal{X}_i$ is a set of nodes, E is a set of links, each representing a binary relation between two objects in V, and $\mathcal{A} = \bigcup_{i=1}^{m} A_i$. If $m = 1$ (i.e., there is only one object type), G reduces to a homogeneous information network.

Figure 1a shows an example of movie AHIN consisting of four types of objects: $T = \{$movies (M), actors (A), genres (G), directors (D)$\}$. The network schema [15] is shown in Fig. 1b. In this AHIN, the M, A and D node types are associated with both numerical and categorical attributes shown in Fig. 1c.

A **meta-path** [15] $\mathcal{P} : T_1 \xrightarrow{R_1} \cdots \xrightarrow{R_l} T_{l+1}$ defines a composite relation $R = R_1 \circ \cdots \circ R_l$ that relates objects of type T_1 to objects of type T_{l+1}. We say $\mathcal{P}$ is symmetric if the defined relation R is symmetric. Giving a meta-path $\mathcal{P}$, the multiple specific paths under the meta-path are called path instances denoted by P. To illustrate, in the AHIN shown in Fig. 1a, the meta-path MDM means two movies are directed by the same director, $M_1 D_1 M_4$ is a path instance of the MDM meta-path. Different meta-paths preserve different semantic meanings in AHIN.

Definition 2. Semi-supervised clustering in an AHIN [9]. Given an AHIN $G = (V, E, \mathcal{A})$, a supervision constraint $(\mathcal{M}, \mathcal{C})$ of must-link set $\mathcal{M}$ and the cannot-link set $\mathcal{C}$, a target object type T_i, the number of clusters k, and a set of meta-paths $\mathcal{PS}$, the problem of semi-supervised clustering of type T_i objects in G is to (1) discover an object similarity measure S that is based on object attributes and meta-paths, and (2) partition the objects in X_i into k disjoint clusters $\mathbf{C} = C_1, \ldots, C_k$ based on the similarity measure S such that the clustering results best agree with the constraint $(\mathcal{M}, \mathcal{C})$.

2.2 Model Overview

In this section, we will take a brief look at the proposed model SCAN. As introduced in Definition 1, the task can be decomposed into two steps, namely node similarity calculation and clustering. In the following, we will detail the two steps.

To begin with, our model calculates the similarity of every node pair based on their attribute similarity and the nodes' network connectedness: The former is obtained by an attribute similarity measure considering coupling relationship amongst attributes, while the latter is derived based on the meta-paths connecting the object pair. Moreover, SCAN assigns a weight to each object attribute and meta-path to compute the overall node similarity S, thus enabling the model to learn different weights for different clustering objectives. As for the supervision constraint, SCAN proposes a flexible penalty function to leverage clustering quality and supervision. Finally, SCAN employs the iterative, 2-step learning process of SCHAIN [9] to determine the optimal weights and cluster assignments as output.

The rest of the paper is organized as follows: The calculation of coupled similarity between nodes is explained in Sect. 3, clustering and weight learning process are explained in Sect. 4, extensive experiments along with parameter discussion are illustrated in Sect. 5. Finally, we conclude the whole paper in Sect. 6.

3 Node Similarity with Coupled Attributes

In this section, we introduce how to calculate the similarity between two nodes in an AHIN. Given a node pair $<x_a, x_b>$ of the same node type T_i in an AHIN, the node similarity between a node pair in an AHIN is twofold: On the one hand, similarity can be obtained from attribute similarity; on the other hand, similarity can be measured by the connectedness of the given node pair, namely link similarity. The overall node similarity matrix S can be obtained by aggregating the coupled attribute similarity matrix S_A and the link similarity matrix S_L. To mine the complex coupling relationships between node attributes, we take one step further to calculate the coupled similarity between node attributes. We will detail this concept in the following section.

3.1 Coupled Node Attribute Similarity

In this section we explain how to capture the coupling relations between node attributes. Given a node pair $<x_a, x_b>$ of type T_i, let $\boldsymbol{f}_a$ and $\boldsymbol{f}_b$ be the attribute vectors of x_a and x_b, respectively (see Definition 1). Recall that A_i is the set of attributes associated with type T_i objects. We define an attribute weight vector $\boldsymbol{\omega}$, whose j-th component, ω_j, captures the importance of the j-th attribute in A_i for the clustering task. The entries of S_A are calculated by:

$$S_A(x_a, x_b) = \sum_{j=1}^{|A_i|} \omega_j \cdot valSim(f_{aj}, f_{bj}). \tag{1}$$

The $valSim()$ function calculates the similarity between the attribute value f_{aj} and f_{bj}. Since in real world, many nodes are assigned with mixed type of attributes. We define A_{ij} as the j-th attribute type of node A_i and calculate the value similarity of different attribute types separately.

For numerical attributes, we firstly normalize numerical attribute to $[0,1]$, and use Eq. (2) to convert distance to similarity. Euclidean distance is adopted as distance metric in this paper.

$$valSim(f_{aj}, f_{bj}) = 1 - d(f_{aj}, f_{bj}), \; if \, A_{ij} \, is \, numerical. \tag{2}$$

However, categorical attributes are associated with each other in terms of certain coupling relationships, ignoring attribute value similarity will lead to coarse similarity [17]. This motivates us to design a coupled similarity measure of categorical attributes. Thus, we go one step further to mine the coupled value similarity between categorical attribute values:

$$valSim(f_{aj}, f_{bj}) = (1 - \eta)S_p(f_{aj}, f_{bj}) \\ + \eta S_c(f_{aj}, f_{bj}), \; if \, A_{ij} \, is \, categorical. \tag{3}$$

We capture two levels of categorical value similarity: The plain similarity S_p to measure the value difference of the categorical value, and the coupled similarity S_c to measure the value similarity in terms of couplings. The overall categorical attribute similarity is obtained by a weighted fusion of S_p and S_c using coupled coefficient η.

For plain similarity, we use SMS [7], which uses 0s and 1s to distinguish the similarity between distinct and identical categorical values, to calculate the value difference. As for the coupled similarity, following IeASV [17], we measure the co-occurrence of the measured attribute to other attributes. To elaborate, we bring forward the inter-coupled attribute value similarity.

Definition 3. Inter-coupled attribute value similarity (IAVS). Given two attribute values f_{aj} and f_{bj} from attribute A_{ij} in an AHIN. The inter-coupled attribute value similarity between attribute values f_{aj} and f_{bj} in terms of attribute A_{ik} is denoted by $\delta_{j|k}(f_{aj}, f_{bj})$.

Although the definition of IAVS is similar to IeASV, IeASV only handles categorical data. Since we are dealing with mixed types of data, we generalize the inter-coupled relative similarity candidate A_{ik} to numerical data. Therefore, the coupled similarity of two attribute values $S_c(f_{aj}, f_{bj})$ can be obtained by an aggregated similarity of other attributes:

$$S_c(f_{aj}, f_{bj}) = \frac{\sum_{k=1, k \neq j}^{|A_i|-1} \delta_{j|k}(f_{aj}, f_{bj})}{|A_i| - 1}. \tag{4}$$

Let us explain the intuition behind the equation by a toy example. In the movie AHIN shown in Fig. 1, suppose we are calculating the coupled similarity $S_c(f_{aj}, f_{bj})$ between attribute value f_{aj} = "USA" and f_{bj} = "UK" of attribute A_{ij} = "country". The coupled similarity of "USA" and "UK" will be the aggregated similarity of USA movies and UK movies in terms of other attributes $A_{ik} \in$ { "language", "budget", "gross", "IMDB score"} calculated by Eq. (4).

For categorical attribute similarity, we adopt the inter-coupled relative similarity based on intersection set (IRSI) [17] to capture the co-occurrence relationship between categorical attributes. However, IRSI is designed to measure the dependencies between categorical variables, while the coupling relationships between categorical and numerical attributes remain undiscussed.

To capture the dependency between a categorical attribute and a numerical attribute, we define $Val(f_{aj}, A_{ik})$ as the values sets of attribute A_{ik} generated by the objects with value f_{aj}. In our example, if $A_{ik} =$ "$gross$", then $\delta_{j|k}(f_{aj}, f_{bj})$ stands for the similarity between USA and UK movies in terms of gross respectively. Thus, $Val(f_{aj}, A_{ik})$ and $Val(f_{bj}, A_{ik})$ stands for the gross values of USA and UK movies. The task is to judge the similarity between the two distributions of $Val(f_{aj}, A_{ik})$ and $Val(f_{bj}, A_{ik})$, the closer the two distributions, the more similar the two categorical values are. This is a problem of measuring two finite and continuous distributions.

Inspired by the histogram, we simplify the problem by discretizing the two distributions to finite discrete distributions. The discretization method can be any of the methods described in [5]. We define $\Psi(V_{ajk}, P_{ajk})$ as the discretized distribution of $Val(f_{aj}, A_{ik})$, where V_{ajk} and P_{ajk} are the discretized value sets (known as bins in the histogram) of $Val(f_{aj}, A_{ik})$ and its corresponding value occurrence frequency. For simplicity, we use uniform quantization as our discretization method in the experiments.

Till now, the dependency between a categorical attribute and a numerical attribute has been simplified to the similarity between two discrete distributions. We first calculate the distance between two distributions and then we convert the distance to similarity. Thus, $\delta_{j|k}(f_{aj}, f_{bj})$ can be calculated by the following equation:

$$\delta_{j|k}(f_{aj}, f_{bj}) = \begin{cases} IRSI(f_{aj}, f_{bj}, A_{ik}), & if\ A_{ik}\ is\ categorical. \\ 1 - d(\Psi(V_{ajk}, P_{ajk}), \Psi(V_{bjk}, P_{bjk})), & if\ A_{ik}\ is\ numerical. \end{cases}$$
(5)

Since A_{ik} is numerical, the difference between different discretized sets in V_{ajk} and V_{bjk} can be measured by the distance between the corresponding discretized numerical attribute values (bins in the histogram). To capture this cross-bin relationship, we adopt the earth mover distance (EMD) [12] to measure the distance between the two value distributions.

3.2   Link Similarity and Similarity Aggregation

In this section, we calculate the similarity between a node pair in terms of network connectedness. We adopt the widely used meta-path to calculate link similarity. Given a meta-path set $\mathcal{PS}$, we define a weight vector $\boldsymbol{\lambda}$. Each meta-path $\mathcal{P}_j \in \mathcal{PS}$ is assigned to a weight λ_j, the overall link similarity S_L is obtained by a weighting scheme:

$$S_L(x_a, x_b) = \sum_{j=1}^{|\mathcal{PS}|} \lambda_j \cdot S_{\mathcal{P}_j}(x_a, x_b),$$
(6)

SCHAIN [9] uses PathSim [15] to measure the meta-path based similarity $S_{\mathcal{P}_j}$ for a given path $\mathcal{P}_j$. However, PathSim is designed for computing the similarity between symmetric meta-paths. In real applications, asymmetric meta-paths are also useful and cannot be ignored. To illustrate, in movie networks, MAMDM and MDMAM are asymmetric yet useful meta-paths which preserve the cooperating relationship between actors and directors. Thus, we use HeteSim [13] to calculate $S_{\mathcal{P}_j}$, so that SCAN is capable of handling both symmetric and asymmetric meta-paths.

$$S_{\mathcal{P}_j}(x_a, x_b) = HeteSim(x_a, x_b), \tag{7}$$

The overall node similarity matrix S represents the aggregated similarity of link similarity and coupled attribute similarity, which can be calculated by a weighted sum of S_A and S_L:

$$S = \alpha S_A + (1-\alpha)S_L, \tag{8}$$

where α is a weighting factor that controls the relative importance of the two similarity matrices.

4 Clustering and Weight Learning

4.1 Leveraging the Supervision

The penalty function is introduced in this part. Recall that our objective is to get $\mathbf{C} = C_1, \ldots, C_k$ based on the similarity matrix S such that the clustering results best agree with the constraint $(\mathcal{M}, \mathcal{C})$. After generating the similarity matrix S, we adopt the clustering and optimization framework of SCHAIN. Different from SCHAIN, we propose a more flexible penalty function to better leverage the supervised information.

We use a semi-supervised normalized cut [14] to measure the similarity between clusters. The penalty function of clustering is:

$$\mathcal{J}(\boldsymbol{\lambda}, \boldsymbol{\omega}, \{C_r\}_{r=1}^k) = \sum_{r=1}^k \frac{links(C_r, \mathcal{X}_i \backslash C_r)}{links(C_r, \mathcal{X}_i)}$$
$$- \beta \sum_{r=1}^k \sum_{\substack{(x_a,x_b)\in\mathcal{M} \\ L(x_a)=L(x_b)=r}} \frac{S(x_a,x_b)}{links(C_r,\mathcal{X}_i)} \tag{9}$$
$$+ \rho \sum_{r=1}^k \sum_{\substack{(x_a,x_b)\in\mathcal{C} \\ L(x_a)=L(x_b)=r}} \frac{S(x_a,x_b)}{links(C_r,\mathcal{X}_i)}.$$

This penalty function is composed of two parts:

1. Clustering quality based on the similarity matrix S: Normalized cut [14] is used to define the similarity between clusters defined as: $NC = \sum_{r=1}^k \frac{links(C_r,\mathcal{X}_i\backslash C_r)}{links(C_r,\mathcal{X}_i)}$, where $links(C_p,C_q) = \sum_{x_a\in C_p, x_b\in C_q} S(x_a,x_b)$. The larger the similarity between different clusters, the worse the clustering quality.

2. Supervision constraints: For an object pair $<x_a, x_b>$ belongs to the same cluster C_r, the presence of $<x_a, x_b>$ in $\mathcal{M}$ indicates good clustering quality; the presence of $<x_a, x_b>$ in $\mathcal{C}$ indicates bad clustering quality.

For flexibility, we define two supervision importance parameters β and ρ to balance the influence of supervised information set. The penalty function of SCHAIN for clustering is equal to Eq. (9) with β and ρ assigned to 1. Thus, SCHAIN regards clustering objective and supervision of $\mathcal{M}$ and $\mathcal{C}$ as equal importance. However, while handling different datasets and different clustering objectives, it is flexible to cluster with different β and ρ. In this paper, we set ρ to a fixed value 1, and use β as a hyper-parameter. The larger β indicates more penalty considered for the object pairs in must-link set $\mathcal{M}$.

By defining $\{z_r\}_{r=1}^k$, where each z_r is a binary indicator vector of length $n = |\mathcal{X}_i|$, $z_r(a) = 1$ represents x_a is assigned to cluster C_r and 0 otherwise, we obtain the overall penalty function for clustering:

$$\mathcal{J}(\lambda, \omega, \{z_r\}_{r=1}^k) = \sum_{r=1}^k \frac{z_r^T(D - S - W \circ S)z_r}{z_r^T D z_r}. \tag{10}$$

$\circ$ is the Hadamard product for two matrices. W is a constraint matrix of $W \in \mathbb{R}^{n \times n}$, where $W(x_a, x_b) = \beta$ for $<x_a, x_b> \in \mathcal{M}$, $W(x_a, x_b) = -1$ for $<x_a, x_b> \in \mathcal{C}$ and 0 otherwise. Furthermore, we add a regularization term to Eq. (10) and get the overall penalty function:

$$\mathcal{J}(\lambda, \omega, \{z_r\}_{r=1}^k) = \sum_{r=1}^k \frac{z_r^T(D - S - W \circ S)z_r}{z_r^T D z_r} + \gamma(\|\lambda\|^2 + \|\omega\|^2). \tag{11}$$

Finally, to find the best clustering, we minimize the penalty function subject to the following constraints: $\sum_{r=1}^k z_r(a) = 1; z_r(a) \in \{0,1\}; \sum_{l=1}^{|A_i|} \omega_l = 1; \omega_l > 0$ and $\lambda_j > 0$. Note that α, β, η and γ are hyper-parameters in the function.

4.2   Model Optimization

There are two objectives to be optimized in the learning process: the clustering results $\{z_r\}_{r=1}^k$ and the weighting vectors λ and ω. In this section, we introduce the optimization of our model. Following SCHAIN [9], we use a mutual updating optimization method. Firstly, given the weights λ and ω, we find the optimal clustering $\{z_r\}_{r=1}^k$. Secondly, given $\{z_r\}_{r=1}^k$, we find the optimal λ and ω. We iterate until $\mathcal{J}$ is smaller than a given threshold ϵ. We will briefly explain how the two update steps are performed.

Step 1: Optimize $\{z_r\}_{r=1}^k$ given λ and ω
In this step, the task is to find the best clustering $\{z_r\}_{r=1}^k$. We define a matrix $\hat{Z}$ with the r-th column of $\hat{Z}$ equals to $D^{\frac{1}{2}} z_r / (z_r^T D z_r)^{\frac{1}{2}}$, s.t. $\hat{Z}^T \hat{Z} = I_k$, where I_k is the identity matrix of $\mathbb{R}^{k \times k}$. Since the weights of meta-paths and attributes, namely λ and ω, are given, the objective function $\mathcal{J}(\lambda, \omega, \{z_r\}_{r=1}^k)$ becomes a

function of clustering indicator matrix $\{z_r\}_{r=1}^k$. The minimization of $\mathcal{J}$ can be derived to a trace maximization problem:

$$\max_{\hat{Z}^T \hat{Z}=I_k} trace(\hat{Z}^T D^{-\frac{1}{2}}(S + W \circ S)D^{-\frac{1}{2}}\hat{Z}), \tag{12}$$

which has a closed form solution by calculating top k eigenvectors [2]. K-means is then adopted to obtain hard clustering results $\{z_r\}_{r=1}^k$.

Step 2: Optimize λ and ω given $\{z_r\}_{r=1}^k$
In this step, the model finds the best λ and ω given fixed cluster $\{z_r\}_{r=1}^k$. As proved in SCHAIN [9], minimizing $\mathcal{J}$ is equivalent to maximizing:

$$\max_{\lambda,\omega} \sum_{r=1}^k \frac{z_r^T(S + W \circ S)z_r}{z_r^T D z_r} - \gamma(\|\lambda\|^2 + \|\omega\|^2), \tag{13}$$

which can be rewritten as:

$$H(\lambda,\omega) = \max_{\lambda,\omega} \frac{f(\lambda,\omega)}{g(\lambda,\omega)}, \tag{14}$$

where $f(\lambda,\omega)$ and $g(\lambda,\omega)$ are two nonlinear multivariate polynomial function. This can be optimized by solving the following non-linear parametric programming problem: Let $f(\lambda,\omega)$ and $g(\lambda,\omega)$ be two multivariate polynomial functions. For a given μ, find $F(\mu) = f(\lambda,\omega) - \mu g(\lambda,\omega)$, s.t. $\sum_{r=1}^k z_r(a) = 1$; $z_r(a) \in \{0,1\}$; $\sum_{l=1}^{|A_i|} \omega_l = 1$; $\omega_l > 0$ and $\lambda_j > 0$. Readers may refer to SCHAIN [9] for proving and other details.

To sum up the whole model, SCAN firstly calculates the coupled attribute node similarity and link similarity between nodes. Secondly, SCAN computes the overall similarity by aggregating the similarity matrices with weighting vectors λ and ω and a balance factor α. The constraint matrix is generated by supervision using supervision importance β. Finally, SCAN adopts an iterative mutual update process to learn the clustering and weighting vectors.

5   Experiments

In this section, we firstly introduce the datasets used in the experiments. Then we discuss the attribute similarity matrix of the IMDB dataset. Further, we show the effectiveness of the SCAN model against 6 representative algorithms. Finally, we explain the influence of different hyper-parameters and the weight learning process.

5.1   Datasets

Yelp[1]. This experiment is similar to the experiment of Yelp-Restaurant dataset used in SCHAIN [9]. The clustering task is to cluster restaurants by three subcategories: "Fast Food", "Sushi Bars" and "American (New) Food". The AHIN

---

[1] http://www.yelp.com/academic_dataset.

is composed of 2,614 business objects (B); 33,360 review objects (R); 1,286 user objects (U) and 82 food relevant keyword objects (K). 5 attributes are considered: 3 categorical attributes (reservation, service, and parking) and 2 numerical attributes (review count and quality star). For meta-paths: we choose BRURB (businesses reviewed by the same user) and BRKRB (businesses receive the same keyword in the reviews).

IMDB5k[2]. We extracted an AHIN network from the IMDB5k dataset. The AHIN is composed of 4 node types: 4,140 movies (M), 4,907 actors (A), 1,867 directors (D) and 24 Genres (G). The movies are extracted into three types by their IMDB scores. Each movie has 2 categorical attributes (content rating and country) and 3 numerical attributes (gross, critic reviews and movie Facebook likes). Three meta-paths: MAM (movies with the same actor), MGM (movies with the same genre) and MDM (movies with the same director) are chosen in our experiments for link-based algorithms. Note that, as mentioned before, Path-Sim based clustering algorithms cannot handle asymmetric meta-paths, though SCAN is able to handle asymmetric meta-paths, we choose symmetric meta-paths for the sake of fairness.

5.2 Discussion of Attribute Value Similarity

The SCAN model captures the coupling relationships between attributes and therefore obtains better node attribute similarity. We will take a closer look at the attribute value similarity of the IMDB dataset. There are two categorical attributes in the IMDB dataset, namely "country" and "content ratings". The similarity between the attribute values are visualized in Fig. 2. Attribute value pairs with higher similarity are visualized with darker colors.

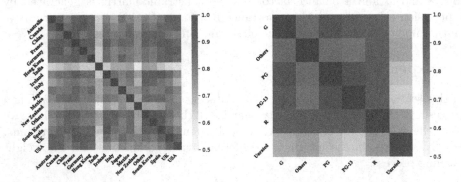

(a) Attribute similarity of IMDB-country   (b) Attribute similarity of IMDB-ratings

Fig. 2. Attribute value similarity of IMDB (Color figure online)

---

[2] https://www.kaggle.com/carolzhangdc/imdb-5000-movie-dataset.

Table 1. The MPAA rating system.

Content rating	Description
G	General Audiences: All ages admitted
PG	Parental Guidance Suggested: May not be suitable for children
PG-13	Parents Strongly Cautioned: Inappropriate for children under 13
R	Restricted: Children under 17 requires adult guardian
Unrated	Not Rated or Unrated: The film has not been submitted for a rating

The attribute similarity of "country" attribute of node type "movie" is shown in Fig. 2a. Top 15 countries in the dataset are analyzed, other countries are combined to the "Others" category. As known, the USA, UK, France, Canada and Australia are all developed countries while India is a developing country. By mining the dependency between movie attributes, our model successfully discovers that USA, UK, France, Canada, and Australia are similar countries, while India is dissimilar to any of the other countries shown in the table. Interestingly, our algorithm also reveals that New Zealand and India are the most dissimilar attribute values in the selected countries.

The Motion Picture Association of America (MPAA) film rating system[3] is used to rate a film's suitability for certain audiences based on its content The attribute similarity of "content rating" attribute of node type "movie" is shown in Fig. 2b. Top 5 genres in the dataset are analyzed ("G", "PG", "PG-13", "R", and "Unrated"), other genres are combined to the "Others" category. The meanings of the rating levels are shown in the Table 1. Generally speaking, our attribute value similarity obtained is reasonable. To illustrate, our model finds that "G" is most similar to "PG" and "G" is more similar to "PG-13" than "R". Meanwhile, "R" is most similar to "PG-13", which is its neighbor level in Table 1; "Others" is somehow similar to other rating levels since its a combined categorical value. "Unrated" is relatively dissimilar to other rating levels.

5.3 Clustering Performance

We conduct experiments of three groups of comparison algorithms: attribute-only, link-only and attribute+link. For each group of comparison algorithms, we consider two algorithms listed as follows.

- **Attribute-only:** The clustering algorithms in the first group consider only object attributes. These are traditional methods which ignore the network structure of an AHIN. We choose Spectral-Learning [6] and a semi-supervised version of normalized cuts [8] as representatives, which are denoted SL and SNcuts, respectively.
- **Link-only:** These methods utilize only the link information of the network and they ignore object attribute values. GNetMine [4], PathSelClus [16] are chosen as representative methods of this category.

---

[3] https://www.mpaa.org/film-ratings/.

- **Attribute+Link:** Methods of this group use both attribute and link information. We consider FocusCO [11], SCHAIN [9], and SCAN-C. Since FocusCO does not consider the heterogeneity of networks, the AHINs are constructed as homogeneous networks for experiments of FocusCO. The SCAN-C is another version of our model using CASV [17] as the categorical attribute value similarity measure. Since CASV only considers categorical data, in the SCAN-C related experiments, numerical attributes are converted to categorical attributes.

Table 2 shows the clustering performance in terms of NMI for Yelp-Restaurant and IMDB5k respectively. We run every experiment 10 times and calculate the average NMI. As can be observed, for both datasets, the clustering performance of SCAN outperforms other algorithms.

For Yelp-Restaurant, the attribute-only algorithms perform relatively worse than link-only algorithms, suggesting that link similarity is more significant of this clustering objective. Since our improvement is mostly on attribute similarity, SCAN turns out to be slightly better than SCHAIN.

For IMDB, due to the sparsity of the network, the attribute-only algorithms perform much better than link-only algorithms, indicating that attribute-based similarity is more significant to the clustering objective. Taking into account the coupling relationships between attributes, SCAN outperforms other algorithms significantly on IMDB dataset. Additionally, SCAN generally outperforms SCAN-C, which proves the importance of mining the coupling relationships between mixed type attributes.

Table 2. NMI comparison on Yelp-Restaurant and IMDB5k

Dataset	Seeds	SL	SNCuts	GNetMine	PathSelClus	FocusCO	SCHAIN	SCAN-C	SCAN
Yelp	5%	0.156	0.190	0.278	0.564	0.088	0.681	0.683	**0.708**
	10%	0.179	0.192	0.288	0.612	0.087	0.703	0.692	**0.727**
	15%	0.289	0.194	0.364	0.632	0.093	0.709	0.714	**0.746**
	20%	0.295	0.198	0.380	0.631	0.090	0.736	0.736	**0.754**
	25%	0.303	0.253	0.399	0.637	0.090	0.742	0.752	**0.764**
IMDB5k	5%	0.118	0.102	0.083	0.085	0.072	0.132	0.180	**0.325**
	10%	0.108	0.191	0.127	0.135	0.075	0.137	0.165	**0.335**
	15%	0.149	0.175	0.157	0.168	0.082	0.191	0.187	**0.370**
	20%	0.156	0.220	0.157	0.203	0.092	0.223	0.240	**0.450**
	25%	0.415	0.259	0.246	0.247	0.090	0.361	0.371	**0.493**

5.4   Weight Learning

SCAN retains the ability of SCHAIN [9] to learn the weights of attributes and meta-paths. We will take clustering of IMDB dataset for illustration. In the following discussion, we assume 25% seed objects. Figure 3a and b exhibit how

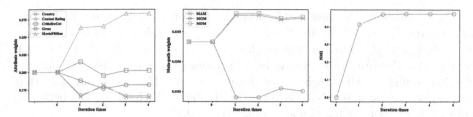

(a) Attribute weight learn- (b) Meta-path weight learn- (c) Clustering performance
ing   ing

Fig. 3. Weight learning on IMDB

our algorithm learns the weight through iteration for attributes and meta-paths. From Fig. 3c, we see that SCAN identifies that the "Facebook likes" attribute to be the most useful attribute to cluster movies. As shown in Fig. 3b, the weights of meta-paths remain almost unchanged through optimization, since the network is really sparse and link similarity does not contribute much to the clustering task. In Fig. 3c, we observe that the overall clustering performance is improving through iterations and finally to an optimal value.

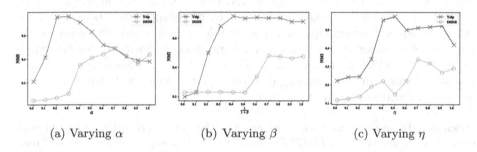

(a) Varying α (b) Varying β (c) Varying η

Fig. 4. Influence of hyper-parameters on Yelp

5.5   Parameters Discussion

There are three hyper-parameters to be tuned in our model: The balance coefficient α for link similarity and attribute similarity, the supervision importance β to leverage constraints, and the coupled coefficient η to balance plain and coupled similarity. The influence of hyper-parameters for Yelp and IMDB are shown in Fig. 4.

Figure 4a shows the relationship between clustering performance and the balance coefficient α. If $\alpha = 0$, only link similarity is considered; if $\alpha = 1$, only attribute similarity is considered. We can observe that the attribute similarity is more important for IMDB and link similarity is more important for Yelp in terms of clustering performance. Therefore, it is better to balance between link similarity and attribute similarity.

Figure 4b shows the relationship between clustering performance and supervision importance β. For visualization, we map the value of β from $[0, \infty]$ to $[0, 1]$, and plot $\frac{1}{\beta+1}$. We can observe that, in both datasets, focusing too much (large β) on the supervision always lead to inferior results. In Yelp, the optimal $\beta = 1.5$, whilst in IMDB, it is better to set $\beta = 0.43$. This result indicates that the importance of supervision in terms of clustering is different, it's better to leverage the supervision and node similarity flexibly.

Figure 4c shows the relationship between clustering performance and the coupled coefficient η. If $\eta = 0$, plain similarity (SMS [7]) is used as our categorical value similarity; if $\eta = 1$, the plain similarity is ignored and only coupled similarity is concerned. We can observe that, for both datasets, it is better to balance the plain similarity and the coupled similarity. The optimal values of η for Yelp and IMDB are 0.5 and 0.7 respectively, which indicates that the attributes from different datasets have different degree of coupling relationships. For highly correlated datasets, high coupled coefficients are recommended.

6   Conclusion

In this paper, we study semi-supervised clustering in attributed heterogeneous information networks. To mine the coupling relationships between node attributes, we propose a new model SCAN. In our model, We present a coupled attribute node similarity measure to capture the dependency between mixed attribute types. Furthermore, we use a flexible approach to better leverage the importance of supervision information and network similarity. Extensive experiments are conducted on two real-world datasets to prove that our model is capable of capturing the coupling relationships between attributes and outperforms other state-of-the-art algorithms.

Acknowledgement. This work is supported by the National Key Research and Development Program of China (2017YFB0803304), the National Natural Science Foundation of China (No. 61772082, 61806020, 61702296) the Beijing Municipal Natural Science Foundation (4182043), the CCF-Tencent Open Fund and the Fundamental Research Funds for the Central Universities.

References

1. Basu, S., Banerjee, A., Mooney, R.: Semi-supervised clustering by seeding. In: In Proceedings of 19th International Conference on Machine Learning, ICML 2002. Citeseer (2002)
2. Bhatia, R.: Matrix Analysis. Graduate Texts in Mathematics, vol. 169. Springer, New York (1997). https://doi.org/10.1007/978-1-4612-0653-8
3. Cao, L.: Coupling learning of complex interactions. Inf. Process. Manag. **51**(2), 167–186 (2015)
4. Ji, M., Sun, Y., Danilevsky, M., Han, J., Gao, J.: Graph regularized transductive classification on heterogeneous information networks. In: Balcázar, J.L., Bonchi, F., Gionis, A., Sebag, M. (eds.) ECML PKDD 2010. LNCS, vol. 6321, pp. 570–586. Springer, Heidelberg (2010). https://doi.org/10.1007/978-3-642-15880-3_42

5. Jin, R., Breitbart, Y., Muoh, C.: Data discretization unification. Knowl. Inf. Syst. **19**(1), 1 (2009)
6. Kamvar, K., Sepandar, S., Klein, K., Dan, D., Manning, M., Christopher, C.: Spectral learning. In: International Joint Conference of Artificial Intelligence. Stanford InfoLab (2003)
7. Kaufman, L., Rousseeuw, P.J.: Finding Groups in Data: An Introduction to Cluster Analysis, vol. 344. Wiley, Hoboken (2009)
8. Kulis, B., Basu, S., Dhillon, I., Mooney, R.: Semi-supervised graph clustering: a kernel approach. Mach. Learn. **74**(1), 1–22 (2009)
9. Li, X., Wu, Y., Ester, M., Kao, B., Wang, X., Zheng, Y.: Semi-supervised clustering in attributed heterogeneous information networks. In: Proceedings of the 26th International Conference on World Wide Web, pp. 1621–1629. International World Wide Web Conferences Steering Committee (2017)
10. Luo, C., Pang, W., Wang, Z.: Semi-supervised clustering on heterogeneous information networks. In: Tseng, V.S., Ho, T.B., Zhou, Z.-H., Chen, A.L.P., Kao, H.-Y. (eds.) PAKDD 2014. LNCS, vol. 8444, pp. 548–559. Springer, Cham (2014). https://doi.org/10.1007/978-3-319-06605-9_45
11. Perozzi, B., Akoglu, L., Iglesias Sánchez, P., Müller, E.: Focused clustering and outlier detection in large attributed graphs. In: Proceedings of the 20th ACM SIGKDD International Conference on Knowledge Discovery and Data Mining, pp. 1346–1355. ACM (2014)
12. Rubner, Y., Tomasi, C.: The earth mover's distance. In: Perceptual Metrics for Image Database Navigation. SECS, vol. 594, pp. 13–28. Springer, Boston (2001). https://doi.org/10.1007/978-1-4757-3343-3_2
13. Shi, C., Kong, X., Huang, Y., Philip, S.Y., Wu, B.: HeteSim: a general framework for relevance measure in heterogeneous networks. IEEE Trans. Knowl. Data Eng. **26**(10), 2479–2492 (2014)
14. Shi, J., Malik, J.: Normalized cuts and image segmentation. IEEE Trans. Pattern Anal. Mach. Intell. **22**(8), 888–905 (2000)
15. Sun, Y., Han, J., Yan, X., Yu, P.S., Wu, T.: PathSim: meta path-based top-k similarity search in heterogeneous information networks. Proc. VLDB Endow. **4**(11), 992–1003 (2011)
16. Sun, Y., Norick, B., Han, J., Yan, X., Yu, P.S., Yu, X.: Pathselclus: integrating meta-path selection with user-guided object clustering in heterogeneous information networks. ACM Trans. Knowl. Discov. Data (TKDD) **7**(3), 11 (2013)
17. Wang, C., Dong, X., Zhou, F., Cao, L., Chi, C.H.: Coupled attribute similarity learning on categorical data. IEEE Trans. Neural Netw. Learn. Syst. **26**(4), 781–797 (2015)

PrivBUD-Wise: Differentially Private Frequent Itemsets Mining in High-Dimensional Databases

Jingxin Xu[1], Kai Han[1($\boxtimes$)], Pingping Song[2], Chaoting Xu[1], and Fei Gui[1]

[1] School of Computer Science and Technology,
University of Science and Technology of China, Hefei, China
jxxu1996@gmail.com, hankai@ustc.edu.cn, {xct1996,guifei}@mail.ustc.edu.cn
[2] School of Arts, Anhui University, Hefei, China
1582922817@qq.com

Abstract. In this paper, we study the problem of mining frequent itemsets in high-dimensional databases with differential privacy, and propose a novel algorithm, *PrivBUD-Wise*, which achieves high result utility as well as a high privacy level. Instead of limiting the cardinality of transactions by truncating or splitting approaches, which causes extra information loss and result in unsatisfactory performance in utility, *PrivBUD-Wise* doesn't make any preprocessing on original database and guarantees high result utility by reducing extra *privacy budget* consumption on irrelevant itemsets as much as possible. To achieve that, we first propose a Report Noisy mechanism with optional number of reported itemsets: SRNM, and what is more important is that we give a strict proof for SRNM in the appendix. Moreover, *PrivBUD-Wise* first proposes a biased *privacy budget* allocation strategy and no assumption or estimation on the maximal cardinality needs to be made. The good performance in utility and efficiency of *PrivBUD-Wise* is shown by experiments on three real-world datasets.

Keywords: Frequent itemsets mining · Differential privacy ·
Smart Report Noisy Max (SRNM) · Privacy budget allocations

1 Introduction

With the rapid growth in data volume and variety, exploring useful information is becoming more and more important. Frequent itemsets mining (FIM) [2] is a basic problem which has been widely studied for finding most frequent itemsets in a transactional database, where each record consisting of a set of items is called a *transaction* [16]. The *support* of an itemset is defined to be the number of transactions containing it. For example, at an online bookstore, each transaction contains a set of purchased books, and the result of an FIM algorithm consists of itemsets with a higher *support* than a given threshold. However, releasing these itemsets may bring privacy issues.

© Springer Nature Switzerland AG 2019
J. Shao et al. (Eds.): APWeb-WAIM 2019, LNCS 11641, pp. 110–124, 2019.
https://doi.org/10.1007/978-3-030-26072-9_8

Differential privacy (DP) [5] is a robust, meaningful, and mathematically rigorous definition of privacy [7]. DP has gradually become a standard concept in privacy preserving areas because it provides strong guarantees for individual data security regardless any background knowledge of an adversary.

We study the algorithms of mining frequent itemsets in high-dimensional databases with differential privacy (DP), and our goal is to achieve both high result utility and time efficiency under a given privacy constraint, i.e., a given *privacy budget* ϵ. FIM algorithms are well-studied in [2] and [12], where two most popular algorithms: Apriori and Fp-growth are proposed respectively. Although Fp-growth is much faster than Apriori, it is not quite suitable for differentially private mechanisms due to its static tree structure. In Apriori-based algorithms, to overcome the challenge of high-dimension, Zeng et al. [17] propose a truncating approach by throwing away some items in long transactions, however, such method could cause too much information loss. Inspired from that, transaction splitting approach that splits large transactions into shorter sub-transactions is adopted [4], but this method leads to the waste of privacy budget and faces the problem of information loss as well.

Different from those two methods, we recompose some details in Apriori and then combine it to our new proposed DP mechanism: SRNM, which lets our algorithm do not change any information of original datasets. Although SRNM is based on RNM mechanism, an existing method proposed in [7], it does offers optional number of reported itemsets in each call and greatly improve time efficiency as shown in experiment Sect. 5.2. As we can see from our experiments, RNM is not computable in real world because of its poor time efficiency. In fact, RNM could be a special case of SRNM (when $z = 1$). What's more, we first give a strict proof of the correctness of SRNM. which is very important in privacy preserving areas.

Another observation is that in Apriori, for algorithm's result, the degrees of importance of different frequent itemsets are different. It is because that Apriori always generates frequent itemsets based on shorter ones.

When combining Apriori with a DP mechanism, to get a higher result utility, more budget should be applied to more important itemsets. As far as we know, there is no research adopting this idea. Privsuper [16], the state-of-the-art algorithm for mining frequent itemsets with DP, also allocates privacy budget averagely on each frequent itemset. However, we do not know the true number of frequent itemsets in advance, a possible way is to pre-compute that number and use the noisy version of it [16,17]. We argue that this way has many disadvantages in Sect. 4.3. Instead, we proposed a novel biased privacy allocation, which allocates more budget on more important itemsets.

Contributions. To sum up, our main contributions are:

- We propose *PrivBUD-Wise*, a novel Apriori-based algorithm for differentially private frequent itemsets mining. To overcome the high-dimensional challenge, based on RNM, we propose SRNM in our algorithm and makes RNM as a special case of SRNM. SRNM overcomes low time efficiency of RNM as shown in our experiment, which means SRNM, instead of RNM, is computable in real world. The proof of SRNM is non-trivial, we give a novel and strict proof of the correctness of SRNM, which is very important in the field of privacy preserving.

– In our algorithm, we first propose a biased budget allocation strategy, which is based on a reasonable intuition. Using this strategy, our algorithm doesn't need to pre-compute the true number of frequent itemsets, which improves the efficiency, and allocates privacy budget in a more reasonable way.

– We prove that *PrivBUD-Wise* satisfies ϵ-differential privacy, and conduct experiments on real-world datasets which show the better performance of our algorithm compared with three state-of-the-art approaches.

2   Related Work

Differential Privacy (DP) was proposed by Dwork et al. [5] and is getting more and more attention and applications such as Google Chrome [8] and Apple iOS [1] in privacy related researches. The most commonly used mechanisms to promise DP are LM mechanism [6] and GM mechanism [11].

In the area of data mining, frequent itemsets mining (FIM) has been widely studied in many applications [10]. The most two popular algorithms to solve FIM is Apriori [2] and Fp-growth [12], but FP-growth is not suitable for DP mechanism because it is based on fp-tree, which is pre-built and remain static. Several researches aim at mining frequent itemsets with differential privacy [3,14, 16,17]. To tackle the high-dimensional challenge, Privbasis is proposed in [14], it transforms the dataset from high dimensional space to lower one. Among apriori-based algorithms, Zeng et al. [17] propose a truncation method to tackle the high-dimensional challenge, and Cheng et al. [4] propose the splitting approach; but they all suffer huge information loss in the worst case [15].

Sparse Vector Technique (SVT) [7] is an effective differential privacy mechanism with no extra privacy consumption on irrelevant answer queries, but it is also well-known by its improper uses. Specifically, SVT is available only for the interactive setting [9], where the queries are unknown to users ahead of time. Unfortunately, FIM is under non-interactive setting. So the Noisycut algorithm [13], which adopts SVT in FIM, has been proved to violate differential privacy [18], with no fix solution until now.

PrivSuper [16] builds a new block for FIM with differential privacy, it searches for the longest frequent itemsets directly and applies a customized mechanism(SEM) to avoid budget wastes. Different from that algorithm, instead of finding frequent top-k itemsets in a given database, we aim at mining the frequent itemsets under a given threshold, which is much more troublesome. As far as we know, there is no research considering privacy budget allocation strategies.

3   Background

3.1   Differential Privacy

Intuitively, differential privacy promises that the change of a single record in a database has little influence on the outputs of an algorithm.

Definition 1 *(ε-differential privacy [5]). for any input database D, let $\mathcal{M}$ be a randomized algorithm, $\mathcal{M}$ is said to satisfy ϵ-differential privacy iff for any $D' \in neb(D)$, and any $\mathcal{O} \subseteq Range(\mathcal{M})$, where $Range(\mathcal{M})$ denotes the set of all possible outputs of $\mathcal{M}$ on D, and Pr denotes probability, it satisfies:*

$$Pr(\mathcal{M}(D) \in \mathcal{O}) \leq e^{\epsilon} \times Pr(\mathcal{M}(D') \in \mathcal{O}) \tag{1}$$

where $neb(D)$ is the neighboring databases set of D, in $neb(D)$, each database differs from D at most one transaction. ϵ is called the *privacy budget*. The smaller ϵ is, the stronger privacy guarantee that a DP algorithm promises.

Definition 2 *(Sensitivity [6]). The Sensitivity of a count query q with numerical outputs is: $S_q = max_{D,D'} \| q(D) - q(D') \|_1$, where $D' \in neb(D)$ and $\| q(D) - q(D') \|_1$ denotes the $\mathcal{L}_1$ distance between $q(D)$ and $q(D')$.*

To perturb the precise results, an effective DP approach is the Laplace Mechanism(LM), which adds random noise subjecting to Laplace Distribution. The probability density function of standard Laplace Distribution with scale b is: $Lap(x|b) = \frac{1}{2b}exp(-\frac{|x|}{b})$. We write Lap($b$) instead of $X \sim$ Lap(b) for convenience.

Theorem 1 *(Laplace Mechanism [7]). Given a count query set $q = (q_1, \ldots, q_k)$, for a given database D, the Laplace Mechanism (LM):*

$$\mathcal{M}_L(D, q, \epsilon) = (q_1(D), \ldots, q_k(D)) + (Y_1, \ldots, Y_k) \tag{2}$$

satisfies ϵ-differential privacy. where Y_i are i.i.d., and $Y_i \sim$ Lap($\frac{S_q}{\epsilon}$)

Lemma 1 *(Sequential Composition [16]). Given a set of n randomized mechanisms $\mathcal{M}_1, \ldots, \mathcal{M}_n$, if $\mathcal{M}_i$ $(1 \leq i \leq n)$ satisfies ϵ_i-differential privacy, then the composite mechanism $\mathcal{M} = \mathcal{M}_1 \circ \mathcal{M}_2 \circ \cdots \circ \mathcal{M}_n$ satisfies $(\sum_i \epsilon_i)$-differential privacy.*

3.2   Frequent Itemsets Mining

In a transactional database D, each *transaction* $t \in D$ is comprised of a set of items, the *support* of an arbitrary itemset s is defined to be the number of transactions that contain s as a subset. $s.supp = |\{t|t \in D, s \subseteq t\}|$. We say s is frequent iff $s.supp$ exceeds a given threshold λ.

For a predefined threshold λ, the aim of an FIM algorithm is to find all itemsets that have a bigger *support* than λ. An itemset *query* means a query for the *support* of the itemset. More explanation of notations can be found in Table 1.

Table 1. Notations

Symbol	Description
D, n	A transactional dataset, $|D|$
t, $t.supp$	A *transaction*, the *support* of t
I	The set of items domain of D
ϵ	Privacy budget
k-itemsets	Itemsets with length equal to k
ϵ_k	Budget used to generate frequent k-itemsets
q, q^k	Query set, query set in k-th round
S_q, d_k	Sensitivity of q, $|q^k|$
λ, λ'	Given threshold of *support*, $\lambda/|D|$
l_{max}	The maximal length of transactions
Lap(b)	The standard Laplace Distribution with scale b
z	The optional parameter in SRNM algorithm

3.3   Apriori

the Apriori algorithm is proposed by [2]. For completeness, we quote the Apriori algorithm in Algorithm 1. Apriori is based on a simple observation: for any itemsets s_1 and s_2, if $s_1 \subset s_2$, then $s_1.supp \geq s_2.supp$. Namely, an itemset will never be frequent if any of its proper subsets isn't frequent. Specifically, it generates all frequent 1-itemsets (with cardinality equal to 1) in the first round, then in the k-th round ($k \geqslant 2$), it generates candidate k-itemsets based on frequent $(k-1)$-itemsets in the apriori-gen [2] step (in line 3 of Algorithm 1), and chooses frequent ones based on their *supports* to generate frequent k-itemsets (line 4–10).

Algorithm 1. Apriori

Input: transactional dataset D, given lower threshold λ
Output: frequent itemsets with *support* over λ.
1: F_1 = frequent 1-itemsets
2: **for** ($k = 2; F_{k-1} \neq \emptyset; ++k$) **do**
3:      C_k = apriori-gen(F_{k-1});
4:      **for** transactions $t \in D$ **do**
5:          C_t = subset(C_k, t);//Candidates contained in t
6:          **for** candidates $c \in C_t$ **do**
7:              $c.count$++;
8:          **end for**
9:      **end for**
10:     $F_k = \{c \in C_k | c.count \geq \lambda\}$;
11: **end for**
        return $\bigcup_k F_k$

---

4 PrivBUG-Wise

4.1 Sensitivity

When applying the LM mechanism to Apriori, in each round of determining frequent itemsets, we need to consider the sensitivity of queries for choosing proper scale of added noises. Formally, at round k, we denote the query set by $q^k = (q_1^k, \ldots, q_{d_k}^k)$, where d_k represents the amount of candidate k-itemsets. we use l_{max} to denote the maximal length of transactions in original dataset. Then we have the following lemma for S_{q^k}:

Lemma 2 [17]. $S_{q^k} \leq min\{\binom{l_{max}}{k}, d_k\}$

In order for insurance, Cheng et al. [4] take that upper bound as S_{q^k}. It is clear that when the dimension of the dataset is too high (i.e., too big l_{max}) and the number of candidate itemsets (i.e., d_k) is too big in large databases, too much noises would be added according to Theorem 1. We find RNM mechanism outperforms LM in Apriori-based cases:

Definition 3 *(RNM [7]). Given a transaction database D, "Report Noisy Max(RNM)" is the following algorithm which determines the highest value of a counting queries set $q = (q_1, \ldots, q_{|q|})$:*

Firstly: add noise: $q_i(D) \to q_i(D) + X_i$; $X_i \sim Lap(1/\epsilon), 1 \leqslant i \leqslant |q|$;
Then: return the itemset s with the largest noisy count.

Lemma 3 [7]. *The RNM algorithm is ϵ-differentially private.*

4.2 SRNM: Smart Report Noisy Max

We notice that the noise amplitude in RNM is independent of S_q compared with LM. In fact, the query sensitivity in RNM mechanism is equal to 1 because any change on one single transaction could affects the output *support* by at most 1. So, regardless of how big l_{max} and d_k are, RNM achieves the same high data utility. A weakness of RNM is that it could be time consuming (shown in experiment Sect. 5.2) because only one itemset can be generated after one call.

To balance the utility and efficiency, Algorithm 2 proposes SRNM which could generate several itemsets with optional number in each call. We first add Laplace noise to all possible queries (line 2 of Algorithm 2), then choose at most z itemsets with a noisy query result over λ (line 3–9).

Theorem 2. SRNM *satisfies ϵ-differential privacy*

Proof. See in Appendix B.

In k-th ($k \geq 1$) round of Apriori, we repeatedly call SRNM to determine all frequent k-itemsets. In line 2 of Algorithm 2, we notice that the noise amplitude is z/ϵ, which is independent of l_{max}. When z is equal to 1, SRNM is equivalent to RNM. As z increases, the amplitude of added noise increases while the number of calls decreases, vice versa. So, SRNM provides a way to balance result utility and efficiency by setting a proper value of z, which is shown in Sect. 5.2.

Algorithm 2. SRNM: Smart Report Noisy Max.

Input: input database D; privacy parameter ϵ; threshold λ; parameter z; candidate k-itemsets queries set q^k

Output: top-z frequent k-itemsets with noisy *support* over λ

1: $F = \emptyset$;
2: add noise: $q_i^k(D) \rightarrow q_i^k(D) + X_i$; $X_i \sim Lap(z/\epsilon), 1 \leqslant i \leqslant |q^k|$
3: **for** $(i = 1; i \leqslant z; + + i)$ **do**
4:     **if** the i-th biggest noisy query result is bigger than λ **then**
5:         add the corresponding itemset with the i-th biggest noisy query result to F;
6:     **else**
7:         break;
8:     **end if**
9: **end for**
        return F

4.3 *Privacy Budget* Allocations

Among All Rounds. The total *privacy budget* is ϵ, and at the i-th round of Apriori, frequent i-itemsets are generated; according to the *sequential composition* property in Lemma 1, we have $\sum_i \epsilon_i = \epsilon$, where ϵ_i represents the budget consumption in round i of Apriori. Traditional methods firstly compute the total round number k_{max}, then get the noisy version of k_{max} in consideration of privacy and use it to distribute budget equally among all rounds.

We argue that the method above has at least 4 drawbacks:

(i) Time consuming: It computes true frequent itemsets to get k_{max} in advance. (ii) Waste of budget: It uses some budget to perturb k_{max}, which could be completely avoidable. (iii) The randomized(perturbed)value of k_{max} will have significant influence on final outputs, which makes the outputs instable, and thus inaccurate. (iv) The allocation strategy is "equally allocation", which doesn't make good use of budget.

An Apriori-based algorithm suffers **propagated error**, which means that if a frequent itemset $\{a\}$ is misidentified to be infrequent, then all supersets of $\{a\}$ will never be identified to be frequent.

To overcome (iv), our biased allocation strategy is $\frac{6}{\pi^2} \cdot \{1, \frac{1}{2^2}, \frac{1}{3^2}, \ldots\} \cdot \epsilon$, which allocates $\frac{6}{\pi^2} \cdot \frac{1}{k^2}\epsilon$ budget to the k-th round, assuming the total budget is ϵ. Our strategy also overcomes (i), (ii) and (iii) since we won't pre-compute k_{max} in advance and ensures the privacy constraint because that: $\frac{6}{\pi^2} \sum_{k=1}^{\infty} \frac{1}{k^2} = 1$.

The convergence speed of $\{\frac{1}{k^2}\}$ is slow, so we won't allocate too little budget in a certain round, thus avoids too high amplitude of noise in SRNM.

In a Certain Round. In k-th round, we iteratively call SRNM to determine frequent k-itemsets. Since we don't know how many calls are needed in advance, we adopt the same mathematical allocation sequence, i.e., $\frac{6}{\pi^2} \cdot \frac{1}{i^2}\epsilon_k$ for i-th call of SRNM in k-th round.

4.4   Our Algorithm

Using SRNM and the allocation strategies, we propose algorithm GenFI in Algorithm 3 for generating k-itemsets $(k = 1, 2, \ldots)$. If $k = 1$, the candidate itemsets is set to be the items set of D (line 2–3). Else, we use apriori-gen of Apriori to generate candidates (line 4–5). Then we use SRNM to generate at most z frequent k-itemsets (line 9), and repeated call SRNM until no more itemset could be generated (line 11–15). At the end, we return frequent k-itemsets F_k.

Algorithm 3. GenFI

Input: parameter:k; input database D; privacy parameter ϵ; threshold λ; parameter z in SRNM

Output: frequent k-itemsets

1: $\epsilon_k = \frac{6}{\pi^2} \cdot \frac{\epsilon}{k^2}$; $F_k = \emptyset$;
2: **if** $k == 1$ **then**
3:      C_k=the items set of D,
4: **else**
5:      $C_k = apriori - gen(F_{k-1})$;
6: **end if**
7: q_k: query set for all itemsets in C_k;
8: $i = 1$;
9: $F_k' = \text{SRNM}(D, \frac{6}{\pi^2} \cdot \frac{\epsilon_k}{i^2}, \lambda, z, q_k)$;
10: **while** $F_k' \neq \emptyset$ **do**
11:      $++i$;
12:      $F_k = F_k \cup F_k'$;
13:      q_k: delete all queries corresponding to F_k';
14:      $F_k' = \text{SRNM}(D, \frac{6}{\pi^2} \cdot \frac{\epsilon_k}{i^2}, \lambda, z, q_k)$;
15: **end while**
          return F_k

Theorem 3. *Algorithm 3 guarantees ϵ_k-differential privacy.*

Proof. By Theorem 2, each call of SRNM in line 14 of Algorithm 3 protects $\frac{6}{\pi^2} \cdot \frac{\epsilon_k}{i^2}$-differential privacy. So, by sequential composition in Lemma 1, Algorithm 3 protects differential privacy for budget less than: $\sum_{i=1}^{\infty} \frac{6}{\pi^2} \cdot \frac{\epsilon_k}{i^2} = \epsilon_k$. Then the theorem follows.

Putting the above discussions together, we propose algorithm PrivBUD-Wise in Algorithm 4 for frequent itemsets mining with differential privacy. We generate frequent 1-itemsets (line 1–2), then repeated call GenFI to generate itemsets until no more itemset can be generated (line 3–5). We return the union of all frequent itemsets an the end.

Theorem 4. *Algorithm 4 guarantees ϵ-differential privacy.*

Proof. By Theorem 3, GenFI protects differential privacy, according to the sequential composition in Lemma 1, Algorithm 4 protects differential privacy and the privacy budget is less than: $\sum_{k=1}^{\infty} \epsilon_k = \sum_{k=1}^{\infty} \frac{6}{\pi^2} \cdot \frac{\epsilon}{k^2} = \epsilon$ Hence, the theorem follows.

Algorithm 4. PrivBUD-Wise
***

Input: input database D; privacy parameter ϵ; threshold λ; parameter z
Output: The set of frequent itemsets
1: $k = 1$;
2: $F_1 =$ GenFI($k, D, \epsilon, \lambda, z$);
3: **for** $(k = 2; F_{k-1} \neq \emptyset; ++k)$ **do**
4:     $F_k =$ GenFI($k, D, \epsilon, \lambda, z$);
5: **end for**
    return $\bigcup_k F_k$

---

5 Experiments

5.1 Experimental Setup

We compare *PrivBUD-Wise* with three state-of-the-art algorithms: truncating approach in [17], splitting approach in [4] and a superset-first approach in [16]. For convenience, we denote our algorithm as PBW and those three algorithms by TA, SA and SFA respectively. The experiments are implemented in Python and performed on a computer with Intel Core i7-6700 3.40 GHz CPU and 32 GB memory.

Datasets. Three real-world datasets' *Mushroom*, *Pumsb-star* and *Retail* are used in our experiments and shown in the Table 2. These three datasets are widely used in most related works.

Table 2. Datasets

Dataset D	$|D|$	$-I-$	$max_{t \in D}|t|$	$\frac{1}{|D|}\sum_{t \in D}|t|$	# frequent itemsets	Proportion
Mushroom [1]	8124	119	23	23.0	23	0.28%
Pumsb-star [1]	49046	2088	63	50.5	20	0.041%
Retail [1]	88162	16470	76	10.3	16	0.018%

Threshold is set to be $0.8|D|$, $0.85|D|$, $0.05|D|$ respectively.

Parameter Settings. When varying z, we set $\epsilon = 1$ and $\lambda' = 0.5$ (on Mushroom), 0.54 (on Pumsb-star), 0.035 (on Retail). When varying ϵ, we set $(z, \lambda') = (4, 0.5)$ for Mushroom, $(5, 0.54)$ for Pumsb-star and $(5, 0.035)$ for Retail. When varying λ', we set $(z, \epsilon) = (4, 1)$ for Mushroom, $(5, 1)$ for both Pumsb-star and Retail.

Evaluation Indicators. *F-score* [4,16,17] is the most popular measuring criteria among all state-of-the-art works. we use U_p and U_c to denote the outputs of a proposed mining algorithm and the correct frequent itemsets respectively. $U = U_p \bigcap U_c$. Precision $= |U|/|U_p|$ and a higher precision means the proportion of un-frequent itemsets in U_p is smaller, recall $= |U|/|U_c|$ and a higher recall means a larger proportion of frequent itemsets are mined.

$$F\text{-}score = 2 * \frac{precision * recall}{precision + recall}$$

A high value of $F\text{-}score$ indicates precision and recall are both high, which means the difference between U_p and U_c is small and means a higher utility.

5.2   Result Evaluation

Effects of the Value of z in SRNM. As we discuss in Sect. 4.2, the best number of reported itemsets (denoted by z) for each call of SRNM is influenced by many factors, such as the sizes of datasets, hardware condition and so on. In general, we can achieve the best utility but the poorest time efficiency by setting $z = 1$. Sometimes one would like to set a bigger value of z (with some loss on the accuracy of results) to get results faster. We notice that the proportion of frequent itemsets is normally tiny, which is illustrated in Table 2. So we try z from 1 to 10 to find a proper value in our experiment settings. Here we conduct experiments of results accuracy and time efficiency on different value of z to set a proper value of z in following experiments:

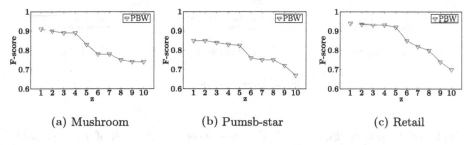

(a) Mushroom   (b) Pumsb-star   (c) Retail

Fig. 1. F-score vs. z

Take Figs. 1(a) and 2(a) for example, we find that the result accuracy (F-score) is stably high when $z \leq 4$, and the time cost is stably small when $z \geq 4$. So for Mushroom dataset, number "4" is a critical point to get both relative high utility and time efficiency under our hardware condition. In Figs. 1(b) and 2(b), we find 4 or 5 are both critical points because they both lead to relative high F-score and little time cost. Moreover, the critical point for Retail is 5 as shown in Figs. 1(c) and 2(c) in the same way. Empirically, the critical point is approxiamately $\lceil \lg |D| \rceil$ and we can set z using that value.

Another important consideration is that when $z = 1$, SRNM is equal to RNM. We can see from Fig. 2 that when z is equal to 1, SRNM gets a poor time efficiency. We find that RNM is not computable in real world. That is why we introduce SRNM rather than applying RNM directly, and that also illustrates the importance of SRNM.

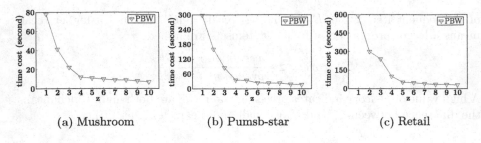

(a) Mushroom   (b) Pumsb-star   (c) Retail

Fig. 2. Time cost vs. z

Efficiency. We conduct experiments on the time efficiency among all algorithms. As shown in Fig. 3, our algorithm outperforms the other three algorithms, and is over 20% more efficient than SFA. One reason is that we apply SRNM, the other reason is that we don't make too much pre-computation (on truncating, estimating k_{max} and so on).

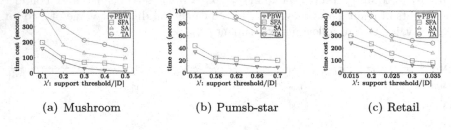

(a) Mushroom   (b) Pumsb-star   (c) Retail

Fig. 3. Time cost vs. λ'

Effects of *Support* Threshold. Figure 4 show the changes of *F-score* with different values of *support* threshold. We can see that PBW outperforms TA, SA and SFA on these three datasets. Especially, we find that our algorithm get higher F-scores with approximate 5% to 10% increase than that of the most state-of-the-art SFA algorithm. By combining Fig. 4 with Fig. 3, we can see that both utility and efficiency of PBW are better than those of SFA and the other two algorithms.

Effects of Privacy Budget (i.e., Added Noise). Figure 5 show the changes of *F-score* when varying *privacy budget*. we can see that a better result utility can be obtained with the increase of ϵ, this is because that a larger value of ϵ means a smaller mount of noise added into original supports. We find that PBW achieves a better utility than TA, SA and SFA for the same privacy constraint, and is about 5% to 10% better than SFA. Moreover, our algorithm shows a more stable performance when ϵ is very small. The reason is that, by applying SRNM, when ϵ decreases, the increasement of noise amount in PBW is much smaller

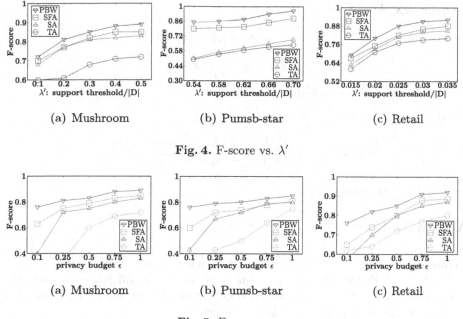

Fig. 4. F-score vs. λ'

Fig. 5. F-score vs. ϵ

than that in the other algorithms, and our budget allocation strategy uses ϵ more effectively in the worse cases.

6 Conclusion

This paper proposes a novel Apriori-based frequent itemsets mining algorithm with differential privacy, and denotes it by *PrivBUD-Wise*. Unlike other proposed methods, we don't make any preprocessing on the original database, or any estimation on the maximal cardinality of frequent itemsets, thus avoid extra information loss. Based on RNM, We first propose a mechanism SRNM with strict proof, SRNM is more efficient than RNM and thus computable in real world. By reducing waste of *privacy budget*, our algorithm achieves a good trade-off between utility, privacy and efficiency. We first propose a reasonable biased budget allocation strategy, which makes our algorithm make better use of privacy budget and achieve better utility under relatively high level privacy constraint (i.e., a small ϵ). Extensive experiments on real datasets verify the higher utility of *PrivBUD-Wise* compared with several state-of-the-art algorithms with same privacy level.

Acknowledgement. This work is partially supported by National Natural Science Foundation of China (NSFC) under Grant No. 61772491, No. U170921, Natural Science Foundation of Jiangsu Province under Grant No. BK20161256, and Anhui Initiative in Quantum Information Technologies AHY150300.

Appendix A

Lemma 4. *For $\forall\ \delta > 0$, and x is a draw from Lap(b), then:*

$$P[x \geq \delta + 1] = e^{-\frac{1}{b}}P[x \geq \delta]$$

where P *denotes the probability.*

Proof.

$$\frac{P[x \geq \delta + 1]}{P[x \geq \delta]} = \frac{\frac{1}{2b}\int_{\delta+1}^{\infty} e^{-\frac{x}{b}}dx}{\frac{1}{2b}\int_{\delta}^{\infty} e^{-\frac{x}{b}}dx} = \frac{e^{-\frac{\delta+1}{b}}}{e^{-\frac{\delta}{b}}} = e^{-\frac{1}{b}}$$

Hence, this lemma follows.

Appendix B

Proof of Theorem 2: Fix $D = D' \cup \{t\}$, where t is a *transaction*. Let v, respectively v', denote the vector of query counts of SRNM when the dataset is D, respectively D'. We use m to denote the number of queries(equal to the number of candidate itemsets). Then we have:

(1) $v_i \geq v_i'$ for $\forall i \in [m]$;
(2) $1 + v_i' \geq v_i$ for $\forall i \in [m]$;

Given an integer z, for every $z' \in [z]$, fix any set $j = (j_1, j_2, \ldots, j_{z'}) \in [m]^{z'}$, to prove differential privacy, we want to bound the ratio(from above and below) of the probabilities that $(j_1, j_2, \ldots, j_{z'})$ is selected with D and with D.

Fix r_{-j}, which is a draw from $[Lap(z/\epsilon)]^{m-z'}$ and is used for all noisy query counts except z' counts corresponding to $j = (j_1, j_2, \ldots, j_{z'})$. We use $P[j|\theta]$ to denote the probability that the outputs of SRNM is j under condition θ.

Firstly, we prove that $P[j|D, r_{-j}] \leq e^{\epsilon}P[j|D', r_{-j}]$: For every $k \in j$, define

$$r_k^* = \min_{r_k} : v_k + r_k > v_i + r_i, \forall i \in [m]\backslash j$$

Then j is the output with D iff for $\forall k \in j$: $r_k \geq r_k^*$.
For all $i \in [m]\backslash j, k \in j$:

$$v_k + r_k^* > v_i + r_i$$
$$\Rightarrow (1 + v_k') + r_k^* \geq v_k + r_k^* > v_i + r_i \geq v_i' + r_i$$
$$\Rightarrow v_k' + (r_k^* + 1) > v_i' + r_i$$

So, if for $\forall k \in j$: $r_k \geq r_k^* + 1$, then the output with D' will be j and the added noise will be (r_j, r_{-j}). So we have:

$$P[j|D', r_{-j}] \geq P[r_k \geq r_k^* + 1 | k \in j] = \prod_{k \in j} P[r_k \geq r_k^* + 1]$$

$$= \prod_{k \in j} e^{-\frac{\epsilon}{z}} P[r_k \geq r_k^*] = e^{-\frac{z'\epsilon}{z}} P[j|D, r_{-j}] \geq e^{-\epsilon} P[j|D, r_{-j}] \tag{3}$$

The second equality is due to Lemma 4. multiply by e^ϵ: $P[j|D, r_{-j}] \leq e^\epsilon P[j|D', r_{-j}]$

We now prove that $P[j|D', r_{-j}] \leq e^\epsilon P[j|D', r_{-j}]$. For every $k \in j$, define: $r_k^* = \min_{r_k} : v_k' + r_k > v_i' + r_i, \forall i \in [m]\backslash j$, then j is the output when the dataset is D' iff for $\forall k \in j$: $r_k \geq r_k^*$.

For all $i \in [m]\backslash j, k \in j$:

$$v_k' + r_k^* > v_i' + r_i$$
$$\Rightarrow 1 + v_k + r_k^* \geq 1 + v_k' + r_k^* > 1 + v_i' + r_i \geq v_i + r_i$$

So, if for $\forall k \in j$: $r_k \geq r_k^* + 1$, then the output with D will be j and the added noise will be (r_j, r_{-j}). So we have:

$$P[j|D, r_{-j}] \geq P[r_k \geq r_k^* + 1|k \in j] = \prod_{k \in j} P[r_k \geq r_k^* + 1]$$

$$= \prod_{k \in j} e^{-\frac{\epsilon}{z}} P[r_k \geq r_k^*] = e^{-\frac{z'\epsilon}{z}} P[j|D', r_{-j}] \geq e^{-\epsilon} P[j|D', r_{-j}]$$

$$(4)$$

multiply by e^ϵ: $P[j|D', r_{-j}] \leq e^\epsilon P[j|D, r_{-j}]$. Hence this theorem follows.

References

1. http://fimi.ua.ac.be/data/
2. Agrawal, R., Srikant, R., et al.: Fast algorithms for mining association rules. In: Proceedings of the 20th International Conference on Very Large Data Bases, VLDB, vol. 1215, pp. 487–499 (1994)
3. Bhaskar, R., Laxman, S., Smith, A., Thakurta, A.: Discovering frequent patterns in sensitive data. In: Proceedings of the 16th ACM SIGKDD International Conference on Knowledge Discovery and Data Mining, pp. 503–512. ACM (2010)
4. Cheng, X., Su, S., Xu, S., Li, Z.: DP-Apriori: a differentially private frequent itemset mining algorithm based on transaction splitting. Comput. Secur. **50**, 74–90 (2015)
5. Dwork, C.: Differential privacy. In: Bugliesi, M., Preneel, B., Sassone, V., Wegener, I. (eds.) ICALP 2006. LNCS, vol. 4052, pp. 1–12. Springer, Heidelberg (2006). https://doi.org/10.1007/11787006_1
6. Dwork, C., McSherry, F., Nissim, K., Smith, A.: Calibrating noise to sensitivity in private data analysis. In: Halevi, S., Rabin, T. (eds.) TCC 2006. LNCS, vol. 3876, pp. 265–284. Springer, Heidelberg (2006). https://doi.org/10.1007/11681878_14
7. Dwork, C., Roth, A., et al.: The algorithmic foundations of differential privacy. Found. Trends® Theoret. Comput. Sci. **9**(3–4), 211–407 (2014)
8. Erlingsson, Ú., Pihur, V., Korolova, A.: RAPPOR: randomized aggregatable privacy-preserving ordinal response. In: Proceedings of the 2014 ACM SIGSAC Conference on Computer and Communications Security, pp. 1054–1067. ACM (2014)
9. Fanaeepour, M., Machanavajjhala, A.: PrivStream: differentially private event detection on data streams. In: Proceedings of the Ninth ACM Conference on Data and Application Security and Privacy, pp. 145–147. ACM (2019)

10. Fournier-Viger, P., Lin, J.C.-W., Vo, B., Chi, T.T., Zhang, J., Le, H.B.: A survey of itemset mining. Wiley Interdisc. Rev.: Data Min. Knowl. Discov. **7**(4), e1207 (2017)
11. Ghosh, A., Roughgarden, T., Sundararajan, M.: Universally utility-maximizing privacy mechanisms. SIAM J. Comput. **41**(6), 1673–1693 (2012)
12. Han, J., Pei, J., Yin, Y.: Mining frequent patterns without candidate generation. ACM SIGMOD Rec. **29**, 1–12 (2000)
13. Lee, J., Clifton, C.W.: Top-k frequent itemsets via differentially private FP-trees. In: Proceedings of the 20th ACM SIGKDD International Conference on Knowledge Discovery and Data Mining, pp. 931–940. ACM (2014)
14. Li, N., Qardaji, W., Su, D., Cao, J.: PrivBasis: frequent itemset mining with differential privacy. Proc. VLDB Endow. **5**(11), 1340–1351 (2012)
15. Li, S., Mu, N., Le, J., Liao, X.: Privacy preserving frequent itemset mining: maximizing data utility based on database reconstruction. Comput. Secur. **84**, 17–34 (2019)
16. Wang, N., Xiao, X., Yang, Y., Zhang, Z., Gu, Y., Yu, G.: PrivSuper: a supersetfirst approach to frequent itemset mining under differential privacy. In: 2017 IEEE 33rd International Conference on Data Engineering (ICDE), pp. 809–820. IEEE (2017)
17. Zeng, C., Naughton, J.F., Cai, J.-Y.: On differentially private frequent itemset mining. Proc. VLDB Endow. **6**(1), 25–36 (2012)
18. Zhang, J., Xiao, X., Xie, X.: PrivTree: a differentially private algorithm for hierarchical decompositions. In: Proceedings of the 2016 International Conference on Management of Data, pp. 155–170. ACM (2016)

A Learning Approach for Topic-Aware Influence Maximization

Shan Tian[1], Ping Zhang[2], Songsong Mo[1], Liwei Wang[1(✉)], and Zhiyong Peng[1]

[1] School of Computer Science, Wuhan University, Wuhan, Hubei, China
{tianshan14,songsong945,liwei.wang,peng}@whu.edu.cn
[2] Huawei, Shenzhen, China
zhangping62@huawei.com

Abstract. Motivated by the application of *viral marketing*, the topic-aware influence maximization (TIM) problem has been proposed to identify the most influential users under given topics. In particular, it aims to find k seeds (users) in social network G, such that the seeds can maximize the influence on users under the specific query topics. This problem has been proved to be NP-hard and most of the proposed techniques suffer from the efficiency issue due to the lack of generalization. Even worse, the design of these algorithms requires significant specialized knowledge which is hard to be understood and implemented. To overcome these issues, this paper aims to learn a generalized heuristic framework to solve TIM problems by meta-learning. To this end, we encode the feature of each node by a vector and introduce a deep learning model, called *deep-influence-evaluation-model* (DIEM), to evaluate users' influence under different circumstances. Based on this model, we can construct the solution according to the influence evaluations efficiently, rather than spending a high cost to compute the exact influence by considering the complex graph structure. We conducted experiments on generated graph instances and real-world social networks. The results show the superiority in performance and comparable quality of our framework.

Keywords: Social network · Influence maximization ·
Graph embedding · Reinforcement learning

1 Introduction

As the popularity of social networks, more and more people tend to share information by the word of mouth in their daily life. As a result, social network has become a new platform for *viral marketing*, which is used to promote products, innovations and opinions. Motivated by these applications, Kempe et al. [14] first formalize the seeds selection in viral marketing as a discrete optimization problem, which is known as Influence Maximization (IM). However, recent studies [7,12,19] have argued that the generic IM is not topic-aware and can not be directly applied to online advertising. For example, imagine that we are looking for spokesmen for new sneakers, we will pay more attention to sports star

© Springer Nature Switzerland AG 2019
J. Shao et al. (Eds.): APWeb-WAIM 2019, LNCS 11641, pp. 125–140, 2019.
https://doi.org/10.1007/978-3-030-26072-9_9

than singing star. Thus there have been many efforts [2,7,12,21,22] extending generic IM to topic-aware IM (TIM) to support more personalized and accurate advertising.

In general, there are two ways to solve TIM [20]. The first way is *IM for topic-relevant targets*, which considers that the user is topic-aware, and wants to maximize the influence on topic-relevant users. These studies mainly focus on differentiating users and compute the influence on activated users by their benefits under specific query topics. The second way is *IM for topic-dependent diffusion*, which formalizes that the *edges* are topic-aware, and wants to maximize the influence under a new diffusion model. These studies [7,9] focus on proposing new diffusion models which can capture the dynamic probabilities under different query topics. The most commonly used model associates each edge with a probability under each topic and calculates the propagation probability under each query as the expected summation of related probabilities.

However, the real scenario is that people have different interests in different topics, and the probability of people influencing each other changes with topics, so neither of these two ways can solve TIM problems solely, not to mention that they are more or less inadequate. For example, the design of all these algorithms requires specialized human knowledge i.e., the dynamic programming method, and trial-and error, which can be very hard to be understood. Moreover, they suffer from the efficiency issues. For example, most current state-of-the-art solutions [22,25] to solve the first branch problems need to generate hundreds of thousands sample sets randomly to select the most influential user online or offline, time-consuming or space-consuming. What's more, regardless of the high cost, problems with same structure but varied in data require new sample processes performed again and again.

In this paper, we establish a greedy heuristic framework to learn the algorithm instead. This data-driven method avoids the traditional complex algorithm designing processes by adopting a combination of graph embedding and reinforcement learning. On the one hand, the embedding method considers both users' preferences and dynamic influence probabilities, which can cover both two branches of TIM and then encode the graph to a vector space. On the other hand, the framework builds a neural network called *deep-influence-evaluation-model* to estimate the influence of each candidate under different circumstances and adopts *reinforcement learning* method to train it. Contrary to the traditional methods, the main advantage of our *deep-influence-evaluation-model* is that it can be trained offline while stored with very little disk space. Based on the model outputs, the framework uses a greedy policy to construct a feasible solution quickly which avoids the heavy online influence computation. This mechanism assures our framework can be generalized to different problem instances without new training processes.

More specifically, our contributions can be summarized as follows:

- To solve the TIM comprehensively, we consider both the difference of users' interests and the dynamics of diffusion probabilities together in our model to solve the two branches of TIM simultaneously, which has not been proposed so far to our limited knowledge.
- We propose a new *graph embedding* network, called *Diffusion2Vec* which can extract features for each user in social network automatically, capturing the properties of each user according to graph structure and his own attributes.
- Based on the embedding network, we define an estimation model called *deep-influence-evaluation-model* (*DIEM*) which is used to calculate the influence of candidate users according to their embeddings, the current partial solution and the query topics. Based on the influence, a feasible solution can be constructed by greedy node selection and addition.
- We adopt an algorithm of reinforcement learning, called *double DQN with prioritized experience replay* to train models. The main advantage of this algorithm is that it can deal with overestimating and delayed reward in a data-efficient way. The training process is set up in such a way that the policy will aim to maximize the targeted influence under each query *directly*.
- We conduct experiments on generated graph instance and real-world social network *Twitter*. The results show the superiority in performance and comparable solution quality of our framework.

2   Problem Definition

In this section, we will define the topic-aware influence maximization (TIM) problem and give a rough sketch our greedy framework.

2.1   TIM Problem

To facilitate our presentation, all frequently used notation are listed in Table 1. TIM introduces *topics* to describe both information characteristics and users' interests, and calculated the influence based on not only the seed set S but also the query topics τ. It focuses on maximizing the *targeted influence* over users who are relevant to the query topics under specific diffusion models.

Let τ denotes the query topics and $\sigma_G(S|\tau)$ denotes the targeted influence spread by the seeds S in an instance of the influence propagation process on graph G. Intuitively, the TIM problem finds a seed set S* with k users to maximize the targeted influence spread by a seed set S over social network G and can be formally defined as follows:

Definition 1. *TIM.* *Given a graph G, a positive integer k and a series of targeted topics τ, TIM selects a set* S* *of k nodes from V as seed set to maximize the* targeted influence $\sigma_G(S^*|\tau)$, *i.e.,* $\sigma_G(S^*|\tau) = \arg\max_{S \subseteq V \wedge |S|=k} \sigma_G(S|\tau)$.

Table 1. Frequent notations used across the paper.

Notation	Meaning
G, V, E	The social network, the vertex set, the edge set
$\tau, k, \widehat{S}, S$	The query topics, the size of seed set, the seed set, current solution set
$u_v^{(t+1)}$	The node embedding of v at $t+1$ iteration
X_v	The properties of node v, including the tag indicted whether it has been selected into S and user profile
p_{uv}^{τ}	The probability of node u successively activate node v under query topics τ
Q	The function to evaluate the influence of each user
$\widehat{Q}$	The estimation model to estimate the influence of each user
$\sigma_G(S^*|\tau)$	The maximum targeted influence spread among all seed set with size k for social graph G under query topics τ
$\sigma_G(S|\tau)$	The targeted influence spread by S for social graph G under query topics τ

2.2   Greedy Framework

We will focus on the popular pattern for designing approximation and heuristic algorithms, namely greedy algorithm. The greedy framework will construct a solution by sequentially adding one node to the partial solution S, based on some evaluation function Q which measures the influence of each candidate in the context of the current partial solution and the query topics. Specifically:

1. A partial solution is represented as an ordered list $S = (v_1, v_2, \ldots, v_{|s|}), v_i \in V$, and $S = \emptyset$ at the start. $\overline{S} = V\backslash S$ denotes the set of candidate nodes for addition, conditional on S. Furthermore, we use the first element of a vector to represent whether the node is selected into S, i.e., $X_{v[0]} = 1$ if $v \in S$ and $= 0$ otherwise.
2. The quality of a partial solution S can be calculated as the targeted influence $\sigma(S|\tau)$ of S.
3. The generic greedy framework selects a node v to be added according to which v maximized the evaluation function, $Q((v, S)|\tau) \in R$. Then the partial solution S will be extended as $S := (S, v^*)$, where $v^* := \arg\max_{v \in \overline{S}} Q((v, S)|\tau)$. This step is repeated until the size of solution set achieved k.

Recently, deep learning has promoted the development of computer vision and speech recognition. Based on feeding sufficient data into the deep neural networks, it's possible to learn better representations than hand-crafted features [16]. Thus, we want to design a powerful deep function Q to estimate the influence of each candidate under certain circumstances.

3   Graph Representation

As mentioned above, the *deep-influence-evaluation-model* $(\widehat{Q})$ must evaluate the influence for each candidate under different circumstances. Intuitively, $\widehat{Q}$ should summarize the state of the partial solution and figure out the influence of each candidate in the context of such a solution and query topics. Here, both the state of the solution and the context of a node v can be very complex, hard to be described in closed form. In order to represent such complex context over social networks, we will leverage a deep learning architecture called *Diffusion2Vec* to parameter model $\widehat{Q}$.

In this section, we first introduce our *topic-aware model* which can capture the dynamics of TIM, then give a thorough explanation about our embedding model: *Diffusion2Vec* and estimation model: $\widehat{Q}$.

3.1   Topic-Aware Model

As we have discussed in the Introduction, the first branch of TIM ignores the dynamic diffusion probabilities dependent on query topics while the second does not consider to differentiate users according to their interests. So neither of these two branches can solve TIM problems completely.

Thus here we propose a model called *topic-aware model* motivated by the user-based collaborative filtering algorithm to merge the two branches together. First, to model a social network as an online advertising platform, we extend each node in G to be associated with a user profile represented by a weighted vector. For example, for a user v who is very interested in sports, music but not in politics, his profile may be described as {<music, 0.7>, <sport, 0.3>, <politics, 0.0>}, where 0.7 called his *benefit* under the *music* topic ($B_v^{\tau_1}$). Then we calculate the sum of benefits related to the query topics to distinguish the targeted users. Next, we define the *similarity* (*sim*) between two users by the *cosine similarity* between their profile vectors. Finally, we can calculate the propagation probability that a user u successfully activate user v with three factors: the initial edge weight w_{uv} which is initialized in $(0, 1]$ and stands for the contact frequencies; the similarity between two users; the sum benefits of user v towards the query topics τ. Formally, the propagation probability between user u and user v under topics τ can be defined as:

$$p_{uv}^{\tau} = (\gamma_1 w_{uv} + \gamma_2 sim(u,v) + \gamma_3 B_v^{\tau})/3, \tag{1}$$

where $\gamma_1, \gamma_2, \gamma_3 \in (0,1)$ and they are used to balance the weights of contact frequency, user similarity and targeted benefits. Since all the value of three factors are in range $(0, 1]$, the propagation probability p_{uv}^{τ} is restricted to range $(0, 1]$, and can then be used in the standard IC model for computing the influence spread. The targeted influence is computed as the summation of benefits of the related users who are activated according to the new diffusion probabilities.

3.2 Embedding: *Diffusion2Vec*

Since graph is an important data representation and traditional graph analysis methods suffer from high computation and space overhead, the graph embedding technique has been put forward to solve the graph analysis problem effectively and efficiently. It related to two aspects of researches, i.e. graph analytics [11] and representation learning [3,4,6]. It converts graph data into low dimensional vectors while the graph structure and other useful information are preserved. For example, *Structure2Vec* [15] embeds each node $v \in V$ into a p-dimensional feature space thus each node v can be easily described as a p-dimensional vector, u_v and the graph can be represented by the set of these vectors.

However due to the lack of consideration about node property and information diffusion characteristics, it can not be directly used in our problem settings. So we proposed a new method named *Diffusion2Vec* which considers both the users' attributes and the dynamic propagation probabilities.

Specifically, the *Diffusion2Vec* will initialize the embedding of each node $u_v^{(0)} = 0$ and for all $v \in V$ update the embeddings synchronously at each iteration as

$$u_v^{(t+1)} \leftarrow F(X_v, \{u_u^{(t)}\}_{u \in N(v)}, \{p(u,v)\}_{u \in N(v)}; \Theta). \tag{2}$$

X_v denotes a vector which contains useful user information. More specifically, the first element of X_v, $X_{v[0]}$ corresponds to a node-specific tag which means whether the node is selected so far, i.e., $X_{v[0]} = 1$ if $v \in S$ and $= 0$ otherwise. The rest part of X_v, named *benefit vector*, correspond to the profile in the *topic-aware model*. For example, for a user v with profile $\{<\text{music}, 0.7>, <\text{sport}, 0.3>, <\text{politics}, 0.0>\}$ and hasn't been selected into S, the X_v will be $\{0, 0.7, 0.3, 0.0\}$. $N_{(v)}$ denotes the neighbors of node v in the graph G. $p(u,v)$ corresponds to probability in the *topic-aware model*. F denotes a nonlinear mapping such as neural network or kernel function.

Based on the update formula, one can see that the embedding of node v can be seen as a combination of its own features and the aggregated impact of other nodes. In other words, the node features can be propagate to other nodes via the nonlinear propagation function F and the more update iterations were carried on, the farther away the features will be propagated and get aggregated at distant nodes. Finally, the update process will terminate until the number of iterations reaches a pre-defined parameter: T. At that time, each node embedding $u_v^{(T)}$ will contain information about its T-hop neighborhoods as determined by graph topology, the involved node features. It is worth nothing that when the X_v degenerates to a binary scalar, this form of node embedding can be used to solve the generic IM problems.

Next, we will design the powerful *deep-influence-evaluation-model* to evaluate the influence of node with these vector representations.

3.3   Deep Influence Evaluation Model ($\widehat{Q}$)

Specifically, we adopted the neural network as the mapping function in *Diffusion2Vec*, then the embedding can be defined as:

$$u_v^{(t+1)} \leftarrow relu\left(\theta_1 X_v + \theta_2 \sum_{u \in N(v)} u_u^{(t)} + \theta_3 \sum_{u \in N(v)} relu(\theta_4 p(u,v))\right), \quad (3)$$

where $\theta_1 \in \Re^{p \times (|\tau|+1)}$, $\theta_2, \theta_3 \in \Re^{p \times p}$ and $\theta_4 \in \Re^p$ are the embedding parameters and relu is the rectified linear unit which is widely used as an activation function in artificial neural networks.

More specifically, we use the embedding $u_v^{(T)}$ for node v and the pooled embedding over the partial solution S, $\sum_{u \in S} u_u^{(T)}$ as the substitution for v and S respectively. Finally the estimation model $\widehat{Q}$ can be defined as:

$$\widehat{Q}(S, v; \Theta, \tau) = \theta_5 relu\left(\left[\theta_6 \sum_{u \in S} u_u^{(T)}, \theta_7 u_v^{(T)}\right]\right), \quad (4)$$

where $\theta_5 \in \Re^{2p}$, $\theta_6, \theta_7 \in \Re^{p \times p}$ and $[.,.]$ is the concatenation operator. The number of iterations T for the graph embedding computation is usually small, such as $T = 4$.

What deserves our attention is the parameter set Θ. Since the model $\widehat{Q}$ is based on the embedding $u_v^{(T)}$, the parameter set Θ is then a collection of 7 parameters actually. That is $\Theta = \{\theta_i\}_{i=1}^7$ and all these parameters must be learned. However, due to the lack of training labels, we will learn these parameters *end-to-end* with reinforcement learning, which will be explained thoroughly in the following section.

4   Parameters Learning

This section shows how to learn the parameters Θ in $\widehat{Q}$. Note that the definition of $\widehat{Q}$ lends itself to an RL formulation and thus can be used as the state-action value in RL directly. Moreover, the state-of-the-art RL algorithms [23,24] will provide strong supports for learning the parameters.

4.1   RL Formulation

Based on the introduction to RL in [27], we show the instantiations of RL components for TIM in Table 2 and define the states, actions, rewards, policy in the RL framework for TIM as follows:

1. *States*: a state is a sequence of actions (nodes) added into the solution set. In other words, the current state equals to the current solution set S. It is easy to see that the *terminal state* comes when the size of S reaches k.

2. *Actions*: an action corresponds to a node in G which have not been added into the solution set S (the tag of $X_{v[0]} = 0$). Since we have represented each node v as a p-dimensional vector, here the action is a p-dimensional vector too.

3. *Rewards*: the reward (after taking action v) is defined as the marginal targeted influence spread by the user v under the query topics. That is, $r(S, v) = \sigma((S \cup \{v\})|\tau) - \sigma(S|\tau)$. As such, the cumulative rewards of a *terminal state* can be described by the seed set $\widehat{S}$, as $R(\widehat{S}) = \sum_{i=1}^{k} r(S_i, v_i)$ which equals to $\sigma(\widehat{S}|\tau)$.

4. *Policy*: based on the *deep-influence-evaluation-model* $\widehat{Q}$, we will adopt a deterministic greedy policy $\pi(v|S, \tau) := \arg\max_{v' \in \overline{S}} \widehat{Q}(v', S|\tau)$. Selecting action v corresponds to add node v of $\overline{S}$ to the partial solution S, which results in a new state S' and a reward $r(S, v)$.

Table 2. Instantiation of reinforcement learning components for TIM.

State	Action	Reward	Termination
Sequence of selected nodes	Add a node to S	The marginal influence	$|S| = k$

4.2   Learning Algorithm

We use *double Deep Q-Networks* (DDQN) [13] with *prioritized experience replay* [26], which can be generalized to work with large-scale function approximation to perform end-to-end learning of the parameters Θ in $\widehat{Q}(S, v; \Theta, \tau)$. We use the term *episode* to represent a series operations of node addition starting from an empty set $\emptyset$, and until termination; a *step* within an episode is a single action (node addition). This approach will update the parameters with a batch of weighted samples from a dataset E, rather than the single sample being currently experienced. The dataset E is populated during previous episodes with an interval, called *n-step*.

The advantage of *prioritized experience replay* is that the model can be learned more effectively. An experience with high expected learning progress measured by the magnitude of its temporal-difference (TD) error will be replayed more frequently to learn more efficiently. And the DDQN has been shown to reduce the overestimations effectively. Both the two methods lead to much better performance when using a neural network as a function approximation [13], a property that also applies for our model $\widehat{Q}$. The *prioritized experience replay* replaces the sampling methods used by DDQN with stochastic prioritization and importance sampling as illustrated in Algorithm 1.

Algorithm 1. Double DQN with prioritized experience replay

Input:
   minibatch b, step-size η, replay period B and capacity N, exponents α and β, budget L.

1: Initialize replay memory H=$\emptyset$, Δ=0, $p_1 = 1$
2: Observe S_0 and choose $A_0 \sim \widehat{Q}(S, v; \Theta, \tau)$
3: **for** $t = 1$ to L **do**
4:     Observe S_t, R_t, γ_t
5:     Store transition $(S_{t-1}, A_{t-1}, R_t, \gamma_t, S_t)$ in H with maximal priority $p_t = \max_{i<t} p_i$
6:     **if** $t \equiv 0 \bmod B$ **then**
7:         **for** $j = 1$ to b **do**
8:             Sample transition $j \sim P(j) = p_j^\alpha / \sum_i p_i^\alpha$
9:             Compute importance-sampling weight $w_j = (N \cdot P(j))^{-\beta}/\max_i w_i$
10:            Compute TD-error $\delta_j = R_j + \gamma_j Q_{t\,\mathrm{arg}\,et}(S_j, \arg\max_a Q(S_j, a)) - Q(S_{j-1}, A_{j-1})$
11:            Update transition priority $p_j \leftarrow |\delta_j|$
12:            Accumulate weight-change $\Delta \leftarrow \Delta + w_j \cdot \delta_j \cdot \nabla_\theta Q(S_{j-1}, A_{j-1})$
13:        **end for**
14:        Update weights $\theta \leftarrow \theta + \eta \cdot \Delta$, reset Δ=0
15:        From time to time copy weights into target network $\theta_{t\,\mathrm{arg}\,et} \leftarrow \theta$
16:    **end if**
17:    Choose action $A_t \sim \pi_\theta(S_t)$
18: **end for**

---

5   Experiment

In this section, we study the performance of *DIEM* on generated instances and real world dataset with different parameter settings. First, we reveal the details of experimental settings. Then, we give an introduction to the baseline solutions [22]. Finally, we report the experimental results and give an explanation.

5.1   Experimental Setup

Instance Generation. We first generate graph instances to evaluate the proposed method against existing state-of-the-art solution WRIS [22]. We generate Barabasi-Albert (BA) [1] graphs which have been used to model many real-world networks. For a given range on the number of nodes, e.g. 1M-10M, we sample the number of nodes uniformly at random in that range, then generate a graph according to BA. Finally we associate each node with a regularized vector randomly to simulate users' profiles.

Real World Dataset. To test the performance on real scenario, we use a dataset collected from *Twitter* for performance evaluation. This dataset contains 41.6 million users with 476 millions tweets [17]. We extract 50 topics from it and user profile is represented by a weighted term vector in the topic space. This term vector is generated by aggregating all the tweets into a giant document and use LDA model[1] to mine users' preferences in the topic space. To test the scalability with increasing number of users, we sampled 10M, 20M, 30M, 40M users from the dataset.

Queries. The two factors that make up a query are topics and seed set size. For the topics, we use queries from AOL search engine[2]. Given 200 topics defined in advance, we first filter the topic queries and retain those only containing our topics manually. For the seed set size, we adopt $k = 30$ for initial estimates on generated instances and vary the number of k from 10 to 50 on *twitter* for further evaluation.

Parameter Settings. The hyperparameters and parameter settings used in our experiments are shown in Table 3. For our method, we simply tune the hyperparameters on small graphs (i.e., the graphs with less than 50 nodes), and fix them for large graphs.

Table 3. Main configurations used in experiments with default values highlighted.

Datasets	Users	Edges	Q Topics	k
Generated	1M–2M, ..., 9M–10M	40–80M, ..., 360–400M	1	30
Twitter	10M, 20M, 30M, **40M**	0.7B, 1.1B, 1.2B, **1.3B**	1, 2, **3**, 4, 5	10, 15, ..., **30**, ..., 50
γ_1	γ_2	γ_3	**Batch size**	**n-step**
0.8	0.8	1	64	3

5.2 Baseline Solutions

To solve the TIM problems, [22] has proposed a weighted sampling technique (WRIS) based on RIS and achieved an approximation ratio of $(1 - 1/e - \varepsilon)$. However due to the method needs to generate hundreds of thousands of random sample sets and requires intensive computation overhead, they further proposed two disk-based solutions, RR and IRR, to support the online query process. The idea is to put the sampling procedure from online to offline and build index on the random sample sets for each topic. During query processing, RR directly loads all the related random sample sets into memory and uses the greedy algorithm to find the top-k seed users. IRR further improves over RR by incrementally loading the most promising RR sets into memory and adopts the top-k aggregation strategy to save computation costs.

Like in [22], we use WRIS as a baseline solution because it uses an online sampling and thus has an effective assurance. Since IRR outperforms RR in the *Twitter* dataset as they reported, we adopt IRR as a baseline to compare the efficiency. That is, we compare the method based on deep learning (*DIEM*) with WRIS and IRR to evaluate the quality by expected targeted influence and efficiency by average running time and disk space respectively.

---

[1] https://github.com/kenneth-orton/TwitterLDATopicModeling.
[2] https://www.aolsearch.com/.

In the following experiments, we evaluate the performance on generated graph instances and *Twitter* respectively.

5.3   Experiments on Generated Instance

The graph embedding framework enable us to train and test on graphs of different sizes, since the same set of model parameters are used. How does the performance of the learned algorithm using small graphs generalize to test graphs of larger sizes? To investigate this, we train *DIEM* on graphs with 0.5M–1M nodes, and test its generalization ability on graphs with up to 10M nodes regarding to one topic and 30 seeds. The results are shown as Table 4.

Table 4. Targeted influence spread when varying test graph size. Values are average *targeted influence* over 10 test instances.

Test size	1M–2M	3M–4M	5M–6M	7M–8M	9M–10M
WRIS	372.1	472.7	548.2	668.4	792.3
DIEM	375.6	478.3	556.3	673.9	798.6

We can see that *DIEM* achieves a very good performance compared with *WRIS*. As the graph size increases to 10M, there are still only subtle differences between the targeted influence they produced, which mainly come from the exploration and utilization in RL. That is to say, our framework can be generalized to larger graph instances without new training processes.

5.4   Experiments on Real-World Dataset

In the following experiments, we evaluate the efficiency and effectiveness of our proposed framework on real-world dataset collected from *Twitter* w.r.t. increasing seed users k, graph size $|V|$ and query topics $|\tau|$ respectively.

Vary the Seed Set Size

Comparison of Efficiency. We first examine the performance with increasing seed set size k. The running time and disk cost are shown in Fig. 1. It is obvious that the methods based on deep learning architecture (*DIEM*) and disk index (*IRR*) are significantly faster than the online sampling method (*WRIS*). The average response time to a query using *DIEM* and *IRR* are 160x and 120x times smaller than *WRIS* respectively. However, *DIEM* takes 100x times less disk space than *IRR*. This is because *IRR* needs to store abundant precomputed sampling sets under each topic while *DIEM* only needs to maintain the model parameters.

As k increases, it takes a slightly longer time for both *DIEM* and *IRR* methods to answer a query. However, *IRR* still spends a little more time than *DIEM*.

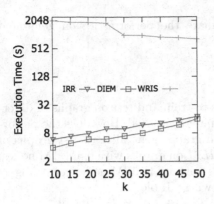 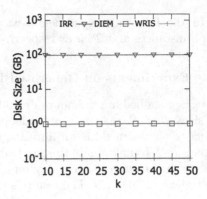

Fig. 1. Varying the seed set size: k

This is because both methods need more iterations to find the seed users, causing more computations but *IRR* costs more disk I/O than *DIEM*. To our surprise, the performance of *WRIS* is slightly faster as k increases. The reason is that the performance of *WRIS* is mainly dependent on the number of RR sets generated, which is inversely proportional to the optimal targeted influence. However, the optimal targeted influence is directly proportional to k.

Comparison of Effectiveness. In addition to the result that *DIEM* and *IRR* are much faster than *WRIS*, the solutions they generated are also not worse than *WRIS*. We report the *targeted influence* of seed sets returned by all the methods in Table 5. There are almost no difference between all the methods. Thus the *targeted influence* spread for all methods will not be presented in the rest of the experiment section, as the results show similar patterns.

Table 5. Targeted influence spread when varying k.

k	WRIS	IRR	DIEM
10	8674.2	8673.7	8673.9
15	10667	10665	10666
20	12089	12099	12093
25	13101	13097	13098
30	13930	13931	13925
35	14619	14616	14617
40	15210	15211	15209
45	15732	15735	15730
50	16212	16214	16215

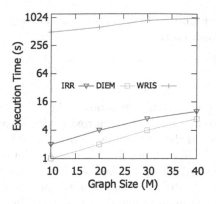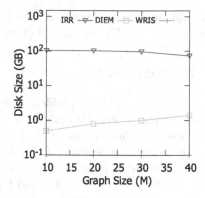

Fig. 2. Varying the graph size: $|V|$

Vary the Graph Size

Comparison of Efficiency. When we increase the graph size, i.e., $|V|$, to test the scalability of our proposed architecture. The results are shown in Fig. 2. All the methods need more time to generate a solution when the graph becomes larger cause more sampling or computation need to be processed. Nevertheless *DIEM* and *IRR* clearly outperform *WRIS* by great margins in all scenarios. However, *DIEM* still takes much less disk space than *IRR*. This is because, as the graph size grows, there are more *RR* sets needed to be built for the *IRR* index while *DIEM* only needs to maintain more parameters in neural networks. It shows that *DIEM* is more effective for larger graphs without compromising its performance superiority.

Vary the Number of Query Topics

Comparison of Efficiency. When we vary the number of query topics $|\tau|$ from 1 to 5, as shown in Fig. 3, the results demonstrate similar pattern: *DIEM* and *IRR* are at least two orders of magnitude better than *WRIS* in a social network

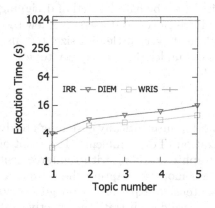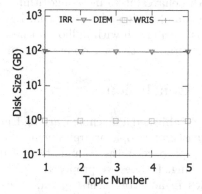

Fig. 3. Varying the query topics number: $|\tau|$

with millions of users. However the solution found by *IRR* takes similar or longer time than our *DIEM*, but is still of worse disk cost.

6   Related Work

IM problem is a NP-Hard problem and has been extensively studied. Approaches to tackling the generic IM problem have mainly two flavors: approximation algorithm ans heuristics. Kempe et al. [14] proposed the first simple greedy algorithm with an approximation ratio of $(1 - 1/e - \varepsilon)$. Since then, there has been a lot of research devoted to improving the efficiency while keeping the theoretical bound [5,10,18]. Although CELF [18] and CELF++ [10] have significantly improved the running time, the methods were only examined in small graphs with thousands of vertexes. RIS [5] uses random sampling technique and is the first method scalable enough to handle graphs with millions of vertexes. The heuristics on IM improved the efficiency by discarding the theoretical bound. For example, Chen et al. [8] used vertex degree as a selection criterion. They propose that fine-tuned heuristics may provide truly scalable solutions to IM with satisfying influence spread and blazingly fast running time.

The above methods cannot be directly applied in online advertising because the same seeds are returned for different advertisements. To solve this problem, the topic-aware IM (TIM) was proposed in [2,22]. Inflex [2] first precomputes a number of top-k seed sets offline and process the online query by finding nearest neighbors among the neighbor approximation. Li et al. [22] adopted a weighted sampling technique based on RIS to make sure the targeted users relevant to the advertisement have a higher probability to be sampled. To meet the real-time processing requirement, they further devise two disk-based index structures to push the sampling procedure from online to offline. However, the solution [22] requires nearly 100 GB disk size to store the sampling sets on handling 5 topics and a graph with 40 million users. Moreover, new sampling RR sets needs to be generated when the method runs on a new graph.

Compared with existing topic-aware IM solutions, our greedy heuristic framework behaves like a meta-algorithm which is driven by data instead of designing the complex algorithm. Our solution not only retrieves seed users in a few seconds with graph with millions of nodes while costs very little disk size, but also is generalized to solve different graph instances under the same query topics.

7   Conclusion

We established a machine learning framework for automatically designing greedy heuristics for topic-aware influence maximization (TIM) problems. The point of our approach is the combination of deep graph embedding with reinforcement learning. Besides we proposed a new diffusion model to capture the characteristics in advertisements spread. Through extensive experiments on generated graph instances and real world social network, we demonstrate the effectiveness of the proposed framework as compared to manually-designed TIM algorithms.

Moreover, the learned heuristics can be generalized to solve different TIM problems, i.e., TIM of larger graph size, TIM on different graph instances under the same topics with all excellent performance.

Acknowledgements. This work is supported by the National Key Research and Development Program of China (Project Number: 2018YFB1003402), key projects of the national natural science foundation of China (Project Number: U1811263) and the Fundamental Research Funds for the Central Universities (Project Number: 2042017kf1017).

References

1. Albert, R., Barabási, A.: Statistical mechanics of complex networks. CoRR cond-mat/0106096 (2001)
2. Aslay, Ç., Barbieri, N., Bonchi, F., Baeza-Yates, R.A.: Online topic-aware influence maximization queries. In: Proceedings of the 17th International Conference on Extending Database Technology, EDBT 2014, Athens, Greece, 24–28 March 2014, pp. 295–306 (2014)
3. Bengio, Y., Courville, A.C., Vincent, P.: Representation learning: a review and new perspectives. IEEE Trans. Pattern Anal. Mach. Intell. **35**(8), 1798–1828 (2013)
4. Bordes, A., Usunier, N., García-Durán, A., Weston, J., Yakhnenko, O.: Translating embeddings for modeling multi-relational data. In: Advances in Neural Information Processing Systems 26: 27th Annual Conference on Neural Information Processing Systems 2013. Proceedings of a Meeting Held at Lake Tahoe, Nevada, United States, 5–8 December 2013, pp. 2787–2795 (2013)
5. Borgs, C., Brautbar, M., Chayes, J.T., Lucier, B.: Influence maximization in social networks: towards an optimal algorithmic solution. CoRR abs/1212.0884 (2012). http://arxiv.org/abs/1212.0884
6. Cai, H., Zheng, V.W., Chang, K.C.: A comprehensive survey of graph embedding: problems, techniques, and applications. IEEE Trans. Knowl. Data Eng. **30**(9), 1616–1637 (2018)
7. Chen, W., Lin, T., Yang, C.: Real-time topic-aware influence maximization using preprocessing. In: Thai, M.T., Nguyen, N.P., Shen, H. (eds.) CSoNet 2015. LNCS, vol. 9197, pp. 1–13. Springer, Cham (2015). https://doi.org/10.1007/978-3-319-21786-4_1
8. Chen, W., Wang, Y., Yang, S.: Efficient influence maximization in social networks. In: Proceedings of the 15th ACM SIGKDD International Conference on Knowledge Discovery and Data Mining, Paris, France, 28 June–1 July 2009, pp. 199–208 (2009)
9. Fan, J., et al.: OCTOPUS: an online topic-aware influence analysis system for social networks. In: 34th IEEE International Conference on Data Engineering, ICDE 2018, Paris, France, 16–19 April 2018, pp. 1569–1572 (2018)
10. Goyal, A., Lu, W., Lakshmanan, L.V.S.: CELF++: optimizing the greedy algorithm for influence maximization in social networks. In: Proceedings of the 20th International Conference on World Wide Web, WWW 2011, Hyderabad, India, 28 March–1 April 2011 (Companion Volume), pp. 47–48 (2011)
11. Grover, A., Leskovec, J.: node2vec: scalable feature learning for networks. In: Proceedings of the 22nd ACM SIGKDD International Conference on Knowledge Discovery and Data Mining, San Francisco, CA, USA, 13–17 August 2016, pp. 855–864 (2016)

12. Guo, J., Zhang, P., Zhou, C., Cao, Y., Guo, L.: Personalized influence maximization on social networks. In: 22nd ACM International Conference on Information and Knowledge Management, CIKM 2013, San Francisco, CA, USA, 27 October–1 November 2013, pp. 199–208 (2013)
13. van Hasselt, H., Guez, A., Silver, D.: Deep reinforcement learning with double Q-learning. In: Proceedings of the Thirtieth AAAI Conference on Artificial Intelligence, Phoenix, Arizona, USA, 12–17 February 2016, pp. 2094–2100 (2016)
14. Kempe, D., Kleinberg, J.M., Tardos, É.: Maximizing the spread of influence through a social network. In: Proceedings of the Ninth ACM SIGKDD International Conference on Knowledge Discovery and Data Mining, Washington, DC, USA, 24–27 August 2003, pp. 137–146 (2003)
15. Khalil, E.B., Dai, H., Zhang, Y., Dilkina, B., Song, L.: Learning combinatorial optimization algorithms over graphs. In: Advances in Neural Information Processing Systems 30: Annual Conference on Neural Information Processing Systems, Long Beach, CA, USA, 4–9 December 2017, pp. 6351–6361 (2017)
16. Krizhevsky, A., Sutskever, I., Hinton, G.E.: ImageNet classification with deep convolutional neural networks. In: Advances in Neural Information Processing Systems 25: 26th Annual Conference on Neural Information Processing Systems. Proceedings of a Meeting Held at Lake Tahoe, Nevada, United States, 3–6 December 2012, pp. 1106–1114 (2012)
17. Kwak, H., Lee, C., Park, H., Moon, S.B.: What is Twitter, a social network or a news media? In: Proceedings of the 19th International Conference on World Wide Web, WWW 2010, Raleigh, North Carolina, USA, 26–30 April 2010, pp. 591–600 (2010)
18. Leskovec, J., Krause, A., Guestrin, C., Faloutsos, C., VanBriesen, J.M., Glance, N.S.: Cost-effective outbreak detection in networks. In: Proceedings of the 13th ACM SIGKDD International Conference on Knowledge Discovery and Data Mining, San Jose, California, USA, 12–15 August 2007, pp. 420–429 (2007)
19. Li, G., Chen, S., Feng, J., Tan, K., Li, W.: Efficient location-aware influence maximization. In: International Conference on Management of Data, SIGMOD 2014, Snowbird, UT, USA, 22–27 June 2014, pp. 87–98 (2014)
20. Li, Y., Fan, J., Wang, Y., Tan, K.: Influence maximization on social graphs: a survey. IEEE Trans. Knowl. Data Eng. 30(10), 1852–1872 (2018)
21. Li, Y., Fan, J., Zhang, D., Tan, K.: Discovering your selling points: personalized social influential tags exploration. In: Proceedings of the 2017 ACM International Conference on Management of Data, SIGMOD Conference 2017, Chicago, IL, USA, 14–19 May 2017, pp. 619–634 (2017)
22. Li, Y., Zhang, D., Tan, K.: Real-time targeted influence maximization for online advertisements. PVLDB 8(10), 1070–1081 (2015)
23. Mnih, V., et al.: Playing Atari with deep reinforcement learning. CoRR abs/1312.5602 (2013)
24. Mnih, V., et al.: Human-level control through deep reinforcement learning. Nature 518(7540), 529–533 (2015)
25. Nguyen, H.T., Dinh, T.N., Thai, M.T.: Cost-aware targeted viral marketing in billion-scale networks. In: 35th Annual IEEE International Conference on Computer Communications, INFOCOM 2016, San Francisco, CA, USA, 10–14 April 2016, pp. 1–9 (2016)
26. Schaul, T., Quan, J., Antonoglou, I., Silver, D.: Prioritized experience replay. CoRR abs/1511.05952 (2015)
27. Sutton, R.S., Barto, A.G.: Reinforcement Learning - An Introduction. Adaptive Computation and Machine Learning. MIT Press, Cambridge (1998)

DeepDial: Passage Completion on Dialogs

Nan Hu, Jianyun Zhou, and Xiaojun Wan[✉]

Institution of Computer Science and Technology,
The MOE Key Laboratory of Computational Linguistics,
Peking University, Beijing, China
{hunan,zhoujianyun,wanxiaojun}@pku.edu.cn

Abstract. Many neural models have been built to carry out reading comprehension tasks. However, these models mainly focus on formal passages like news and book stories. Although human dialog is the most important part of daily life, machine reading comprehension on dialogs (i.e., passage completion on dialogs) has not been sufficiently explored. Existing models show some weaknesses when comprehending dialogs and they are unable to capture global information over a distance and local detailed information at the same time. This paper introduces a neural network model DeepDial that aims at addressing the problems mentioned above. The model explores both word-level and utterance-level information in a dialog, and achieves the state-of-the-art performance on the benchmark dataset constructed from a TV series *Friends*.

Keywords: Dialog · Reading comprehension · Passage completion · Deep learning

1 Introduction

Dialog seems to be the most important part of our daily life, especially on the Internet, which happens among people at anytime, anywhere. And a huge amount of conversational data is generated on the Internet every day. A machine must be able to comprehend an arbitrary daily dialog and give a reasonable response in order to own "intelligence". In fact, the well-recognized indicator of artificial intelligence, the Turing test, is based on dialog [14]. There exist many research fields related to dialog, such as conversational AI [16] and machine reading comprehension on dialog. However, understanding a dialog is never an easy task for machines. Ever since the invention of CNN [6], RNN [7] and other neural networks, deep neural models have been widely utilized to deal with dialogs.

The task of Machine Reading Comprehension is to train models that can read, understand and answer questions about a text [11]. Since [1] and [5] created two benchmarks based on Cloze-style questions [21], people have shown more interest in this task. But CNN Daily Mail [1] and Children's Book Test [5] are not dialogs. They comprise of formal passages, and most models based on

---

N. Hu and J. Zhou are equally contributed.

© Springer Nature Switzerland AG 2019
J. Shao et al. (Eds.): APWeb-WAIM 2019, LNCS 11641, pp. 141–155, 2019.
https://doi.org/10.1007/978-3-030-26072-9_10

these datasets consider the passage as just a sequence of words, and carry out word-level exploration [1,4,5,11,12]. But the dialog has its own structure. A dialog comprises of several utterances spoken by different speakers with different purposes and topics. A speaker can interrupt the current conversation and raises his/her ideas at any time. As a result, information in a dialog tends to be more scattered. If we directly apply existing models mentioned above to our task, the performance on queries that need the information of a few utterances will not be satisfactory.

[2] created a dialog-based reading comprehension (i.e. passage completion on dialogs) dataset in 2018, and proposed a neural approach to this task, treating each utterance as an independent unit. We found that in three situations this model tends to make mistakes: (1) exact match, i.e., the utterance and query shares the same sequence of meaningful words, which includes the answer to the query; (2) when a query needs the model to gather information from several incontinuous words; (3) short-distance coreference resolution, for example, the machine already knows *"you"* in the utterance *"you look beautiful today"* is the answer, and it then needs to know who *"you"* refers to. We suspect this is due to the weakness of the CNN module they used: it is hard for it to locate relevant information accurately and put emphasis on it. Queries in (3) expose another weakness of their model. As it represents each utterance independently, the relations between utterances will be ignored.

To summarize, existing models have the problem that they can not deal with concentrated information (in-utterance cues and word-level information) and scattered information (global information like a person's opinions about marriage and who acts decently in a dating) effectively at the same time. Meanwhile, few models take the dialog structure into consideration. After carefully analyzing the characteristics of dialog structure, we put forward our model – DeepDial, which maintains word-level information and dialog structure information at the same time. Specifically speaking, word-level information include pronoun usage (*he, she, you...*), meaningful words (*married, awkward, chatting...*) and local entity information (like *Mike ate the apple.*). Dialog structure information is the global information of a dialog, for example, who takes part in the dialog, what a person means by making several utterances. In addition, our model carries out a thorough interaction between the query and dialog.

In this paper, we give a quantitative analysis of queries in order to find existing models' weakness and improve them accordingly (Sect. 3). Our neural network model, DeepDial, will be introduced in Sect. 4. We will show that our model achieves better performance on queries that need either concentrated or scattered information (Sect. 5). It also outperforms the state-of-the-art model in [2]. In Sect. 6, we focus on the ability of several models to solve queries that require different amount of information. A thorough result analysis will be given on this aspect.

The contributions of this study are summarized as follows:

– We propose a neural network model DeepDial to address the challenging passage completion task over dialogs, which has two innovations: (1) it employs a novel method to extract information from each utterance and represent the dialog, which maintains word-level information and dialog structure information at the same time; (2) it utilizes a fusion attention module that treats the dialog and query as two modalities and uses an attention mechanism to explore their interaction.

– Our model achieves the state-of-the-art performance. To make a further discussion, we perform a quantitative analysis of the queries of the challenging task in order to measure a model's ability to process concentrated information and scattered information. Experiments show that our model is good at both kinds of information acquisition.

2   Task Description and Data Analysis

2.1   Task Definition

The task of passage completion on dialogs was proposed by [2] and a benchmark dataset was generated from a TV series *Friends*. An entry in the dataset can be defined as a tuple $(\mathcal{D}, \mathcal{Q}, \mathcal{A}, a)$, where $\mathcal{D}$ is a dialog comprised of utterances, $\mathcal{Q}$ is a query that gives a description of some key information in $\mathcal{D}$, $\mathcal{A}$ is a set of possible answers and a the right answer (a needs to be predicted in the development and test sets). A certain entity in $\mathcal{Q}$ is replaced by a placeholder. And the task is to read the dialog ($\mathcal{D}$), comprehend the given question ($\mathcal{Q}$), and decide which entity in $\mathcal{A}$ can be filled into the placeholder. Names of characters are anonymized as @*entxx*, and note that anonymised names are allocated randomly.

Table 1 shows a dialog in *Friends*. Each dialog is a whole scene, which contains a relatively integrated plot. Here the dialog comprises of 16 utterances. An utterance has the sentences spoken by a speaker, along with some description in brackets.

Table 2 contains 3 queries generated manually according to the dialog in Table 1. They are summarizing some information that could be found in the dialog. But the difficulty of these queries varies. For example, query 3 can be answered simply by utterance 10. While for query 1, no direct evidence shows whom the furniture belong to, and thus we have to infer from multiple utterances to get the answer. That is a fairly challenging question which needs global inference. In the following subsection we will introduce a quantitative method to measure the difficulty of a query.

The whole dataset contains 1681 scenes. From these scenes, 13487 queries are generated and wrapped into entries. The entries are split into training set, development set and test set with a ratio of **8:1:1**.

2.2   Data Analysis

Above we discussed the difficulty variety of queries. We will give a clear measurement of a query's difficulty and discuss what makes it difficult.

Table 1. A dialog from *Friends*: Season 1, Episode 1, Scene 4

Index	Speakers	Utterances
1	@ent00	(squatting and reading the instructions) I'm supposed to attach a brackety thing to the side things, using a bunch of these little worm guys. I have no brackety thing, I see no whim guys whatsoever and - I can not feel my legs
2	-	(@ent01 and @ent02 are finishing assembling the bookcase.)
3	@ent01	I'm thinking we've got a bookcase here
4	@ent02	It's a beautiful thing
5	@ent01	(picking up a leftover part) What's this?
6	@ent02	I would have to say that is an 'L' - shaped bracket
7	@ent01	Which goes where?
8	@ent02	I have no idea
9	-	(@ent01 checks that @ent00 is not looking and dumps it in a plant.)
10	@ent01	Done with the bookcase!
11	@ent02	All finished!
12	@ent00	(clutching a beer can and sniffing) This was @ent03's favorite beer. She always drank it out of the can, I should have known
13	@ent01	Hey - hey - hey - hey, if you're gonna start with that stuff we're outta here
14	@ent02	Yes, please don't spoil all this fun
15	@ent01	@ent00, let me ask you a question. She got the furniture, the stereo, the good TV - what did you get?
16	@ent00	You guys

As stated in [1], the difficulty is measured with the depth of inference. Inspired by this, we design a quantitative measurement of "difficulty": we make a quantitative analysis of the queries and classify them according to two factors: how many utterances a query needs to obtain the answer and the maximum span between the needed utterances. The "maximum span" is defined as follows: we index the utterances chronologically, pick out the largest and smallest index i, j of the needed utterances, then $i - j + 1$ is the maximum span. The first factor is to measure a model's ability of synthesizing several utterances and make an effective inference, and the second is to measure whether it could extract information over a long distance.

Figure 1 shows the results based on manual analysis. 200 queries are sampled from the development set, and analyzed manually. From the two figures we can see that 38.8% of the queries need three or more utterances to answer, and 51.7% of the queries have maximum span above two. This shows the difficulty of the dataset, as scattered information is more difficult for a model to gather and effectively utilize.

Table 2. Queries generated from the dialog in Table 1

Index	Query	Answer
1	Meanwhile, the guys attempt to construct @placeholder' new furniture	@ent00
2	@placeholder says the bookcase is finished	@ent02
3	@placeholder is done with the bookcase	@ent01

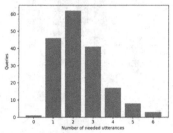
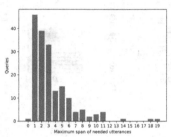

(a) The number of utterances that are needed to answer a query.

(b) The maximum span between the utterances that are needed to answer a query.

Fig. 1. Difficulty analysis of the dataset.

3 Our DeepDial Model

We design a neural network model - DeepDial, which comprises of a dialog representation generator, a dialog-query aligner, a fusion attention module and a classification module. Main innovations lie in the first three modules. For the dialog representation generator, we apply a novel method to extract information from each utterance, which maintains word-level information (pronoun usage, meaningful words and entity information, etc.) and the dialog structure at the same time. Meanwhile, it keeps the structure of dialog inconsideration. Thus, our model solve the problems in existing model in Sect. 1. For the query-dialog aligner and the fusion attention module, we regard the query and the dialog as two modalities and assume the interaction of the two modalities can predict the answer of the reading comprehension task. This model is able to answer queries about word-level information such as exact-match information, pronoun referent and lexical inference or about structural information such as the plot of whole dialog and all the entities that take part in a certain activity.

3.1 Query and Dialog Representation

The approach to deal with the difficulty of dialog structure will be discussed here. We put forward a new module to get the dialog representation. And how the query representation is generated will also be shown here.

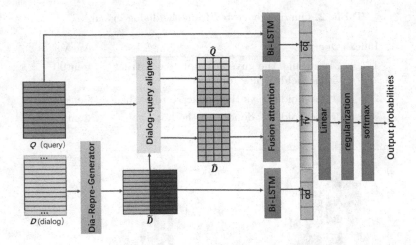

Fig. 2. An overview of DeepDial.

Consider a query of m words $\overline{Q} = \{w_t^Q\}_{t=1}^m$ and a dialog of k words $\overline{D} = \{w_t^D\}_{t=1}^k$. All words of the dialog are stacked one by one. Specially, the speaker of each utterance is regarded as the first word of the utterance. The first layer of the model convert words in Q and D to their respective word-level embeddings ($Q = \{e_t^Q\}_{t=1}^m$ and $D = \{e_t^D\}_{t=1}^k$). We simply use $Q = \{e_t^Q\}_{t=1}^m$ to be the representation of the query (Fig. 2).

The dialog representation generator derives a context vector for each time step in D. We apply Bi-LSTM on D, but only the hidden state at the middle location of every utterance are extracted ($\overrightarrow{h_t^{fwd}}, \overleftarrow{h_t^{bwd}}$). The total pair number is equal to the number of utterances in dialog D. Every pair of ($\overrightarrow{h_t^{fwd}}, \overleftarrow{h_t^{bwd}}$) is concatenated to be the representation of the corresponding utterance $u_t^{\tilde{D}}$.

Then all the vectors extracted are stacked as the new representation of the dialog. Assuming that the dialog has n utterances, its new representation matrix is $\tilde{D} = \{u_t^{\tilde{D}}\}_{t=1}^n$.

Figure 3 shows the process described above. To make the flow chart clear, only one utterance is shown.

3.2 Dialog-Query Aligner

Inspired by multimodal attention ideas and mechanisms [3,10], we design a dialog-query aligner and a fusion attention in DeepDial model.

From now the representation of dialog is $\tilde{D} = \{u_t^{\tilde{D}}\}_{t=1}^n$ and query is $Q = \{e_t^Q\}_{t=1}^m$. Each vector $u_t^{\tilde{D}}$ has the information of an utterance, while each e_t^Q has the information of a word. So alignment should be done first before using query and dialog like two modalities.

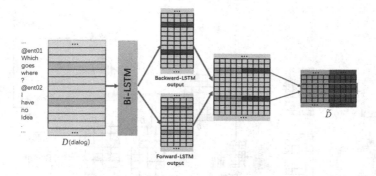

Fig. 3. Dialog representation generator

We make use of the idea in Text-CNN [6] to get respective feature vectors of $u_i^{\widetilde{D}}$ and e_j^Q. Here the window length of each filter is 1. Now we get respective feature matrices $\widehat{D} = \{d_i^{\widehat{D}}\}_{i=1}^n$ and $\widetilde{Q} = \{cw_j^{\widetilde{Q}}\}_{j=1}^m$. Then inner product of every pair of $d_i^{\widehat{D}}$ and $cw_j^{\widetilde{Q}}$ is calculated to estimate the similarity of utterance i in dialog and word j in query. The similarity of utterance i in dialog and word j in query is notated as $s_{i,j}$, and S is the matrix composed by all $s_{i,j}$. Using the similarity matrix S, a new query representation $\widehat{Q} = \{q_t^{\widehat{Q}}\}_{t=1}^n$ is calculated and aligned to $\widehat{D}$. Each $q_t^{\widehat{Q}}$ is a weighted sum of $\widetilde{Q} = \{cw_j^{\widetilde{Q}}\}_{j=1}^m$.

$$cw_j^{\widetilde{Q}} = 1D\_CNN(e_j^Q) \tag{1}$$

$$d_i^{\widehat{D}} = 1D\_CNN(u_i^{\widetilde{D}}) \tag{2}$$

$$s_{i,j} = \; <d_i^{\widehat{D}}, cw_j^{\widetilde{Q}}> \tag{3}$$

$$q_t^{\widehat{Q}} = \sum_{j=1}^m s_{t,j} cw_j^{\widetilde{Q}} \tag{4}$$

"Alignment" here has three meanings. One is that the dialog representation $\widehat{D}$ and the query representation $\widehat{Q}$ has same number of rows. Another is about the granularity. Both every row in $\widehat{D}$ and every row in $\widehat{Q}$ has sentence-level information. Finally, $q_t^{\widehat{Q}}$ has the sentence-level information most related to $d_t^{\widehat{D}}$.

A symmetric process is not conducted because it would only align query and dialog in terms of shape. However, it would increase the difference in granularity and content.

3.3 Fusion Attention

The fusion attention module derives the main part of the final representation vector for prediction. Figure 4b shows how we do the interaction of query and dialog to generate the fusion representation vector, from which the answer could be implied effectively.

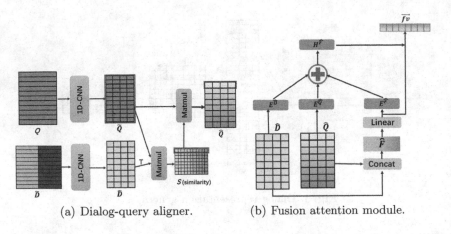

(a) Dialog-query aligner.   (b) Fusion attention module.

Fig. 4. Two modules.

We first obtain (self) attention weights from the query matrix $\widehat{Q}$ and dialog matrix $\widehat{D}$ respectively. The weights notations are $E^{\widehat{Q}} = \{\overline{a}_i^{\widehat{Q}}\}_{i=1}^n$ and $E^{\widehat{D}} = \{\overline{a}_i^{\widehat{D}}\}_{i=1}^n$. Then every pair of $(q_i^{\widehat{Q}}, d_i^{\widehat{D}})$ is concatenated to generate a fusion matrix of $\widehat{Q}$ and $\widehat{D}$, which is $F = \{[q_i^{\widehat{Q}}, d_i^{\widehat{D}}]\}_{i=1}^n$. F is transited to $\widehat{F} = \{f_i^{\widehat{F}}\}_{i=1}^n$ using a fully-connected layer. The attention weight of $\widehat{F}$ is computed in the same way and is notated by $E^{\widehat{F}} = \{\overline{a}_i^{\widehat{F}}\}_{i=1}^n$. The new weights $E^{\widehat{F}}$ are more accurate, so double attention is paid when calculating the final weights ($H^{\widehat{F}} = \{h_i^{\widehat{F}}\}_{i=1}^n$) of the fusion matrix. $\overrightarrow{fv}$ is the weighted sum of $\widehat{F}$ with the weights $\{h_i^{\widehat{F}}\}_{i=1}^n$. ($\{\overline{a}_i^{\widehat{D}}\}_{i=1}^n$ and $\{\overline{a}_i^{\widehat{F}}\}_{i=1}^n$ are calculated using the same process as $\{\overline{a}_i^{\widehat{Q}}\}_{i=1}^n$ with $\widehat{D}$ and $\widehat{F}$ while the parameters are seperate).

$$a_i^{\widehat{Q}} = (V_{\widehat{Q}})^T tanh(W_{\widehat{Q}} q_i^{\widehat{Q}}) \tag{5}$$

$$\overline{a}_i^{\widehat{Q}} = \frac{exp(a_i^{\widehat{Q}})}{\sum_{j=1}^n exp(a_j^{\widehat{Q}})} \tag{6}$$

$$h_i^{\widehat{F}} = (\overline{a}_i^{\widehat{Q}} + \overline{a}_i^{\widehat{D}} + 2 * \overline{a}_i^{\widehat{F}})/4 \tag{7}$$

$$\overrightarrow{fv} = \sum_{i=1}^n h_i^{\widehat{F}} \cdot f_i^{\widehat{F}} \tag{8}$$

$V_{\widehat{Q}}, W_{\widehat{Q}}$ are parameters, while $V_{\widehat{Q}}$ is a vector and $W_{\widehat{Q}}$ is a matrix.

3.4   Classification Module

We add some information from the origin query representation Q and dialog representation $\widetilde{D}$. The complementary information is extracted using standard Bi-LSTM. This idea is kind of similar to Residual Network [9]'s motivation:

getting information from the origin representations to lighten information lost resulted from too complicated model.

$$\vec{cd} = Bi - LSTM(\{u_i^{\tilde{D}}\}_{i=1}^n) \tag{9}$$

$$\vec{cq} = Bi - LSTM(\{e_i^Q\}_{i=1}^n) \tag{10}$$

Then the concatenated vector $\vec{v} = \vec{cd}||\vec{fv}||\vec{cq}$ contains almost all the information needed to answer the query. It is delivered to a stack of fully-connected layer, a regularization layer and a softmax layer. Then we get the probability belonging to each class. Cross entropy loss is adopted as the loss function.

$$\vec{l} = W^{vl}\vec{v} + \vec{b} \tag{11}$$

$$r_i = \frac{l_i - \sum_{j=1}^u l_j/|l|}{\sqrt{\sum_{j=1}^u \left(l_j - \sum_{p=1}^u l_p/|l|\right)^2}} \tag{12}$$

$$\hat{y}_i = \frac{exp(r_i)}{\sum_{j=1}^u exp(r_j)} \tag{13}$$

$$prediction = \arg\max_i(\{\hat{y}_i\}_{i=1}^u) \tag{14}$$

$$loss = -\sum_{i=1}^u y_i * \log\hat{y}_i + (1 - y_i) * log(1 - \hat{y}_i) \tag{15}$$

u is the length of $\vec{v}$, $\widehat{Y} = \{\hat{y}_i\}_{i=1}^u$ is the output probabilities and $Y = \{y_i\}_{i=1}^u$ is the one-hot presentation of the answer.

4   Experiments

4.1   Training Details

The pre-trained 100-dimensional Glove word vectors [15] are used to do the word embedding. The hidden length of all the Bi-LSTM layers is fixed as 64. And 1D-CNN layer has 50 filters. The activation function of the convolutional layer is *ReLU*, and other activation function used in our model is *tanh*. The dropout rate is set to be 0.5 [26]. The Adam optimizer [27] is used to train the model. We also applied pruning strategy to deal with long queries and dialogs, similar to [2].

4.2   Baselines

To make a comprehensive comparison, we pick some typical models, reproduce and evaluate them respectively.

Majority. This model simply predict the dominant entity that appears in one scene as answer.

Table 3. Results on the development set and test set. (Results marked by * are directly taken from [2], DeepDial+ is the model replacing Bi-LSTM in Dialog-Representation-Generator with Transformer)

Index	Model	Development	Test
1	Majority*	28.61	30.08
2	CNN + LSTM	58.64	55.43
3	Bi-LSTM	72.42	70.58
4	Attentive Reader	75.24	72.35
5	Attention-Over-Attention	46.60	49.80
6	CNN+LSTM+UA+DA*	72.21	72.42
7	CNN+LSTM+UA+DA	77.53	76.72
8	DeepDial (CNN as dialog representation generator)	58.41	56.32
9	DeepDial (without fusion attention)	60.56	60.49
10	DeepDial (without $\vec{cd}$ and $\vec{cq}$)	76.40	76.10
11	DeepDial	78.65	79.61
12	DeepDial+	**81.77**	**83.18**

Bi-LSTM. It is the baseline of all kinds of reading comprehension tasks. It views the dialog as a sequence of words, and uses Bi-LSTM to generate the context vectors for the dialog and query [20]. It then concatenates the two context vectors to predict the answer.

CNN+LSTM. This model differs from Bi-LSTM. It first generates a representation of each utterance using CNN, and then feeds them to Bi-LSTM to get the vector for prediction.

Attentive Reader. [1] first used attention mechanism to carry out this kind of task. The dialog is treated as a sequence of words, and the model uses the query to make a utterance-level attention. This method explores the mutual information between the utterance and query, hence outperforms simple Bi-LSTM model by a large margin.

CNN+LSTM+UA+DA. [2] treated each utterance as an independent unit, and used CNN to produce a vector representation for every utterance. An utterance-level attention and a query-level attention were applied in this model. As stated in Sect. 1, this model is not good at picking up relevant information in a long sentence. It also misses some relation between utterances.

4.3 Result

The results of models above are shown in Table 3. CNN+LSTM shows a poor performance (58.64%/55.43%) as the utterance encoder leaves out too much word-level information. While Bi-LSTM performs much better (72.42%/70.58%).

Based on Bi-LSTM, Attentive Reader utilizes attention mechanism to focus on words that are more relevant to the query, and gets a better result (75.24%/72.35%). The result for Attention-Over-Attention, however, seems abnormal, but a similarly poor result is also reported in [2] on the same dataset. Also, we tried the state-of-the-art model GA-Reader on CNN/DailyMail, but it performs badly on this dataset.

Our model outperforms all the other models. It achieves an accuracy of 78.65% on development set and 79.61% on test set before we tune the hyper-parameters on this model.

Three ablation tests are performed. First, we replace the dialog representation generator by Convolutional Neural Network to show that our dialog representation generator can generates a fusion dialog representation of word-level and utterance-level, which is indispensable for the model. Then is the dialog-query attention module, which makes it possible to pick out important cues from the dialog according to the query accurately and efficiently. The last experiment is removing the completion information (namely, $\vec{cd}$ and $\vec{cq}$). The result decreases by 2.2% on the development set and 3.5% on the test set. This shows that $\vec{cd}$ and $\vec{cq}$ adds some useful supplement information of the dialog and query which is left out for the model's complexity.

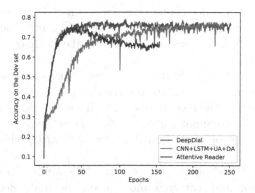

Fig. 5. Training curve of three models (Accuracy on the development set).

Figure 5 shows the accuracy curves of three models on development set. DeepDial convergences as fast as Attentive Reader, while it achieves an accuracy higher than the reproduced CNN+LSTM+UA+DA model with a smoother curve.

5   Analysis

5.1   Running Examples

Table 4 shows some prediction results made by our model. The first query needs only one utterance, and is the simplest exact-match problem.

Table 4. Correct predictions made by our model.

Query	Dialog
@placeholder says @ent01 is "so much the smitten kitten"	(**@ent02**) Oh, you are so much the **smitten kitten**. You should ask him out
@ent02 says that, when **@placeholder** was a kid, @ent02 barely had enough pieces of parents to make one whole one	(**@ent02**) I mean, well, cause **when I was growing up**, you know my dad left, and my mother died, and my stepfather went to jail, so I barely had enough pieces of parents to make one whole one ...
@ent05, @ent03, and @ent05 attend **@placeholder**'s new musical, @ent04!	(@ent00) (reading the program) Ooh! Look! Look! Look! Look, there's **@ent01**'s picture! This is so exciting! (@ent02) You can always spot some-one who's never seen one of **his plays** before ... (@ent03) (Gesturing) Y'know, it's not just **@ent04**, it's @ent04!

The second query also needs one utterance, but it is harder. To answer this query, the machine should be able to make a lexical inference – "when I was growing up" equals "when I was a kid", and make it clear the referent of the pronoun *I*.

The third one is more complicated. The query asks who holds a new musical, *@ent04*. Three utterances are needed to solve this problem. In the first utterance *@ent01* is spoken out. In the second utterance, from "his plays" we know *@ent01* holds a play. But still we cannot make sure *@ent01* is the answer. From the third utterance we know the play mentioned above is *@ent04*, which is sufficient to make the inference.

Those examples listed above show our model's ability to handle both concentrated information and scattered information and make deep inferences.

5.2   More Analysis

The quantitative data analysis mentioned earlier can be a good tool to test a model's ability to acquire concentrated information and scattered information, as shown in Fig. 6. We compared our model with the previous state-of-the-art model CNN+LSTM+UA+DA.

As shown in Fig. 6a, CNN+LSTM+UA+DA endures a decrease of accuracy (only 70%) on queries which could be answered by a single utterance. A similar result to this can be found in [2]. It reveals the model's weakness on processing concentrated information. Our model overcomes that, and achieves the accuracy of over 80% on queries that need 1 or 2 utterances. Also, due to the fact that our model maintains the dialog structure information, and carries out a deep

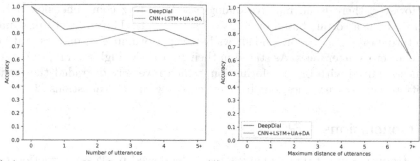

(a) Accuracy vs. Number of needed ut- (b) Accuracy vs. Maximum span of
terances.   needed utterances.

Fig. 6. Two quantitative measurements.

interaction between the dialog and query, it shows an even better result on
queries that need 4 utterances than CNN+LSTM+UA+DA.

Figure 6b measures a model's ability to gather information over different distances. Our model outperforms CNN+LSTM+UA+DA on both short distances
(maximum span of 1, 2, 3) and long distances (maximum span of 5, 6), which
shows the strength of our model.

6   Related Work

[1] gave two reading comprehension benchmarks CNN and Daily Mail. They did
a quantitative analysis of answer difficulty on two datasets, and classified the
queries into six categories, from *Simple* to *Complex*. They also introduced the
Attentive Reader and **Impatient Reader**, which utilized attention mechanism to help better propagate information over long distances.

[5] built another benchmark Children's Book Test, whose queries came from
children's books. And they proposed a structure **MemNN** to deal with this
task.

The datasets mentioned above have similar format. For each question, there
is a *context*, a *query* that misses one word from it, a set of alternative answers C
and an answer a. The standardisation of dataset format promotes the research
of machine reading comprehension a lot.

[4] proposed the **AOA-Reader**, using an "attended attention" to predict
answers, which seemed quite simple but performed pretty well on both datasets.

[11] proposed **Iterative Alternating Neural Attention** to accumulatively
optimize the attention vector for several loops. The number of loops is pre-
determined.

And their are also some other models which have achieve great result in
reading comprehension task, such as AS reader [17], GA reader [18], stanford
AR [19] and EpiReader [28].

However, there is no massive dialog-based reading comprehension dataset until [2] built a corpus based on a TV series *Friends*. Their model takes dialog structure into consideration, and shows better performance on queries that need inference cross utterances. As stated in their paper, the high level representation of cues is mixed with other information, which reversely degraded the model's ability to answer easy questions like "exact string match" questions [2].

7   Conclusions

We introduce a quantitative analysis for dialog-based passage completion tasks and find that existing models are not able to acquire concentrated information and scattered information at the same time. In order to address this problem, we design a neural network model called DeepDial. DeepDial involves a dialog representation generator, which maintains local and global information, a dialog-query aligner and a fusion attention module that pick the relevant information from the dialog. Experimental results show that our model achieves high accuracy on almost all kinds of queries. It also outperforms the current state-of-the-art model. For the future work, we can explore more dialog-relevant information, for example: (1) add a named entity recognition [23] and coreference resolution module; (2) find a better way to make use of speakers; (3) adopt a better method to utilize the dialog structure, such as tree-LSTM [24] or structured attention [29].

Acknowledgements. This work was supported by National Natural Science Foundation of China (61772036) and Key Laboratory of Science, Technology and Standard in Press Industry (Key Laboratory of Intelligent Press Media Technology). We appreciate the anonymous reviewers for their helpful comments.

References

1. Hermann, K.M., et al.: Teaching machines to read and comprehend. In: NIPS 2015, pp. 1693–1701 (2015)
2. Ma, K., Jurczyk, T., Choi, J.D.: Passage completion on multiparty dialog. In: Proceedings of NAACL-HLT 2018, pp. 2039–2048 (2018)
3. Gu, Y., Yang, K., Fu, S., Chen, S., Li, X., Marsic, I.: Multimodal affective analysis using hierarchical attention strategy with word-level alignment. arXiv preprint arXiv:1805.08660 (2018)
4. Cui, Y., Chen, Z., Wei, S., Wang, S., Liu, T., Hu, G.: Attention-over-attention neural networks for reading comprehension. In: ACL 2017, pp. 593–602 (2017)
5. Hill, F., Bordes, A., Chopra, S., Weston, J.: The goldilocks principle: reading children's books with explicit memory representations. In: ICLR 2016 (2016)
6. Kim, Y.: Convolutional neural networks for sentence classification. arXiv preprint arXiv:1408.5882 (2014)
7. Mikolov, T., Karafiát, M., Burget, L., Černocký, J., Khudanpur, S.: INTERSPEECH-2010, pp. 1045–1048 (2010)
8. Wang, W., Yang, N., Wei, F., Chang, B., Zhou, M.: Gated self-matching networks for reading comprehension and question answering. In: ACL 2017, pp. 189–198 (2017)

9. He, K., Zhang, X., Ren, S., Sun, J.: Deep residual learning for image recognition. In: Proceedings of The IEEE Conference on Computer Vision and Pattern Recognition (CVPR), June 2016
10. Pham, H., Manzini, T., Liang, P.P., Poczos, B.: Seq2Seq2Sentiment: multi-modal sequence to sequence models for sentiment analysis. arXiv preprint arXiv:1807.03915 (2018)
11. Sordoni, A., Bachman, P., Bengio, Y.: Iterative alternating neural attention for machine reading. arXiv preprint arXiv:1606.02245 (2016)
12. Cui, Y., Liu, T., Chen, Z., Wang, S., Hu, G.: Consensus attention-based neural networks for Chinese reading comprehension. arXiv preprint arXiv:1607.02250 (2016)
13. Vaswani, A., et al.: Attention is all you need. arXiv preprint arXiv:1706.03762 (2017)
14. Turing, A.M.: Computing machinery and intelligence. Mind **59**(236), 433–460 (1950)
15. Pennington, J., Socher, R., Manning, C.D.: GloVe: global vectors for word representation. In: EMNLP 2014, pp. 1532–1543 (2014)
16. Asghar, N., Poupart, P., Hoey, J., Jiang, X., Mou, L.: Affective neural response generation. arXiv preprint arXiv:1709.03968 (2017)
17. Kadlec, R., Schmid, M., Bajgar, O., Kleindienst, J.: Text understanding with the attention sum reader network. In: ACL 2016, pp. 908–918 (2016)
18. Dhingra, B., Liu, H., Cohen, W.W., Salakhutdinov, R.: Gated-attention readers for text comprehension. CoRR 2016 (2016)
19. Chen, D., Bolton, J., Manning, C.D.: A thorough examination of the CNN/daily mail reading comprehension task. In: ACL 2016, pp. 2358–2367 (2016)
20. Cheng, J., Dong, L., Lapata, M.: Long short-term memory-networks for machine reading. In: Proceedings of the 2016 Conference on Empirical Methods in Natural Language Processing, EMNLP 2016, Austin, Texas, USA, 1–4 November (2016)
21. Taylor, W.L.: Cloze procedure: a new tool for measuring readability. Journal. Q. **30**(4), 415–433 (1953)
22. Bahdanau, D., Cho, K., Bengio, Y.: Neural machine translation by jointly learning to align and translate. CoRR, abs/1409.0473 (2014)
23. Lample, G., Ballesteros, M., Subramanian, S., Kawakami, K., Dyer, C.: Neural architectures for named entity recognition. arXiv preprint arXiv:1603.01360 (2016)
24. Tai, K.S., Socher, R., Manning, C.D.: Improved semantic representations from tree-structured long short-term memory networks. arXiv preprint arXiv:1503.00075 (2015)
25. Hochreiter, S., Schmidhuber, J.: Long short-term memory. Neural Comput. **9**(8), 1735–1780 (1997)
26. Srivastava, N., Hinton, G.E., Krizhevsky, A., Sutskever, I., Salakhutdinov, R.: Dropout: a simple way to prevent neural networks from overfitting. J. Mach. Learn. Res. **15**(1), 1929–1958 (2014)
27. Kingma, D.P., Ba, J.: Adam: a method for stochastic optimization. arXiv preprint arXiv:1412.6980 (2014)
28. Trischler, A., Ye, Z., Yuan, X., Suleman, K.: Natural language comprehension with the EpiReader. arXiv preprint arXiv:1606.02270 (2016)
29. Kim, Y., Denton, C., Hoang, L., Rush, A.M.: Structured attention networks. In: ICLR 2017 (2017)

A Survival Certification Model Based on Active Learning over Medical Insurance Data

Yongjian Ren[1], Kun Zhang[1,2], and Yuliang Shi[1,2(✉)]

[1] School of Software, Shandong University, Jinan, China
ryjsdu@outlook.com, kunzhangcs@126.com, shiyuliang@sdu.edu.cn
[2] Dareway Software Co., Ltd., Jinan, China

Abstract. In China, Survival Certification (SC) is a work carried out for the implementation of Social Insurance (SI) policies, mainly for retirees. If a retiree is dead but his family has not notified the SI institution, then the SI institution will continue to issue pensions to the retiree. This will lead to the loss of pensions. The purpose of SC is to block the "black hole" of pension loss. However, currently, SC work mainly relies on manual services, which leads to two problems. First, due to the large number of retirees, the implementation of SC usually occupies a large amount of manpower. Secondly, at present, SC work requires all retirees to cooperate with the work of local SI institutions, while some retirees have problems with inconvenient movement or distant distances. These phenomena will lead to an increase of social costs and a waste of social resources. Thus, in this paper, a SC model based on active learning is proposed, which helps staff to narrow the scope of attention. First, we extract features from medical insurance data and analyze their effectiveness. Then, we study the effects of kinds of feature selection functions and classifiers on the SC model. The experimental results show that the model can effectively predict death and can greatly reduce the range of high-risk populations.

Keywords: Active learning · Survival Certification · Machine learning · Medical

1 Introduction

SC, short for Retirees Survival Certification, is a demand arising from the implementation of SI policies. At present, the state stipulates that retirees need to go to the local SI institution every year to cooperate with the implementation of SC work, that is, retirees need to prove that they are still alive and in line with the conditions for continuing to receive pensions. The purpose of SC is to prevent or mitigate the loss of pensions due to "pension fraud". Pension fraud refer to the behavior that after the death of the pensioner, the family member fails to report the death to the SI institution timely due to intentional or unintentional

© Springer Nature Switzerland AG 2019
J. Shao et al. (Eds.): APWeb-WAIM 2019, LNCS 11641, pp. 156–170, 2019.
https://doi.org/10.1007/978-3-030-26072-9_11

reasons and continues to receive the pensions. At present, SC mainly relies on manual services, which leads to two problems. First, due to the large number of retirees, SC often consumes a lot of human resources and time. This in turn led to a long period of renewal of retirees' survival information and the inability to detect the loss of pensions in time. For example, at present, the state stipulates that retirees only need to perform SC once a year. Second, at this stage, SC requires all retirees to cooperate with the work of the SI institution, which leads to an increase in social costs and a waste of social resources. Therefore, there is an urgent need for a SC model that can significantly narrow the screening range for high-risk deaths in specific future time periods.

The research on mining the potential information of medical data and discovering its hidden knowledge to provide decision-assisted services for humans has gained wide attention and great progress. However, SC for retirees still faces challenges. First, existing survival prediction studies combining medical data and machine learning are directed at specific diseases or treatments, such as heart disease [2,3,10,13,14,20], various cancers [7,16,18,19,21,27,28], and liver transplants [22]. When focusing on a particular type of diseases or treatments, researchers can take advantage of the detailed information in the medical records to construct features that are tailored to the type of disease. However, SC for retirees will have to face almost all possible diseases. In fact, our medical insurance data set contains tens of thousands of disease IDs. Therefore, we have to construct features from a more macro perspective, and these features do not depend on specific diseases and treatments.

The second challenge of SC is the imbalance of data. Considering the continuous development of living standards and medical technology, the death samples many years ago may not be helpful for the training of the SC model, and may even mislead the model. Therefore, the death samples in recent years are very important and useful. On the one hand, the number of surviving samples is much larger than the number of dead samples we can get. Too few death samples may cause the model to fail to learn enough. On the other hand, for retirees who have recently died, we may not be able to get their death news in time. Therefore, it is very important to effectively find death samples that have recently died but whose family members have not informed us (or the SI institution). This is important for expanding the death sample set and improving the accuracy of the SC model.

To this end, this paper proposes a SC model for retirees. First, based on the fact that the coverage of medical insurance now is broad and the fact that diseases are closely related to death, we use the medical insurance data to construct a feature set for retirees. The feature set covers information about retirees' basic information, economic income, and historical illnesses, and does not depend on specific diseases. Furthermore, we make SC problem as a classification problem taking the advantage of active learning [4,8,12]. The contributions of this article are as follows:

- For the issue of retirees' SC, we have constructed a feature set that covers basic information, career information, economic income, historical illness information, and recent health status of retirees. After the chi-square test and Spearman correlation analysis, the rationality and effectiveness of the feature were preliminarily verified.

- We consider the retirees' SC as a two-classification problem. We apply the active learning model and modify the sample selection function to make it suitable for our application scenarios. The model achieves an effective discovery of undocumented death samples and achieves an effective extension of the death sample set.
- We conducted experiments on a real medical insurance data set containing hundreds of thousands of retirees and thousands of dead retirees. We conducted comparative experiments on a variety of sample selection functions and a variety of classifiers, and verified the validity of the SC model proposed in this paper.

The rest of this paper is as follows. The related work is represented in Sect. 2. Section 3 shows the set of features we constructed and the relevance of each feature to the target. Section 4 describes the retirees SC model proposed in this paper. Section 5 shows the experimental results. Finally, this article is concluded in Sect. 6.

2   Related Work

In recent years, machine learning combined with medical data to describe the health status of patients has continued to be one of the hot spots of survival prediction. Rajkomar et al. proposed a scalable deep learning with good accuracy on multiple predictive topics including in-hospital mortality [23]. The authors point out that because most of the work in the various predictive tasks is to pre-process, merge, customize, and clean up the data set, the scalability of the model is greatly limited. Therefore, the authors suggest using the Fast Healthcare Interoperability Resources (FHIR) format to represent the patient's entire original EHRs and combine it with deep neural networks to create scalable and accurate predictions for a variety of clinical scenarios.

Sanz et al. proposed a prediction system based on a multi-classifier system that targets the survival status of patients with severe trauma [26]. The study turned survival prediction into a classification problem and studied 462 patients who were treated at Navarre Hospital in Spain. In this paper, we also consider the SC problem as a classification problem, but the patient's disease range is more extensive.

Rizzo et al. proposed the use of computational argumentation theory to study the complex relationship between biomarkers and survival [24]. Using a knowledge base of 51 different biomarkers, the authors constructed a parametric model based on biomarkers to infer the risk of death in the elderly. The research and our work are both for the elderly, but the biomarkers used in this study are difficult to obtain for the large population in SC scenarios.

Zhang et al. considered the more general categories of medical data when studying survival prediction problems, including a large number of historical measurements with intermittent changes [29]. The authors proposed a new semi-proportional risk model using local time-varying coefficients and a new data model learning standard for coefficient optimization. In our work, in order to fully characterize the current health status of retirees, we have constructed a number of features that are not dependent on specific diseases.

In addition, Rouzbahman et al. improved the prediction accuracy of three prediction tasks including ICU1 death by clustering before prediction [25]. Karmakar et al. used multitasking transfer learning to study in-hospital mortality in four types of patients [15]. Abbod et al. used neural networks to predict brain death [1]. And Brause et al. studied the death prediction in patients with septic shock [6].

3   Feature Construction and Evaluation

We extracted features and constructed data sets based on the medical insurance data of retirees in X city in Shandong Province. The feature set we constructed contains 27 features, as shown in Fig. 1. These features are divided into five cat-

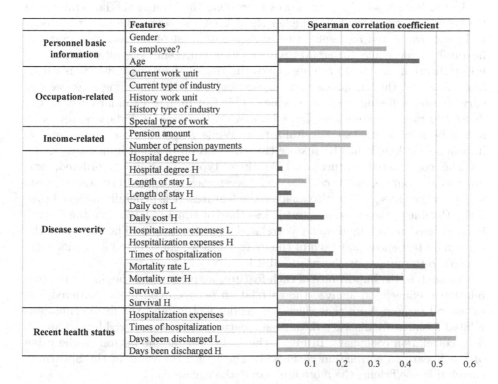

Fig. 1. Features for SC. Note that since the gender feature and the five occupation-related features are disordered, these six features are not analyzed by the Spearman correlation coefficient. (Color figure online)

Table 1. Chi-square test results of features

Feature	P-value
Hospital degree L	0.000004
Hospital degree L	0.014645
The other 25 features	<0.000001

egories, which respectively describe the basic information, occupational information, economic status, historical illness information and the recent (in this paper, the last 3 months) health status of the retirees. From the perspective of the timeline, we not only describe the worst illness in the history of retirees (The feature whose name is followed by the letter "H", for example, *Hospital degree H*), but also describe the most recent illness (the feature whose name is followed by the letter "L", for example, *Hospital degree L*). Since our medical insurance data contains tens of thousands of disease IDs, first, we divided all diseases into 21 categories according to the ICD-10 standard, and then divided all diseases into 251 subcategories.

In the feature set, *Hospital degree* represents the level of the hospital where the retiree is hospitalized. The higher the degree, the better the hospital's medical level. *Daily cost* represents the average daily cost of the retiree during the hospitalization; while *Hospitalization expenses* represents the total cost of the hospitalization. *Mortality rate* describes the risk of the disease, which is calculated based on the 251 subcategories. *Survival* describes how long a retiree can survive after suffering from the diseases. This feature has a similar effect to the *Mortality rate*. For a retiree, *Days have been discharged L* indicates how long it has been since the last hospitalization. While *Days have been discharged H* indicates how long it has been since the worst hospitalization in history.

The constructed features include three types: continuous, ordered, and unordered. To measure the correlation between each feature and the target, first, we apply the chi-square test to analyze the features. It is generally believed that if the P-value of the chi-square test is less than or equal to 0.05, then the feature is considered to be significantly correlated with the target. The results of the chi-square test show that each of the 27 features we constructed is significantly related to the target, as shown in Table 1.

To analyze the importance of each feature, we further use the Spearman correlation coefficient to analyze the correlation between continuous/ordered features and targets. Note that since the gender feature and the five occupation-related features are disordered, these six features are not analyzed by the Spearman correlation coefficient. In Fig. 1, the scale value at the bottom is the value of the Spearman correlation coefficient. The larger the value of the Spearman correlation coefficient, the more important the feature is.

In this paper, the semantics of the classification target is *whether this person will die?*. The blue histogram indicates that the value of the feature is positively correlated with the target, and the orange histogram indicates that the value of the feature is negatively correlated with the target. For example, for the feature *Age*, the blue histogram represents the semantics of *the older the person, the higher the probability of death*. For the feature *Is employee?*, the orange histogram represents the semantics of *workers have a lower probability of death than inhabitants*. Similarly, for the feature *Hospitalization expenses*, the blue histogram represents the semantics of *the higher the total cost of hospitalization in the recent period, the greater the probability of death*. For *Days been discharged L*, the orange histogram represents the semantics of *the greater the time span from the last hospitalization to the present, the lower the probability of death*.

4   SC Model

4.1   SC Model Overall

Figure 2 shows the overall flow of the SC model. Among them, the upper part is the training process of the model, and the lower part is the test/application process of the model. Since the number of survival samples is much larger than

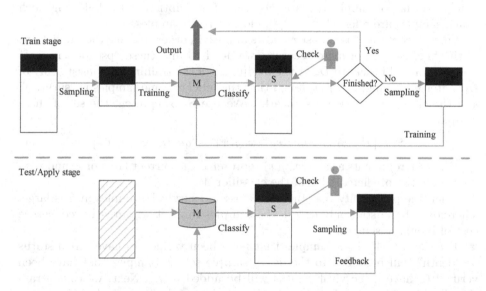

Fig. 2. The SC model. In this figure, the black rectangle indicates the set of death samples marked with the date of death (denoted by D), including samples that actively report death information to the SI institution and samples that have been verified as dead by the staff. The white rectangle represents the set of survival samples (denoted by L). L is the union of L_1 and L_2. L_1 denotes the set of "default survival" samples and L_2 denotes the set of "confirmed survival" samples. The gray rectangle represents samples that have been classified as death but the current label are "default survival", denoted by S, and S is a subset of L_1. It is S that contains the samples we are interested in. In order to minimize the workload of human workers, what really needs staff verification is set $S' \subset S$. S' is obtained by a sample selection function $F(\cdot)$.

the number of dead samples, we use the undersampling method to resample the survival samples to obtain the training set during the training stage.

First, we discuss the labels of the samples.

We temporarily label a sample as "death" or "survival". Samples labeled as "death" include two categories: samples that actively inform SI agencies of death information, and samples that have been found dead by employees of SI agencies. Samples labeled for "survival" also include 2 categories: (a), samples that the SI institution has not received the death notice of and the staff has not checked, which will be defaulted to survival samples; (b), samples that are suspected and verified by staff and confirmed to be survival, which have been confirmed to be survival.

Therefore, in the SC model proposed in this paper, a sample is labeled as "death" or "survival". In order to reduce the workload of the staff, the "survival" sample is further divided into "default survival" and "confirmed survival". The set of "death" samples is denoted by D. The set of "survival" samples is denoted by L. The set of "default survival" samples is denoted by L_1. and the set of "confirmed survival" samples is denoted by L_2. $L = L_1 \bigcup L_2$. Obviously, in the set L_1, the actual state of most samples is survival, and the actual state of a small number (or no) of samples is death. For ease of description, in the following, such death samples are referred to as "pseudo-survival" samples.

Since L contains "pseudo-survival" samples, in order to make the predictions of the final SC model more accurate, we need to find these "pseudo-survival" samples and add them to D. At the beginning of the training, L consists only of L_1. After a training, it will be found that the label of some samples is "survival" and the classification result is "death". We use S to represent the set of these samples:

$$S = \{x | y(x) = 'survival' \ \& \ \hat{y}(x) = 'death' \ \& \ x \in L_1\} \tag{1}$$

Where x represents a sample. $y(x)$ represents the current label of x, and $\hat{y}(x)$ represents the predicted label of the classifier M.

Then the probability that there are "pseudo-survival" samples in S is large. Therefore, if the staff only checks the samples in S, it will greatly reduce the cost of human resources.

For the samples in S, samples that have been verified to have a real status of "death" will be added to the death sample set D. Samples that have been verified to have a "survival" status will be added to L_2. Next, we can retrain the classifier M with a better training set. After a limited number of manual markings and retraining, an ideal classifier is finally obtained.

In the test/application stage of the SC model, the sample set generated by manual verification is fed back to the classifier M. The logic of the feedback mechanism is that, as time goes by, due to the continuous development of living standards and the medical level, the mapping relationship between sample features and sample labels may change. The existence of a feedback mechanism will enable the SC model to capture this change based on the latest data.

It should be noted that the difference of training stage between SC and the active learning is: (1) The labeled samples in the active learning are incremental.

However, in SC, the total number of labeled samples is constant, and the changes are the labels of some samples. (2) The classical active learning training process is also incremental, that is, each time only the newly marked data can be put into the model for training. However, in SC, in order to eliminate or reduce the interference of the "pseudo-survival" samples, each training will use the re-sampling to obtain the training set. Therefore, the training process of the SC model is not incremental training. Nevertheless, the SC model proposed in this paper still maintains the core idea of the classical active learning model, that is, in order to reduce the human workload, only the most worthwhile samples in the data set are selected for manual labeling. The cost of non-incremental training is still secondary to the cost of manual labeling.

4.2 Sample Selection Functions

In active learning, the sample selection function is designed to help researchers select samples that are more useful to the classifier from unlabeled samples. These samples are added to the training set after being labeled by human workers, which can effectively improve the accuracy of the classifier. Although intuitively speaking, we should find all the pseudo-survival samples, but it is costly. Our goal is to find as many pseudo-survival samples as possible that contribute to the classification. Therefore, some existing strategies in active learning will also help us. For example, we can only consider the "uncertainty" of the sample [17]. Another strategy is that we can take into account the "diversity" of the sample together with the "uncertainty".

To measure the "uncertainty" of a sample, the most basic method is Least Confidence (LC), developed by Culotta and McCallum [9]. Another measure of "uncertainty" is the Margin approach. Margin approach is prone to selecting instances with maximum margin between posterior probabilities of the two most likely class labels [5]. Since SC belongs to the 2-classification problem, the Margin measurement method has the same effect as the LC measurement method. Therefore, this paper only considers the LC method to measure the "uncertainty" of the sample. In the classic active learning model, the smaller the probability that the classifier divides the sample into the class, the less confidence the classifier has for the sample to belong to the class. As shown in the following formula:

$$x_{LC}^* = \arg\max_x \ 1 - P_M(\hat{y}(x) = y(x)) \tag{2}$$

Where x_{LC}^* represents a sample selected according to the LC method, and $P_M(\hat{y}(x) = y(x))$ represents the probability of classifying the sample x into class y according to the classifier M. However, in SC problem, the sample that the staff most needs to verify is not the sample that minimizes the confidence of the classifier. The sample that people care about is a sample that the classifier strongly believes to be dead but the existing label is "default survival" (i.e. $x \in S'$). In this paper, we name this method of describing the "uncertainty" of the sample as the "Maximum Suspicion" method (denoted as MS). Therefore, we adjust the above formula to suit our scenario:

$$x_{MS}^* = \arg\max_{x \in S'} P_M(\hat{y} = {'death'}|x \in S') \tag{3}$$

Where $\hat{y}$ is short for $\hat{y}(x)$.

Taking into account the "diversity" of the sample can make the sample selected by the sample selection function more representative, so that the classifier can learn more information. The measurement problem of the sample "diversity" is usually translated into a measurement problem of "similarity" between samples. Typical similarity metrics include "Cosine similarity", "Gaussian similarity", and the like. In order to reduce the computational complexity of the sample similarity, a typical method is to utilize the "center point" $\bar{x}$ of the candidate sample set. The similarity between x and $\bar{x}$, denoted by $sim(x, \bar{x})$, is used to approximate the similarity between the sample x and other samples in the finally selected sample set S'.

Finally, we can get a sample selection function that takes into account the sample "uncertainty" and sample "diversity":

$$x_{MS\&sim}^* = \arg\max_{x \in S'} \lambda \cdot P_M(\hat{y} = {'death'}|x) + (1 - \lambda) \cdot (1 - sim(x, \bar{x})) \tag{4}$$

4.3   Classifier M

In the SC model, the task of the classifier M is to effectively classify the data set according to the selected features. Therefore, the classifier M is not tightly coupled to the SC model. We can apply different classifiers in the SC model. Simple and effective classifiers are our first choice, such as logistic regression (for short, logis), Bayesian classifiers, neural networks, and Random Forests (RF). In active learning, in addition to selecting a single classifier, a mechanism called "Query-By-Committee" (QBC) can also be used [11]. Multiple classifiers give classification results for the same sample, and each classifier then votes on the category of the sample like a "committee" member. Finally, the final classification results are given by the "committee". Different scoring mechanisms are available for us. For example, one option is that the results given by the various members of the "committee" have the same weight. Alternatively, the weight of each member of the "committee" is related to the members previous classification accuracy.

5   Experiments

5.1   Data Set

The research in this paper is based on real medical insurance data from Z City, China. The data set covers about 120,000 pensioners, including 8,109 deaths in 2016. The training set consists of samples who died in 2016 and samples who survived at the end of 2016. There are 3 test sets for January–March 2017. In the training stage of the model, since the number of survivors is far greater than the number of deaths, the training set A' actually used in this paper is constructed according to the resampling with $survivors : deaths = 1 : 1$.

5.2 Influence of Sample Selection Function and Classifier on Sample Set Expansion Process

We assume that only 3178 (about 40%) of the 8109 deaths in 2016 have actively informed the employees of local SC institution, that is, the 3178 deaths formed the initial death sample set D. Now, we can simulate the discovery process of "pseudo-survival" samples based on the real data set. As mentioned earlier, there are a variety of sample selection functions available to us, denoted by $F(\cdot)$. This paper examines three strategies including four sample selection functions, as shown in Table 2. In the strategy that takes into account the sample "uncertainty" and the sample "diversity", the weight λ is set to be 0.5.

Table 2. Strategies of sample selection functions

$F(\cdot)$	Sample selection strategy
F (Random)	Random
F (MS)	Using uncertainty
F (MS & Cos)	Using uncertainty and diversity based on Cosin similarity
F(MS & Gau)	Using uncertainty and diversity based on Gaussian similarity

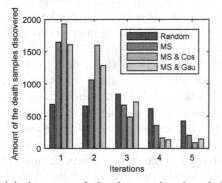

(a) Amount of death samples found in each iterations

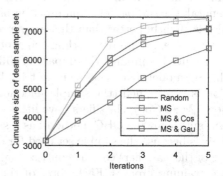

(b) Cumulative size of set D

Fig. 3. The growth of set D with different sample selection functions.

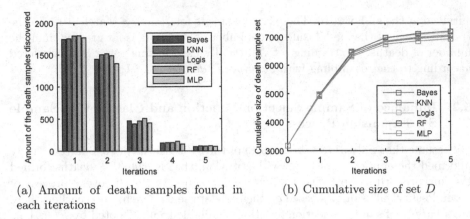

(a) Amount of death samples found in each iterations

(b) Cumulative size of set D

Fig. 4. The growth of set D with different classifiers.

In our experiments, the complete training process with each sample selection function contains 5 iterations. The set S' at each iteration consists of 2000 selected samples (the size of S is greater than 2000).

Taking the QBC classifier as an example, Fig. 3 shows the discovery of pseudo-survival samples under different sample selection functions. Figure 3(a) shows the number of pseudo-survival samples found in each iterations. Figure 3(b) shows the cumulative number of death samples after each iteration (i.e., the size of the death sample set D) under different sample selection functions. As can be seen from the Fig. 3(a), when there are many pseudo-survival samples (e.g., in the first 2 iterations), $F(MS)$, $F(MS\&Cos)$, $F(MS\&Gau)$ could have a much higher hit rate than $F(Random)$. From Fig. 4(b) we can see that $F(MS\&Cos)$ has the best performance in the task of finding pseudo-survival samples. $F(MS\&Gau)$ is not much different from $F(MS)$, but both of them are better than $F(Random)$. In general, the sample selection function using the uncertainty and diversity principles is slightly better than the sample selection function using only uncertainty; both strategies are better than the random selection strategy.

Figure 4 compares the impact of different classifiers on the training stage of the SC model. It can be seen that in the case where the sample selection function is determined (the sample selection function in this experiment is fixed to be $F(MS\&Cos)$, classifiers have little influence on the sample selection process. Among them, the QBC method has a slight advantage over any single classifier. This is because the sample set used for each iteration (including the initial training of the classifier) contains only a small amount of interference data throughout the training process. Each voter of the Committee can give a rational vote to ensure the validity of the QBC method. In addition, we found that random forest classifiers and logistic regression classifier are slightly more specific to other classifiers.

The experimental results in Figs. 3 and 4 show that in the training stage of SC, the role of the sample selection function is more important than that of the classifier. Taking sample uncertainty and sample diversity into account could make a better performance for the task of finding more samples. At the same time, the QBC mechanism will further help to improve the efficiency of sample selection.

5.3 The Effect of Different Classifiers on the Prediction Results

Figure 5 illustrates the impact of different classifiers on the SC model with sample selection function $F(GS\ \&\ Cos)$. According to the needs of SC, we are concerned about the "precision" and "recall" of death samples. Figure 5(a) first shows the proportion of correct classification on the survival samples and on the death samples. Among them, the proportion of correct classification on the death sample is the "recall". Since the number of deaths per month (about a few hundred people) is much smaller than the number of survival samples (more than 100,000 retirees), the number of false positives in the model will be considerable, resulting in a lower precision and F1 score,as shown in Fig. 5(b). Figure 5(a) shows that different classifiers have no significant impact on the classification accuracy of the SC model. The SC model can find most (more than 70%) of the deaths while reducing the large amount of human workload (about 80%). However, Fig. 5(b) shows that a large number of samples predicted to be death are false positives.

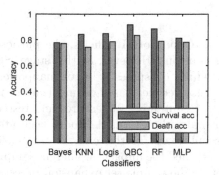 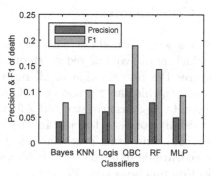

(a) Prediction accuracy on deaths and survivals

(b) Precision and F1 of deaths

Fig. 5. The effect of different classifiers on the prediction results.

5.4 SC Model Prediction Results

Figure 6 shows the classification results of the proposed SC model with sample selection function $F(MS\&Cos)$ and QBC classifier. We use the data of the first three months of 2017 as test sets. The experimental results show that the overall accuracy of the SC model in survival/death classification tasks is above 80%.

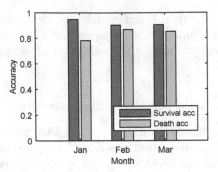

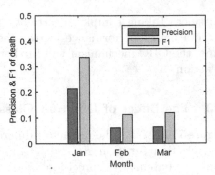

(a) Prediction accuracy on deaths and sur-   (b) Precision and F1 of deaths
vivals

Fig. 6. Prediction results of SC model.

Therefore, the model can effectively reduce the attention of the staff of the SI institution and help the staff identify the majority of the dead. However, due to the large proportion of survivors and deaths per month, the precision and F1 scores of the SC model are not high. At the same time, we observed that the SC model is affected by the monthly changes.

6   Conclusion

SC is a work carried out with the implementation of SI policies, the purpose of which is to avoid or reduce the loss of pensions. However, at this stage, SC requires the participation of all retirees, which consumes a lot of manpower. Therefore, this paper proposes a SC model for retirees based on medical insurance data. We first constructed a feature set containing 27 features and tested the validity of each feature. In order to make the SC model more accurate, we recommend using death samples in recent years. To this end, we combined active learning to build the SC model, and studied the impact of different sample selection functions and different classifiers on the sample set expansion process. Finally, we conducted a lot of experiments on the sample set of more than 100,000 retirees. The experimental results show that our model can effectively predict the survival status of retirees, thus can help staff to improve work efficiency.

Acknowledgments. This work was supported by the National Key Research and Development Plan of China (No. 2018YFC0114709), the Natural Science Foundation of Shandong Province of China for Major Basic Research Projects (No. ZR2017ZB0419), the TaiShan Industrial Experts Program of Shandong Province of China (No. tscy20150305).

References

1. Abbod, M.F., Cheng, K.Y., Cui, X.R., Huang, S.J., Han, Y.Y., Shieh, J.S.: Ensembled neural networks for brain death prediction for patients with severe head injury. Biomed. Sig. Process. Control **6**(4), 414–421 (2011)
2. Acharya, U.R., et al.: An integrated index for detection of sudden cardiac death using discrete wavelet transform and nonlinear features. Knowl.-Based Syst. **83**(C), 149–158 (2015)
3. Acharya, U.R., Fujita, H., Sudarshan, K.V., Ghista, D.N., Lim, W.J.E., Koh, J.E.W.: Automated prediction of sudden cardiac death risk using Kolmogorov complexity and recurrence quantification analysis features extracted from HRV signals. In: International Conference on Systems, Man, and Cybernetics, pp. 1110–1115 (2015)
4. Beygelzimer, A., Dasgupta, S., Langford, J.: Importance weighted active learning. In: Proceedings of the 26th Annual International Conference on Machine Learning (ICML), pp. 49–56 (2009)
5. Bottou, L.: Une approche thorique de l'apprentissage connexioniste; applications la reconnaissance de la parole (1991)
6. Brause, R., Hanisch, E.: An alarm system for death prediction. Int. J. Monit. Surveill. Technol. Res. **1**(2), 29–39 (2013)
7. Chaddad, A., Sabri, S., Niazi, T., Abdulkarim, B.: Prediction of survival with multi-scale radiomic analysis in glioblastoma patients. Med. Biol. Eng. Comput. **56**(12), 2287–2300 (2018)
8. Cohn, D.A., Ghahramani, Z., Jordan, M.I.: Active learning with statistical models. J. Artif. Intell. Res. **4**, 129–145 (1996)
9. Culotta, A., McCallum, A.: Reducing labeling effort for structured prediction tasks. In: National Conference on Artificial Intelligence, pp. 746–751 (2005)
10. Ebrahimzadeh, E., Manuchehri, M.S., Amoozegar, S., Araabi, B.N., Soltanian-Zadeh, H.: A time local subset feature selection for prediction of sudden cardiac death from ECG signal. Med. Biol. Eng. Comput. **56**(7), 1253–1270 (2018)
11. Freund, Y., Seung, H.S., Shamir, E., Tishby, N.: Selective sampling using the query by committee algorithm. Mach. Learn. **28**(2–3), 133–168 (1997)
12. Fu, Y., Zhu, X., Li, B.: A survey on instance selection for active learning. Knowl. Inf. Syst. **35**(2), 249–283 (2013)
13. Fujita, H., et al.: Sudden cardiac death (SCD) prediction based on nonlinear heart rate variability features and SCD index. Appl. Soft Comput. **43**(C), 510–519 (2016)
14. Garca, D.I., Roquei, N.G., De, F.C., Calvo, D.: Analysis of the high-frequency content in human qrs complexes by the continuous wavelet transform: an automatized analysis for the prediction of sudden cardiac death. Sensors **18**(2), 560 (2018)
15. Karmakar, C.K., Saha, B., Palaniswami, M., Venkatesh, S.: Multi-task transfer learning for in-hospital-death prediction of ICU patients. In: Annual International Conference of the IEEE Engineering in Medicine and Biology Society (EMBC), pp. 3321–3324 (2016)
16. Lee, Y.J., Mangasarian, O.L., Wolberg, W.H.: Survival-time classification of breast cancer patients. Comput. Optim. Appl. **25**(1–3), 151–166 (2003)
17. Lewis, D.D., Gale, W.A.: A sequential algorithm for training text classifiers. In: Croft, B.W., van Rijsbergen, C.J. (eds.) SIGIR 1994, pp. 3–12. Springer, London (1994). https://doi.org/10.1007/978-1-4471-2099-5_1

18. Li, S., Wang, J., Liu, Y., Sun, X.: Prediction of death rate of breast cancer induced from average microelement absorption with neural network. In: Li, K., Li, X., Irwin, G.W., He, G. (eds.) LSMS 2007. LNCS, vol. 4689, pp. 414–421. Springer, Heidelberg (2007). https://doi.org/10.1007/978-3-540-74771-0_47

19. Malhotra, K., Navathe, S.B., Chau, D.H., Hadjipanayis, C., Sun, J.: Constraint based temporal event sequence mining for glioblastoma survival prediction. J. Biomed. Inform. **61**(C), 267–275 (2016)

20. Miao, F., Cai, Y., Zhang, Y., Fan, X., Li, Y.: Predictive modeling of hospital mortality for patients with heart failure by using an improved random survival forest. IEEE Access **6**, 7244–7253 (2018)

21. Montazeri, M., Montazeri, M., Montazeri, M., Beigzadeh, A.: Machine learning models in breast cancer survival prediction. Technol. Health Care Off. J. Eur. Soc. Eng. Med. **24**(1), 31 (2015)

22. Raji, C.G., Chandra, S.S.V.: Graft survival prediction in liver transplantation using artificial neural network models. J. Comput. Sci. **16**, 72–78 (2016)

23. Rajkomar, A., et al.: Scalable and accurate deep learning with electronic health records. NPJ Digit. Med. **1**(1), 1–10 (2018)

24. Rizzo, L., Majnaric, L., Dondio, P., Longo, L.: An investigation of argumentation theory for the prediction of survival in elderly using biomarkers. In: Iliadis, L., Maglogiannis, I., Plagianakos, V. (eds.) AIAI 2018. IAICT, vol. 519, pp. 385–397. Springer, Cham (2018). https://doi.org/10.1007/978-3-319-92007-8_33

25. Rouzbahman, M., Jovicic, A., Chignell, M.H.: Can cluster-boosted regression improve prediction of death and length of stay in the ICU? IEEE J. Biomed. Health Inform. **21**(3), 851–858 (2017)

26. Sanz, J., Paternain, D., Galar, M., Fernandez, J., Reyero, D., Belzunegui, T.: A new survival status prediction system for severe trauma patients based on a multiple classifier system. Comput. Methods Programs Biomed. **142**, 1–8 (2017)

27. Vanneschi, L., Farinaccio, A., Giacobini, M., Mauri, G., Antoniotti, M., Provero, P.: Identification of individualized feature combinations for survival prediction in breast cancer: a comparison of machine learning techniques. In: Pizzuti, C., Ritchie, M.D., Giacobini, M. (eds.) EvoBIO 2010. LNCS, vol. 6023, pp. 110–121. Springer, Heidelberg (2010). https://doi.org/10.1007/978-3-642-12211-8_10

28. Vanneschi, L., Farinaccio, A., Mauri, G., Antoniotti, M., Provero, P., Giacobini, M.: A comparison of machine learning techniques for survival prediction in breast cancer. BioData Min. **4**(1), 12 (2011)

29. Zhang, J., Chen, L., Vanasse, A., Courteau, J., Wang, S.: Survival prediction by an integrated learning criterion on intermittently varying healthcare data. In: AAAI Conference on Artificial Intelligence, pp. 72–78 (2016)

A Novel Approach for Air Quality Inference and Prediction Based on DBU-LSTM

Liang Ge[1,2(✉)], Aoli Zhou[1,2], Junling Liu[1,2], and Hang Li[1,2]

[1] College of Computer Science, Chongqing University, Chongqing, China
{geliang,aolizhou,liujunling,hangli}@cqu.edu.cn
[2] Chongqing Key Laboratory of Software Theory and Technology, Chongqing, China

Abstract. The inference and prediction of fine-grained air quality are two important directions in urban air computing. Solving these two problems can provide useful information for urban environmental governance and residents' health improvement. In this paper, we propose a general approach to solve these two problems with one model, while most other existing works use different models to solve them. Our model is based on deep bidirectional and unidirectional long short-term memory (DBU-LSTM) neural network, which can capture bidirectional temporal dependencies and spatial correlation from time series data containing spatial information. To infer and predict the air quality of the target region, we use the historical meteorological data of the target region and the historical air quality data of regions which are similar to the target. Urban heterogeneous data such as point of interest (POI) and road network are used to evaluate the similarities between urban regions. We also use a tensor decomposition method to complete the missing historical air quality data onto monitoring stations, which reduces the error of our model. We evaluated our approach on real data sources obtained in Beijing, and the results show its advantages over recent literature.

Keywords: Air quality · Inference · Prediction · RNN · LSTM

1 Introduction

With the growth of the economy and society, air pollution problems are getting more and more attention, especially for rapidly developing countries (e.g., China and Brazil). The high concentration of air pollutants may cause much significant heart disease and lung disease. The information about air quality, such as the concentration of PM_{25} is important to urban environmental governance and the health of residents. The specific monitoring station in different regions collect the air quality information, but the number of monitoring stations in cities is limited due to the expensive cost of building and maintaining, and there is no air quality monitoring station in most urban regions. Considering the equipment

© Springer Nature Switzerland AG 2019
J. Shao et al. (Eds.): APWeb-WAIM 2019, LNCS 11641, pp. 171–186, 2019.
https://doi.org/10.1007/978-3-030-26072-9_12

maintenance, recalibration, and any other problems, the data of air quality collected by the monitoring station may also be inaccurate or missing. Hence there are three main problems in urban air computing:

1. Infer the current time air quality in regions without monitoring station.
2. Predict the future time air quality in regions with monitoring station.
3. Complete the missing historical data of monitoring stations.

This paper aims to solve the above three problems by using urban heterogeneous data, such as air quality, meteorology, POI, road networks. Air quality is affected by many factors, such as pollutant emissions, meteorological factors, traffic flow, and human activities [21]. The POI in the urban region provides information about sources of air pollutants. The strong correlation between meteorology and air quality in the urban region provide information about the trends of air quality. Vehicle emissions are a major source of air pollutants. Figure 1 shows that the blue and orange lines represent the trend of individual air quality index (IAQI) of two monitoring stations, the X-axis represents the time sequence, and the Y-axis represents the concentration of air pollutant. This figure exposes the air quality trends of monitoring stations are similar in the same city. We can use the information about known regions to derive information about those unknown regions, based on the correlation between regions in the same city. The correlation between regions is not only related to distance but also related to similarity. Those regions with similar topographical distribution, building distribution, and road structure have more similar trends in air quality. We used heterogeneous spatial data to calculate the similarities.

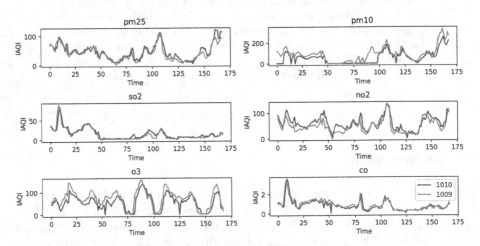

Fig. 1. The trend of IAQI of two monitoring stations within one week. (Color figure online)

We propose a general approach to infer and predict air quality with one model, based on the DBU-LSTM neural network. The bidirectional long short-term memory (BI-LSTM) layer process sequence data in both forward and backward directions with two separate hidden layers, capturing bidirectional temporal dependencies and spatial features from historical series data which contains spatial information. The unidirectional long short-term memory (LSTM) layer before the output layer achieves a better inference or prediction value. Deep learning models require a lot of training data to prevent overfitting, we use the tensor decomposition method to complete the historical air quality data of the monitoring station that provide a large number of training samples. Experiments show that DBU-LSTM is superior to the recent literature when solving the topics of inference, prediction of fine-grained air quality. In summary, our contributions state as follows:

1. We apply urban heterogeneous data to a general efficient approach, that trains DBU-LSTM neural network to infer current time air quality of no-monitoring regions and predict the future time air quality of all regions.
2. We use the tensor decomposition method to complete the missing data of monitoring stations, which increases the training samples and improves the accuracy of the model.
3. We analyzed the correlation between regions in the same city and proposed the concept and calculation method of the regional similarity matrix.

2   Related Work

2.1   Traditional Methods

Traditional methods use different models, e.g., spatial interpolation methods, and statistical models. The main idea of spatial interpolation methods is based on the hypothesis of spatial continuity of air pollutant distribution. Wong et al. [17] evaluated the performance of most spatial interpolation methods, such as spatial averaging, nearest neighbor, inverse distance weighting (IDW), kriging, etc. But the accuracy of spatial interpolation methods is not satisfactory, due to the geographically sparse original datasets. The main idea of statistical models is modeling the relationship between air quality and some influencing factors. Hasenfratz et al. [8] proposed Generalized Additive Models (GAMs) to infer air quality by land-use features and traffic data. Statistical models need labeled data to build data-driven models, but the labeled data collected from stations is limited and the problem of missing data is also very serious.

2.2   Urban Computing

Urban computing aims to tackle urban challenges by using the heterogeneous big data that has been generated in cities (e.g., traffic flow, human mobility, and geographical data) [20]. Zheng et al. [21] developed a co-training approach for air quality inference based on joint learning of temporal-spatial data. Zheng et al. [22] proposed a approach for air quality prediction based on hybrid predictive models. Zhu et al. [23] used a spatial-temporal granger causality model for air quality prediction. These methods both use urban heterogeneous data to solve air quality inference or prediction problems in different models. With growing heterogeneous data is acquired, these models are becoming more and more difficult to fit these data.

2.3   Deep Learning

In recent years, some literatures have begun to address these air pollution problems through deep learning. Deep learning method used to deal with high dimensional data problem, which has the capability of capturing non-linear relationship.

Qi et al. [16] developed a deep neural network including three stages (feature extraction, self-coding training, semi-supervised regression) to infer and predict air quality. Kök, İbrahim et al. [9] proposed a method using the LSTM to predict air quality. Le and Duc [11] combined the Convolutional Neural Network (CNN) and LSTM to predict the real air pollution. However, these deep learning methods proposed in above literatures are overly dependent on the capabilities of the model, and there is no effective processing of the data which entered into the model.

3   Design of Approach

3.1   Framework

The framework of our approach is shown in Fig. 2, including three main processes: Data Acquisition and Preprocessing, Data Completion and Feature Extraction, and Air Quality Inference and Prediction.

In the process of Data Acquisition and Preprocessing, we acquire a variety of original urban heterogeneous data, formatting these data and provide it to the next stage.

In the process of Data Completion and Feature Extraction, we first use the tensor decomposition method to complete the missing historical air quality data of the monitoring stations. We extract spatial features from POI data, road network data and geographic location data, and then calculate the regional similarity matrix. For a target inference or prediction region, we select k similar monitoring station regions through the regional similarity matrix and then combine the historical meteorological data of the target region and the historical air quality data of similar regions into time series data. Hence the time series data contains the temporal-spatial information.

In the process of Air Quality Inference and Prediction, we train the DBU-LSTM network to model time series data and then use the network to infer or predict the air quality after the specified time step in the target region. When the time step is zero, the model is used to infer the air quality of no-monitoring station region. Otherwise, the model is used to predict the air quality of monitoring station region.

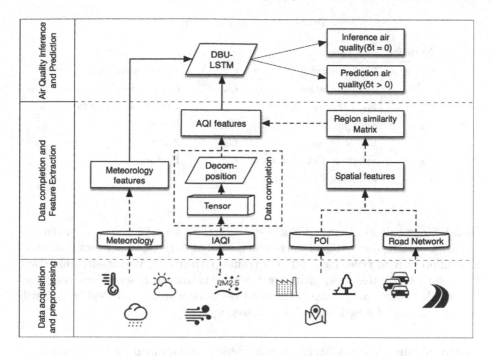

Fig. 2. The framework of our approach.

3.2   Extraction of Features

The fine-grained region is represented by one-kilometer grid cell, we divide a city into disjointed grids. Symbol G denotes the collection of all regions, G_u denotes the collection of regions without monitoring station, and G_v denotes the collection of regions with monitoring station.

Meteorological Features. Local meteorological conditions strongly affect the daily concentration of air pollutants. [15] We extract six meteorological features, i.e., weather, temperature, humidity, pressure, wind direction, wind speed. The weather and wind direction are not numeric. We use the numbers one to five to indicate the five values of the weather and use the numbers one to eight indicate the eight values of the wind direction.

POI Features. Different POI categories have different effects on the concentration of air pollutants. We only select those POI categories that have an impact on air quality. For example, the category of Vehicle service including the gas station, toll, etc. which reflect the density of cars in the region. We classify important POI into 13 categories, which is shown in Table 1. Symbol $g.F_p$ denotes the POI feature extracted from region g, which contains the total number of each POI category.

Table 1. Category of POIs.

Notation	POI category	Notation	POI category
C_1	Vehicle service	C_8	Social institution
C_2	Transportation	C_9	Companies
C_3	Food & Beverage	C_{10}	Culture & Education
C_4	Shopping	C_{11}	Parks
C_5	Life & Hospital	C_{12}	Nature
C_6	Sports & Entertainment	C_{13}	Factory
C_7	Hotels & Real estates		

Road Network Features. Vehicle emissions are a major source of urban air pollution. Traffic flow in the urban region is related to the road network structure, and different road types have different traffic patterns [2]. We classify important roads into four categories, including freeway, trunk road, secondary road and link road. Symbol $g.F_r$ denotes the road network feature from region g, which contains the total length of each road category.

Regional Similarity Matrix. Topographical variations in urban regions lead to complex spatiotemporal variations in pollutant concentration [18]. We propose the concept of the regional similarity matrix to describe the similarity of spatial features between regions. $g.F_c$ denotes the geographical coordinates of region g. $g.F_{spa}$ denotes the spatial feature of a region, which combine the POI features $g.F_p$, road features $g.F_r$ and geographical coordinates of the region $g.F_c$ into one feature.

We normalize the spatial features of each region to $[0, 1]$ and then calculate the distance between no-monitoring station regions and monitoring station regions by Eq. 1. Mat_S denotes the regional similarity matrix, where rows represent all the region G and the columns represent all regions containing monitoring station G_v. $Mat_S(i,j)$ represents the similarity between the region g_i and the region g_j in the similarity matrix, where $g_i \in G$, and $g_j \in G_v$. The smaller the value, the more similar they are.

$$Mat_S(i,j) = Mat_S(g_i, g_j) = \sqrt{\|g_i.F_{spa} - g_j.F_{spa}\|^2} \tag{1}$$

For any region, we can use the region similarity matrix to select those regions with monitoring station.

3.3 Data Completion

Completing the missing historical data of the monitoring station can provide more training data for deep learning model. We use the tensor decomposition method to complete the missing values.

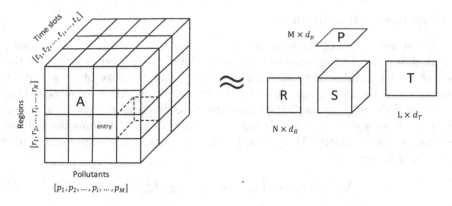

Fig. 3. Structure of the air quality tensor decomposition.

The tensor $A \in \mathbb{R}^{N \times M \times L}$ represents the temporal-spatial air quality values of the monitoring stations in different time slots is shown in Fig. 3, which has three dimensions denoting N regions, M air pollutants categories, and L time slots, respectively. An entry A_{ijk} represents the jth air pollutant concentration value (IAQI) for region i at time slot k.

We use tucker decomposition [10] method to decomposes tensor A into four parts including a core tensor $S \in \mathbb{R}^{d_R \times d_P \times d_T}$, space related matrix $R \in \mathbb{R}^{N \times d_R}$, air pollutant categories related matrix $P \in \mathbb{R}^{M \times d_P}$ and time related matrix $T \in \mathbb{R}^{L \times d_T}$. The loss function to control the error of decomposition defined by Eq. 2.

$$Loss(S, R, P, T) = \frac{1}{2} \|A - S \times_R R \times_P P \times_T T\|^2 \\ + \frac{\lambda}{2}(\|S\|^2 + \|R\|^2 + \|P\|^2 + \|T\|^2) \tag{2}$$

The symbol $\|\bullet\|^2$ denotes the $l2$ norm, the $\times_*$ is the tensor-matrix multiplication, and * denotes for the mode of a tensor. The first term is to control the decomposition error and the second term is a regularization penalty to avoid overfitting. d_R, d_P, d_T usually very small, denoting the number of latent factors. λ is a parameter controlling the contribution of the regularization penalty. We adopt the gradient descent algorithm to the optimization problem. By minimizing the objective function, we can get optimized R, P, and T. Then we can recover missing values in A by Eq. 3.

$$A_r = S \times_R R \times_P P \times_T T \tag{3}$$

Each entry in recover tensor A_r is not empty. We consider those entries that are vacant in the original tensor A but have a value in the recover tensor A_r as filling values.

3.4   Problem Description

We define some symbols to describe the problem of the research. We consider one kind of individual air quality index from the target region as a target value for our model to infer or predict, and y denotes the category of air pollutant. Symbol t denotes the current time, $t + \delta t$ denotes the future time after δt time steps, and T denotes the historical time steps. One time step represents one hour. For a region of $g \in G$, we select k regions with monitoring stations through the regional similarity matrix Mat_S. The input data in the time t represented by a vector as follows:

$$X_t = [g.F_m^t, g_1.F_a^t, ..., g_i.F_a^t, ..., g_k.F_a^t], i \in k \tag{4}$$

which connect $k+1$ vectors into one vector, the $g.F_m^t$ denotes the meteorological feature vector of the target region, the $g_i.F_a^t$ denotes the air quality value vector of the similar region g_i. Finally we merge each historical time step input data vector into one vector as follows:

$$X = [X_{t-T}, X_{t-T+1}, ..., X_{t-1}] \tag{5}$$

The vector X also represents the time series data contains spatial information. The vector Y denotes the true value of the y air pollutant of the target region at the time step $t + \delta t$ as follows:

$$Y = [y^{t+\delta t}] \tag{6}$$

We use (X,Y) as a training sample of the model. When the $\delta t = 0$, the model is used for inference. Moreover, when the $\delta t > 0$, the model is used for prediction.

3.5   Deep BU-LSTM Network

The deep bidirectional and unidirectional LSTM neural network includes four components: BI-LSTM layer, LSTM layer, input layer and output layer. The architecture of DBU-LSTM is shown in Fig. 4, the BI-LSTM layer is behind the input layer and the LSTM layer is before the output layer. When feeding the time series data with temporal-spatial information to the BI-LSTM layer, both the spatial correlation of the AQI in different regions and the temporal dependencies of the AQI captured during the feature learning process. The BI-LSTM is behind the input layer which can capture more useful information from temporal-spatial series data, learning more useful features.

When the output layer generates the inferred or predicted value, the LSTM layer in front of the output layer only needs to utilize features learned from BI-LSTM layer, calculate iteratively along the forward direction and generate the predicted values, which can capture forward dependency. The detail about LSTM, BI-LSTM and training process is explained as follow.

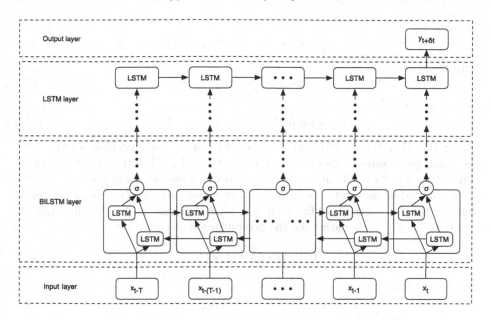

Fig. 4. The deep bidirectional and unidirectional LSTM architecture

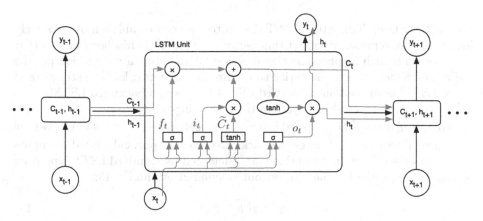

Fig. 5. Standard LSTM unit architecture.

LSTM. LSTM is a kind of gated recurrent neural networks. It works well on sequence-based tasks with long-term dependencies, which handle the gradient exploding and vanishing in traditional RNNs. The LSTM cell is shown in Fig. 5, which has three gates namely input gate, forget gate, and output gate, that control the information flow through the cell and the neural network. At time t, the input gate, the forget gate, the output gate and the cell input state are denoted as i_t, f_t, o_t, $\widetilde{C}_t$ respectively. The formula as follows, input gate:

$$i_t = \sigma(W_i x_t + U_i h_{t-1} + b_i) \tag{7}$$

forget gate:

$$f_t = \sigma(W_f x_t + U_f h_{t-1} + b_f) \qquad (8)$$

output gate:

$$o_t = \sigma(W_o x_t + U_o h_{t-1} + b_o) \qquad (9)$$

cell input state:

$$\widetilde{C}_t = tanh(W_C x_t + U_C h_{t-1} + b_C) \qquad (10)$$

where W_i, W_f, W_o and W_C are the weight matrices connecting input x_t to the three gates and cell input state, Ui, U_f, U_o and U_C are the weight matrices connecting previous cell output state h_{t-1} to the three gates and cell input state, b_i, b_f, b_o and b_C are the bais vectors of the three gates and cell input state. σ denotes the gate activation function and $tanh$ denotes the hyperbolic tangent function. The formula calculates the cell output state:

$$C_t = i_t * \widetilde{C}_t + f_t * C_{t-1} \qquad (11)$$

hidden layer output:

$$h_t = o_t * tanh(C_t) \qquad (12)$$

where i_t, f_t, $\widetilde{C}_t$ and C_t have the same dimension.

BI-LSTM. Deep bidirectional LSTM architecture can establish a progressively higher level of representations of time series data [5,12]. It has been proved that in many fields such as phoneme classification [6] and speech recognition [5], the bidirectional networks are superior to unidirectional networks. The structure of an BI-LSTM layer contains a forward LSTM layer and a backward LSTM layer, and connect the two hidden layers to the same output layer.

We use $\overrightarrow{h}$ denotes the forward layer output, calculated by inputs in a positive sequence. Moreover, $\overleftarrow{h}$ denotes the backward layer output, calculated by inputs in a reversed sequence. Both of them are calculated by standard LSTM updating equations, respectively. The unit output calculated by the Eq. 13:

$$y_t = \sigma(\overrightarrow{h}_t, \overleftarrow{h}_t) \qquad (13)$$

where the σ is the function combined the two outputs, such as concatenating, summation, average, and multiplication.

Training. We use the mean square error formula as the loss function of the model. For the training of the DBU-LSTM model, we adopt the backpropagation through time algorithm to solve it. The training process of DBU-LSTM model is shown in Algorithm 1. The Θ denotes the all model parameter include weight W, U and bais b of each gate and cell state. The line 1–7 represents the training samples generated process, and the line 8–13 represents the training process.

Algorithm 1. Train algorithm of DBU-LSTM

Input:
    Collection of monitoring station region G_v
    Historical air quality of G_v: $\{g.F_a^0, g.F_a^1, ..., g.F_a^{n-1}\}, g \in G_v$;
    Historical meteorology of G_v: $\{g.F_m^0, g.F_m^1, ..., g.F_m^{n-1}\}, g \in G_v$;
    The future time offset δt, historical time steps T;
    The target air pollutant type y;
Output:
    DBU-LSTM model $\mathcal{M}$
1:   $\mathcal{D} \leftarrow \varnothing$
2: **for all** available time step for region g $(T \leq t \leq n-1)$ **do**
3:     select k similar monitoring stations
4:     $X = [g.F_m^{t-T}, g_1.F_a^{t-T}, ..., g_k.F_a^{t-T}, ..., g.F_m^{t-1}, g_1.F_a^{t-1}, ..., g_k.F_a^{t-1}]$
5:     $Y = [g.y^{t+\delta t}]$
6:     put an training sample (X, Y) into $\mathcal{D}$
7: **end for**
8: initialize the model parameter Θ
9: **while** stopping criteria is not met **do**
10:    randomly select a batch of samples $\mathcal{D}_b$ from $\mathcal{D}$
11:    update Θ by minimizing the objective function with $\mathcal{D}_b$
12: **end while**
13: **return** the learned DBU-LSTM model $\mathcal{M}$

---

4 Experiments

4.1 Data and Parameter Setting

In the evaluation, we apply our model to the topics of inference and prediction of fine-grained air quality based on real data sources obtained from Beijing, China. The real urban data sources used in experiments as follows.

1. Air quality data: We obtained the data sets from Microsoft Research Asia (MSRA) website contributed by Zheng et al [19]. It consists of $PM_{2.5}$, PM_{10}, SO_2, NO_2, CO and O_3 every hour from 36 monitoring stations in Beijing, which spanned a whole year, from May. 2014 to May. 2015.
2. Meteorological data: We collect fine-grained meteorological data from Open-WeatherMap [14], It consists of weather, temperature, humidity, pressure, wind direction, wind speed from 16 districts in Beijing, which spanned a whole year, from May. 2014 to May. 2015.
3. POI data: We collect the fine-grained POI data in Beijing from Amap [1] with its APIs.
4. Road network data: We collect the road network data in Beijing from Open-StreetMap [13] with its APIs.

In the air quality data, data is completely missing at some time slots, which accounts for 11.8% of the total data. We use the tensor decomposition method to complete this part of the missing data. We divide the Beijing city into disjoint one-kilometre grids, the total number of grids is 3600. The size of G_u is 3564, and the size of G_v is 36. We set the length of time steps $T = 12$, the number of similar stations $k = 3$, and the offset time steps $\delta \in [0, 48]$. The dimensionality of the input vector is $6 \cdot T + 6 \cdot T \cdot k = 288$, the dimensionality of output is 1. For each experiment, we select 6400 h of data for training data and 2100 h of data for testing data. There are 36 air quality monitoring stations in Beijing. Hence, the number of train set is $6400 * 36 = 230400$, and the number of the test set is $2100 * 36 = 75600$, that constructing a large-scale data set.

We adopt the root mean square error (RMSE) to evaluate the approaches, which defined by Eq. 14:

$$RMSE = \sqrt{\frac{\sum_{i=1}^{n}(y_i - \widehat{y_i})^2}{n}} \tag{14}$$

where y_i is a ground-truth IAQI from the air quality stations, $\widehat{y_i}$ is the inferred or predicted IAQI value, and n is the total number of test data.

We compare our model with several closely related methods as follow:

- SVR [3]: Support vector regression, a supervised learning models with associated learning algorithms that analyze data used for regression analysis.
- RNN [7]: Recurrent neural network, a class of artificial neural network where connections between nodes form a directed graph along a sequence.
- LSTM [9]: LSTM network, an RNN composed of LSTM units, which were developed to deal with the exploding and vanishing gradient problems when training traditional RNNs.
- GRU [4]: Gate recursive unit, a gating mechanism in RNN. It is similar to LSTM unit, but lack an output gate, which having fewer parameters and exhibits better performance on smaller datasets.
- BI-LSTM [5]: Bidirectional LSTM network, a special LSTM that processes sequence in both forward and backward directions with two separate hidden layers.

4.2   Results and Discussions

As results are shown in Table 2, we summarize the RMSE on the test set using SVR, RNN, LSTM, GRU, BI-LSTM, and DBU-LSTM, respectively. We observe that DBU-LSTM achieve the best performance in air quality inference and the best performance in air quality prediction, except in 1–12 prediction. For all the methods, the infer RMSE on the test set is lowest, and the predict RMSE on test set increases as the prediction hour increases. GRU performs slightly better than DBU-LSTM in the 1–12 h prediction but much worse than DBU-LSTM with the prediction hour increases. GRU has fewer parameters than LSTM, so it is easier to converge and train faster. When the size of the data is large, the performance of the LSTM may be better than the GRU. DBU-LSTM performs better than LSTM and GRU, because GRU and LSTM only find relevant information from the past, not find relevant information from the future. DBU-LSTM performs slightly better than BI-LSTM. Deep bidirectional LSTM acquires more temporal-spatial features from entire time series data, which processes time series

Table 2. The RMSE on the test set using SVR, RNN, LSTM, GRU, BI-LSTM and DBU-LSTM, respectively.

	Inference	Prediction			
	Hour 0	Hour 1–12	Hour 12–24	Hour 25–36	Hour 37–48
SVR	0.0671	0.0736	0.0784	0.0872	0.1091
RNN	0.0645	0.0693	0.0745	0.0813	0.0934
LSTM	0.0639	0.0683	0.0738	0.0792	0.0906
GRU	0.0633	**0.0667**	0.0730	0.0789	0.0891
BI-LSTM	0.0631	0.0671	0.0726	0.0773	0.0879
DBU-LSTM	**0.0628**	0.0669	**0.0718**	**0.0767**	**0.0863**

Table 3. The RMSE on the same model with and without data completion.

	Inference	Prediction			
	Hour 0	Hour 1–12	Hour 12–24	Hour 25–36	Hour 37–48
LSTM without DC	0.0663	0.0718	0.0766	0.0827	0.0953
LSTM with DC	0.0639	0.0683	0.0738	0.0792	0.0906
DBU-LSTM without DC	0.0641	0.0698	0.0756	0.0798	0.0898
DBU-LSTM with DC	0.0628	0.0669	0.0718	0.0767	0.0863

data in both forward and backward directions with two separate hidden layers. DBU-LSTM based on BI-LSTM, adding a unidirectional LSTM layer in front of the output layer, which can process learned features well to infer and predict. The traditional SVR performs worse than recurrent neural networks because they do not have the ability to modeling the higher level of representations of time series data well.

As results are shown in Table 3, regardless of the LSTM model or the DBU-LSTM model, the RMSE obtained on the completed dataset is smaller than the RMSE obtained on the missing dataset. Our DBU-LSTM model has improved accuracy in predicting and inferring air quality, after complete the missing historical data by Data Complete (DC). The data completion based on tensor decomposition is effective for improving the experimental results, in the approach framework.

To evaluate the prediction accuracy of DBU-LSTM model, we use DBU-LSTM to predict 1 h interval pm25 IAQI of two days (January 1 to 2, 2015) in one station region. As results are shown in Fig. 6, the blue line shows the real IAQI flow, and the orange line shows the predicted IAQI flow. We can notice that the performance of the DBU-LSTM model is quite good during the whole time.

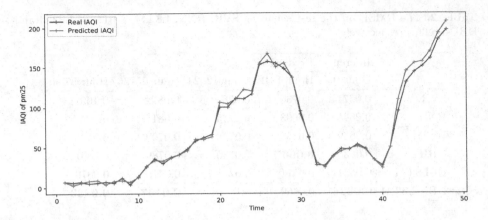

Fig. 6. Comparison of predict and real IAQI flows of pm25. (Color figure online)

The reason why our approach achieves the best performance as follow:

1. We use heterogeneous urban data to provide temporal-spatial information for the model, which combine the similar region historical air quality and historical meteorological into time series data.
2. We complete the missing historical data of monitoring station regions and increase the sample of training.
3. The DBU-LSTM we proposed can learn the temporal-spatial information from the time series data and use learned features to predict better.

5   Conclusion

In this paper, we have studied two important problems in urban air computing: the inference and prediction of fine-grained air quality. The solutions to these two problems can provide crucial information to support air pollution control, and consequently generate great humanities, societal, and technical impacts. In order to solve these two problems, we proposed a general approach based on one model, while most other existing works are to establish different models. We proposed the concept of the regional similar matrix to describe the spatial similarity between urban regions. We combined the meteorological data of the target region and the air quality data of similar regions into time series data. Our model based on bidirectional and unidirectional LSTM network, which can capture bidirectional temporal dependencies and spatial features from time series data. In order to increase the training sample, improve the accuracy of the model, we proposed a tensor decomposition method to complete the missing historical air quality data in monitoring stations. Extensive experiments evaluated on real data sources collected from Beijing, results showing our approach achieves superior performance than comparable rivals.

References

1. Amap. https://lbs.amap.com/api/webservice/guide/api/search. Accessed 4 June 2018
2. Bagieński, Z.: Traffic air quality index. Sci. Total Environ. **505**, 606–614 (2015)
3. Basak, D., Pal, S., Patranabis, D.C.: Support vector regression. Neural Inf. Process. Lett. Rev. **11**(10), 203–224 (2007)
4. Cho, K., et al.: Learning phrase representations using rnn encoder-decoder for statistical machine translation. arXiv preprint arXiv:1406.1078 (2014)
5. Graves, A., Jaitly, N., Mohamed, A.R.: Hybrid speech recognition with deep bidirectional lstm. In: 2013 IEEE Workshop on Automatic Speech Recognition and Understanding (ASRU), pp. 273–278. IEEE (2013)
6. Graves, A., Schmidhuber, J.: Framewise phoneme classification with bidirectional LSTM and other neural network architectures. Neural Netw. **18**(5–6), 602–610 (2005)
7. Grossberg, S.: Recurrent neural networks. Scholarpedia **8**(2), 1888 (2013)
8. Hasenfratz, D., Saukh, O., Walser, C., Hueglin, C., Fierz, M., Thiele, L.: Pushing the spatio-temporal resolution limit of urban air pollution maps. In: 2014 IEEE International Conference on Pervasive Computing and Communications (PerCom), pp. 69–77. IEEE (2014)
9. Kök, İ., Şimşek, M.U., Özdemir, S.: A deep learning model for air quality prediction in smart cities. In: 2017 IEEE International Conference on Big Data (Big Data), pp. 1983–1990. IEEE (2017)
10. Kolda, T.G., Bader, B.W.: Tensor decompositions and applications. SIAM Rev. **51**(3), 455–500 (2009)
11. Le, D.: Real-time air pollution prediction model based on spatiotemporal big data. arXiv preprint arXiv:1805.00432 (2018)
12. LeCun, Y., Bengio, Y., Hinton, G.: Deep learning. Nature **521**(7553), 436 (2015)
13. OpenStreetMap. http://www.openstreetmap.org/. Accessed 4 June 2018
14. OpenWeatherMap. https://openweathermap.org/. Accessed 4 June 2018
15. Pearce, J.L., Beringer, J., Nicholls, N., Hyndman, R.J., Tapper, N.J.: Quantifying the influence of local meteorology on air quality using generalized additive models. Atmos. Environ. **45**(6), 1328–1336 (2011)
16. Qi, Z., Wang, T., Song, G., Hu, W., Li, X., Zhang, Z.M.: Deep air learning: interpolation, prediction, and feature analysis of fine-grained air quality. IEEE Transact. Knowl. Data Eng. **30**, 2285–2297 (2018)
17. Wong, D.W., Yuan, L., Perlin, S.A.: Comparison of spatial interpolation methods for the estimation of air quality data. J. Expo. Sci. Environ. Epidemiol. **14**(5), 404 (2004)
18. Zhang, X., Gong, Z.: Spatiotemporal characteristics of urban air quality in china and geographic detection of their determinants. J. Geog. Sci. **28**(5), 563–578 (2018)
19. Zheng, Y.: Forecasting fine-grained air quality based on big data (2015). https://www.microsoft.com/en-us/research/project/urban-air/. Accessed 4 June 2018
20. Zheng, Y., Capra, L., Wolfson, O., Yang, H.: Urban computing: concepts, methodologies, and applications. ACM Transact. Intell. Syst. Technol. (TIST) **5**(3), 38 (2014)
21. Zheng, Y., Liu, F., Hsieh, H.P.: U-air: When urban air quality inference meets big data. In: Proceedings of the 19th ACM SIGKDD international conference on Knowledge discovery and data mining, pp. 1436–1444. ACM (2013)

22. Zheng, Y., Yi, X., Li, M., Li, R., Shan, Z., Chang, E., Li, T.: Forecasting fine-grained air quality based on big data. In: Proceedings of the 21th ACM SIGKDD International Conference on Knowledge Discovery and Data Mining, pp. 2267–2276. ACM (2015)
23. Zhu, J.Y., Sun, C., Li, V.O.: An extended spatio-temporal granger causality model for air quality estimation with heterogeneous urban big data. IEEE Transact. Big Data **3**(3), 307–319 (2017)

Predictive Role Discovery of Research Teams Using Ordinal Factorization Machines

Tong Liu[1], Weijian Ni[1(✉)], Qingtian Zeng[1], and Nengfu Xie[2]

[1] College of Computer Science and Engineering,
Shandong University of Science and Technology, Qingdao, China
liu_tongtong@foxmail.com, niweijian@gmail.com, qtzeng@163.com
[2] Agricultural Information Institute, Chinese Academy of Agricultural Sciences,
Beijing, China
xienengfu@caas.cn

Abstract. In this paper, we address the problem of research role discovery, especially for large research institutes where similar yet separated teams co-exist. The roles that researchers play in a research team, i.e., principal investigator, sub-investigator and research staff, typically exhibit an ordinal relationship. In order to better incorporate the ordinal relationship into a role discovery model, we approach research role discovery as an ordinal regression problem. In the proposed approach, we represent a research team as a heterogeneous teamwork network and propose OrdinalFM, short for Ordinal Factorization Machines, to learn the role prediction function. OrdinalFM extends the traditional Factorization Machines (FM) in an effort to handle the ordinal relationship among learning targets. Experiments with a real-world research team dataset verify the advantages of OrdinalFM over state-of-the-art ordinal regression methods.

Keywords: Role discovery · Research team · Ordinal regression · Factorization Machines

1 Introduction

Research teams, instead of sole scientists, have become increasingly crucial to the success of modern scientific production in recent years. Typically, each member in a well organized research team can be categorized as: Principal Investigator (PI), the one who takes overall responsibility for the research team and is the head of the team; Sub-Investigator (SI), the one who takes charge of specified sub-tasks delegated by the PI; Research Staff (RS), the one who is led by PI or SI and can be viewed as common members in the research team. Role analysis of research team members is an important issue in research team management. In fact, the role of each member represents the degree of status in the research team, i.e., the PI is the most important figure in the research team, and the SIs

© Springer Nature Switzerland AG 2019
J. Shao et al. (Eds.): APWeb-WAIM 2019, LNCS 11641, pp. 187–195, 2019.
https://doi.org/10.1007/978-3-030-26072-9_13

are more salient than the RSs. Formally speaking, the research roles exhibit an ordinal relationship: PI $\succ$ SI $\succ$ RS. Obviously, the ordinal relationship provides valuable prior information for the research role discovery task.

In this paper, we aim to propose a predictive research role discovery approach under the scenario where "parallel" research teams – that operate separately and have comparable compositions – co-exist. More specifically, there are two phases in our proposed approach: the training phase and the predicting phase. During the training phase, a number of exemplary research teams are taken as training samples and used to train a research role discovery model. Practically, the exemplary research teams, which are composed of members with prior-assigned roles, are the teams with outstanding performance or considered as representative ones in a research institute. During the predicting phase, the research role discovery model is employed to predict the roles of researchers in research teams that are either newly-formed or to-be-assessed.

Rather than employing traditional classification or regression approaches, we formulate the problem of research role discovery as ordinal regression (OR), so as to handle the ordinal relationship among research roles in a more principled way. Furthermore, we develop a novel ordinal regression approach named **Ordinal Factorization Machines (OrdinalFM)**. The proposed OrdinalFM is built upon the state-of-the-art factorization based predictive model – Factorization Machines (FM), with new extensions to handle outputs with natural orders.

2   Relation Work

Role discovery involves partition nodes in a network into subsets (also called positions or roles) such that the nodes in a subset share similar traits. In summary, existing work on role discovery can be roughly categorized into three groups: (i) metric-based, (ii) unsupervised, and (iii) semi-supervised. Rule-based approaches basically resort to designing some metrics to identify the roles of nodes in a network. Unsupervised approaches compute the role membership based on some pre-defined notion of graph equivalence such as stochastic equivalence [11] and structural similarity [2]. Besides, Non-negative Matrix Factorization (NMF) was found to be a viable method for computing role membership in an unsupervised context [5,12]. Recently, semi-supervised learning techniques, e.g., graph-based transductive learning [3] and Markov random fields [15], have been leveraged for role discovery.

Ordinal Regression (OR, sometimes called ordinal classification) is an important type of supervised learning problem. OR is unique in that there is an ordinal relationship among the target variables whereas the distances between target variables are not defined. Traditional machine learning algorithms have been redesigned for the OR problem. Neural networks are one of the most widely explored models for its ability in naturally handling multiple classes. The basic idea is to design a new output layer to preserve the ordinal relationship among outputs [7,10]. The adaptations of support vector machines to OR have also received much attention in the literature. The most representative one is Support Vector Ordinal Regression (SVOR) [4], in which parallel discrimination hyperplanes are learned, one for each ordinal target. A number of variants of SVOR have been proposed in a various context such as online learning [6] and multiple instance learning [14].

3   The Proposed Approach

In this section, we first describe how to generate the feature representation of research team members, and then elaborate on the proposed OrdinalFM.

3.1   Research Team Representation

Typically, a research team is composed of a number of cooperated researchers, each of them participating in different collaboration activities such as co-authoring an academic paper or co-developing a software. In our solution, we construct a heterogeneous graph representation, named heterogeneous teamwork network, for each research team.

Formally, the heterogeneous teamwork network of a research team is an undirected edge-weighted graph $G = (M \cup A \cup P, E^{(\mathrm{MA})} \cup E^{(\mathrm{AP})})$ where M, A and P denote the set of team members, collaboration activities and property tags, respectively. An edge in $E^{(\mathrm{MA})}$ takes the form of $e = \langle m, a \rangle \in M \times A$ and indicates that the member m takes part in the collaboration activity a. Likewise, an edge in $E^{(\mathrm{AP})}$ takes the form of $e = \langle a, p \rangle \in A \times P$ and indicates that the collaboration activity a is attached with the property tag p. Besides, $h^{(\mathrm{MA})} : E^{(\mathrm{MA})} \to \mathbb{R}^+$ and $h^{(\mathrm{AP})} : E^{(\mathrm{AP})} \to \mathbb{R}^+$ are functions that assign a weight to each edge in $E^{(\mathrm{MA})}$ and $E^{(\mathrm{AP})}$, respectively, according to the affinity between the two nodes.

After representing a research team as a heterogeneous teamwork network, the role of a team member can be reflected by the position of the corresponding node in the network. In order to characterize the positions of nodes in a network, we generate feature representations of team members using a variety of network metrics. Intuitively, the PI of a research team tends to reside at the center part of the network and the SIs are more likely to be the centers of sub-groups of the network. Therefore, centrality metrics can be useful indicators for research roles. In this work, we exploit the following centrality metrics [8] as features, each of which measures how important a node is in a network from different perspectives:

1. *Degree centrality*: The fraction of nodes in a network that the given node is connected to.
2. *Average neighbor degree*: The average degree of the neighborhood of the given node.
3. *Closeness centrality*: The reciprocal of the sum of the shortest path distances from the given node to others.

4. *Betweenness centrality*: The sum of the fraction of all-pairs shortest paths that pass through the given node.
5. *Communicability centrality*: The sum of closed walks of all lengths starting and ending at the node.
6. *Closeness vitality*: The change in the sum of distances between all node pairs when excluding the node.
7. *Eigenvector centrality*: A relative score of the node based on the corresponding component of the eigenvector of the network's adjacency matrix.
8. *Katz centrality*: A relative score of the node based on the total number of walks between a pair of nodes.
9. *Pagerank*: The ranking score of Google's PageRank algorithm.
10. *Local clustering*: The fraction of possible triangles through the node.

3.2 Research Role Prediction Model

Traditional Factorization Machines. Factorization Machines (FM) [13] are a class of predictive models that are capable of modeling pairwise feature interaction by learning a latent vector for each feature. Formally, FM predicts the output associated with a real-valued feature vector $\mathbf{x} = (x_1, \cdots, x_d)^\top \in \mathbb{R}^d$ by:

$$f_{\mathrm{FM}}(\mathbf{x}) = w_0 + \sum_{j=1}^{d} w_j x_j + \sum_{j=1}^{d} \sum_{j'=j+1}^{d} \langle \mathbf{v}_j, \mathbf{v}_{j'} \rangle x_j x_{j'} \qquad (1)$$

where w_0 is the global bias, w_j the weight associated with the j-th feature, and $\mathbf{v}_j \in \mathbb{R}^p$ the latent vector associated with the j-th feature. $p \in \mathbb{N}^+$ is the hyper-parameter that defines the dimensionality of factor vectors.

FM is initially proposed for binary classification and regression, and thus cannot be directly applied to ordinal regression, such as the research role discovery problem addressed in the paper.

Ordinal Factorization Machines. One key to ordinal regression approaches is to deal with the ordering relationship among learning targets. In essence, the ordering relationship imposes an unimodal constraint on the distribution of posterior predictions. Figure 1 shows three posterior probability distributions in which the left two are unimodal while the right one is multimodal. As for an ordinal regression problem, a multimodal posterior probability would be unreasonable. Taking Fig. 1(c) as an example, a team member is predicted with the highest probability being PI, but with higher probability being RS than SI. This is counterintuitive because the posterior probabilities are expected to coincide with the ordering relationship, i.e., PI $\succ$ SI $\succ$ RS.

With the above considerations, we propose OrdinalFM by introducing a postprocessing strategy on the scalar output of FM which enforces unimodal ordinal probabilistic predictions. As depicted in Fig. 2, OrdinalFM consists of two components, FM component and ordinal postprocessing component. The FM component is a traditional FM which models pairwise feature interactions through

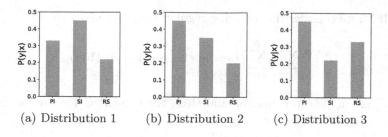

(a) Distribution 1   (b) Distribution 2   (c) Distribution 3

Fig. 1. Three posterior probability distributions

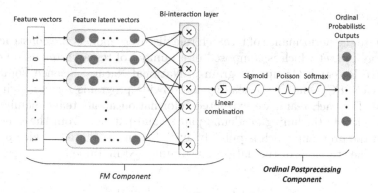

Fig. 2. The architecture of OrdinalFM

inner product between feature latent vectors, as well as a linear combination with each features.

The Ordinal postprocessing component aims to generate ordinal probabilistic predictions based on the raw output of FM. We first feed the raw output $f_{\text{FM}}(\mathbf{x})$ into a sigmoid unit: $f'(\mathbf{x}) = \frac{1}{1+e^{-f_{\text{FM}}(\mathbf{x})}}$. Then, inspired by the recent work on deep ordinal classification [1], we make use of the Poisson distribution to enforce discrete unimodal probability distributions over the ordinal targets. The probability mass function of Poisson distribution is defined as $p(k; \lambda) = \frac{\lambda^k \cdot e^{-\lambda}}{k!}$, where $k \in \mathbb{N}^*$ denotes the number of events occurring in a particular interval of time, and $\lambda \in \mathbb{N}^+$ is the parameter of the Poisson distribution that can be regarded as the mean number of events occurring in a particular interval of time.

On basis of the Poisson distribution, we view $f'(\mathbf{x})$ as the distribution parameter and calculate the prediction for the r-th class ($1 \leq r \leq R$) as follows:

$$h_r(\mathbf{x}) = \log(p(r; \lceil f'(\mathbf{x}) \cdot R \rceil)) = r \log(\lceil f'(\mathbf{x}) \cdot R \rceil) - \lceil f'(\mathbf{x}) \cdot R \rceil - \log(r!) \quad (2)$$

Finally, the unimodal probability distribution can be calculated by a softmax over $h_r(\mathbf{x})$: $p(r|\mathbf{x}) = \frac{e^{-h_r(\mathbf{x})}}{\sum_{r'=1}^{R} e^{-h_r(\mathbf{x})}} (1 \leq r \leq R)$.

Parameter Learning. It is easy to see that the proposed OrdinalFM has the same model complexity with traditional FM. To learn the model parameters in

OrdinalFM, we consider the following regularization framework:

$$\mathcal{L} = \sum_{(\mathbf{x},y)\in D} \sum_{r=1}^{R} (\llbracket y = r \rrbracket \cdot \ln p(r|\mathbf{x})) + \alpha \cdot \Omega(\Theta) \tag{3}$$

where $\llbracket \cdot \rrbracket$ is the indicator function and $\Omega(\Theta)$ the regularizer for parameters.

4   Experiments

4.1   Dataset

To evaluate the performance of research role discovery models, we constructed an evaluation dataset which is composed of a number of research teams funded by Innovative Research Groups Program of National Natural Science Foundation of China (NSFC), one of the leading nationwide programs for research teams in China. For each team, we extracted information about team members and research outputs (including academic papers and patents) from the concluding report of the program which is publicly available on the shared service portal of NSFC[1]. The statistics of the dataset is summarized in Table 1.

Table 1. Statistics of dataset

	Total number	Ave. number per team
Team	35	–
Team member	7428	212.2
Team member - PI	35	1
Team member - SI	269	7.7
Team member - RS	7124	203.5
Collaboration activity	9950	284.3
Property tags	18212	520.3

4.2   Performance Comparison

In order to verify the proposed OrdinalFM method, we compare it with several traditional ordinal regression methods, i.e., ORBoost [9], POMNN [7], SVOREX and SVORIM [4]. The hyper-parameters of each method are chosen from a certain range (shown in Table 2) using 5-fold cross-validation within the training set. For other parameters, we use default settings provided by the implementations.

First, we evaluate the prediction performance of OrdinalFM and baselines. Mean Absolute Error (MAE), marco-Precision, marco-Recall and macro-F1 are used as the evaluation measures. In this experiment, the evaluation results are

---

[1] http://npd.nsfc.gov.cn.

Table 2. The ranges of hyper-parameter selection

Method	Hyper-parameter	Range
SVOR	Trade-off parameter C	{0.001, 0.01, 0.1, 1, 10}
POMNN	Neurons in the hidden layer M	{10, 25, 50, 75, 100}
OrdinalFM	Dimensionality of factor vector p	{5, 10, 15}
	Trade-off parameter λ	{0.001, 0.01, 0.1, 1, 10}

calculated by team-level leave-one-out cross validation: we first evaluate the prediction performance of each team by taking the team as the testing data and the remaining ones in the data set as the training data, then report the average performance of all the teams. Table 3 shows the prediction performance of all the methods. It is clear that OrdinalFM outperforms all the counterparts in terms of both MAE and marco-F1, validating the importance of introducing feature interactions to learn an effective ordinal regression model, especially in the case of research role identification where the network metric based features are correlated. Among all the baselines, SVOR methods consistently perform better than others. We also notice that POMNN does not perform as well as expected. It may be that traditional neural network can be easily stuck in a local minimum. Thus advanced training techniques need to be employed to learn a better POMNN model.

Table 3. Comparison with ordinal regression methods

	MAE	Macro-Pre	Macro-Rec	Macro-F1
ORBoost-LR	0.0664	0.6676	0.6938	0.6772
ORBoost-All	0.0690	0.6642	0.6881	0.6732
SVORIM-Gau	0.0539	0.6941	0.6824	0.6865
SVORIM-Lin	0.0518	**0.7253**	0.6720	0.6910
SVOREX-Gau	0.0608	0.6545	0.6855	0.6661
SVOREX-Lin	0.0653	0.6706	**0.6974**	0.6805
POMNN	0.0686	0.6009	0.6271	0.6105
OrdinalFM	**0.0502**	0.7015	0.6878	**0.6926**

Then, we compare the training time of different methods by varying the size of training data. We randomly select 1, 2, ⋯, and 35 teams from the evaluation dataset and learn research role discovery models using SVOREX-Gau, SVOREX-Lin, SVORIM-Gau, SVORIM-Lin and OridinalFM, respectively. Each model is learned with the best parameter setting. Figure 3 shows the plots of training time of all the methods. As expected, OrdinalFM has an approximately linear training time cost w.r.t. the size of the training data and is much more efficient than SVOR. In many situations, especially when the size of training data is large, the differences in training efficiency are significant due to the near-quadratic scalability of SVOR. Among all the SVOR methods, linear kernel needs less training time than Gaussian kernel, and SVOREX is slightly faster than SVORIM.

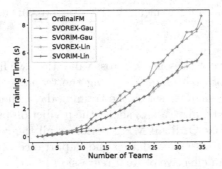

Fig. 3. Training time with varying dataset size

5   Conclusion

In this study, we consider predictive role analysis of research teams and place more emphasis on the ordinality within research roles. We formulate the problem as ordinal regression and propose a novel model – Ordinal Factorization Machines (OrdinalFM). The proposed OrdinalFM extends FM by enforcing discrete unimodal probability distribution on the real outputs of FM. Extensive empirical evaluations, conducted over a dataset composed of a number of real-world research teams, demonstrate the effectiveness of the proposed OrdinalFM as it achieves better prediction performance compared with several baseline methods, as well as keeps high computational efficiency.

Acknowledgement. This work is partially supported by Natural Science Foundation of China (61602278, 71704096 and 31671588) and SDUST Higher Education Research Project (2015ZD03).

References

1. Beckham, C., Pal, C.: Unimodal probability distributions for deep ordinal classification, pp. 411–419 (2017)
2. Brandes, U., Lerner, J.: Structural similarity: spectral methods for relaxed block-modeling. J. Classif. **27**(3), 279–306 (2010)
3. Cheng, Y., Agrawal, A., et al.: Social role identification via dual uncertainty minimization regularization. In: Proceedings of the IEEE ICDM, pp. 767–772 (2014)
4. Chu, W., Keerthi, S.S.: New approaches to support vector ordinal regression. In: Proceedings of the 22nd ICML, pp. 145–152 (2005)

5. Gilpin, S., Eliassi-Rad, T., Davidson, I.: Guided learning for role discovery (GLRD). In: Proceedings of the 19th ACM SIGKDD, pp. 113–121 (2013)
6. Gu, B., Sheng, V.S., Tay, K.Y., Romano, W., Li, S.: Incremental support vector learning for ordinal regression. IEEE TNNLS **26**(7), 1403–1416 (2015)
7. Gutiérrez, P.A., Tiňo, P., Hervás-Martínez, C.: Ordinal regression neural networks based on concentric hyperspheres. Neural Netw. **59**, 51–60 (2014)
8. Koschützki, D., Lehmann, K.A., Peeters, L., Richter, S., Tenfelde-Podehl, D., Zlo-towski, O.: Centrality indices. In: Brandes, U., Erlebach, T. (eds.) Network Analysis. LNCS, vol. 3418, pp. 16–61. Springer, Heidelberg (2005). https://doi.org/10.1007/978-3-540-31955-9_3
9. Lin, H.-T., Li, L.: Large-margin thresholded ensembles for ordinal regression: theory and practice. In: Balcázar, J.L., Long, P.M., Stephan, F. (eds.) ALT 2006. LNCS, vol. 4264, pp. 319–333. Springer, Heidelberg (2006). https://doi.org/10.1007/11894841_26
10. Liu, X., Zou, Y., Song, Y., Yang, C., You, J., Kumar, B.V.K.V.: Ordinal regression with neuron stick-breaking for medical diagnosis. In: Leal-Taixé, L., Roth, S. (eds.) ECCV 2018. LNCS, vol. 11134, pp. 335–344. Springer, Cham (2019). https://doi.org/10.1007/978-3-030-11024-6_23
11. Nowicki, K., Snijders, T.A.B.: Estimation and prediction for stochastic blockstructures. J. Am. Stat. Assoc. **96**(455), 1077–1087 (2001)
12. Pei, Y., Zhang, J., Fletcher, G.H.: DyNMF: role analytics in dynamic social networks. In: Proceedings of the 27th IJCAI, pp. 3818–3824 (2018)
13. Rendle, S.: Factorization machines with libFM. ACM Trans. Intell. Syst. Technol. (TIST) **3**(3), 57 (2012)
14. Xiao, Y., Liu, B., Hao, Z.: Multiple-instance ordinal regression. IEEE Trans. Neural Netw. Learn. Syst. (2018)
15. Zhao, Y., Wang, G., Yu, P.S., Liu, S., Zhang, S.: Inferring social roles and statuses in social networks. In: Proceedings of the 19th ACM SIGKDD, pp. 695–703 (2013)

Deep Learning for Online Display Advertising User Clicks and Interests Prediction

Zhabiz Gharibshah[1]($\boxtimes$), Xingquan Zhu[1], Arthur Hainline[2],
and Michael Conway[2]

[1] Department of Computer and Electrical Engineering and Computer Science,
Florida Atlantic University, Boca Raton, FL 33431, USA
{zgharibshah2017,xzhu3}@fau.edu
[2] Bidtellect Inc., Delray Beach, FL 33483, USA
{arthur,mike}@bidtellect.com

Abstract. In this paper, we propose a deep learning based framework for user interest modeling and click prediction. Our goal is to accurately predict (1) the probability that a user clicks on an ad, and (2) the probability that a user clicks a specify type of campaign ad. To achieve the goal, we collect page information displayed to users as a temporal sequence, and use long-term-short-term memory (LSTM) network to learn latent features representing user interests. Experiments and comparisons on real-world data shows that, compared to existing static set based approaches, considering sequences and temporal variance of user requests results in an improvement in performance ad click prediction and campaign specific ad click prediction.

1 Introduction

Computational advertising is mainly concerned about using computational approaches to deliver/display advertisement (Ad) to audiences (*i.e.* users) who might be interested in the Ad, at the right time [1]. The direct goal is to draw users' attention, and once the Ads are served/displayed on the users' device, they might take actions on the Ads and become potential buying customers. Due to the sheer volumes of online users and the advertisements, and users have different background and interests, not to mention their changing habits and interests, finding users interests is often the key to determine whether a user is interested in an Ad.

In display advertising, because AdExchange often only passes very limited information about the user [2], such as user device type, user agent, page domain name and URL, etc., in order to predict user interests, the industry commonly relies on generative modeling. Historical data is used to build tree-structured

---

This research is sponsored by Bidtellect Inc. and by US National Science Foundation through Grant No. CNS-1828181.

J. Shao et al. (Eds.): APWeb-WAIM 2019, LNCS 11641, pp. 196–204, 2019.
https://doi.org/10.1007/978-3-030-26072-9_14

models whose parameters are used to derive the CTR value of the new impression. Common generative models include CTR hierarchy trees [3] or hierarchical Bayesian frameworks [4]. One inherent advantage of the generative model is that the model provides transparent interpretability for business to understand which factor(s) contribute the most to the CTR values. However, due to the limitations of the models, such methods can normally estimate only a handful of parameters (*e.g.* using a number of selected factors to split the tree hierarchy), and are unable to consider many rich information from users, publishers, and websites for accurate CTR estimation.

Different from generative models, the increasing popularity of machine learning, particularly deep learning, has driven a set of predictive modeling methods, which treat user clicks as binary events, and uses supervised learning to train a classifier to predict the likelihood of an impression being clicked by users [5], including some deep neural networks based CTR estimation methods [6]. Such methods normally work on tens of thousands of features, and are often more powerful than generative models.

In this paper, we propose to consider temporal user information to estimate user clicks and user interests. We generalize these two problems as a binary classification task (for user click prediction) and a multi-class classification task (for user interest prediction). More specifically, we collect users' page visits as a temporal sequence, and train deep LSTM (long-term short-term memory) networks to make predictions.

2 LSTM Network for User Clicks and Interests Modeling

2.1 Problem Definition

let U be the a set of users $\{u_1, u_2, u_3, ..., u_n\}$ and R be a set of events. Each event denoted by $r_{u_j}^{t_i}$ represents the occurrence that an advertisement is displayed to a user u_j in a specific context at time t_i. In this case, the event is encoded as a real-valued vector ($r_{u_i}^{t_i} \in \mathbb{R}^d$). The context in display advertising industry is the page visited by the user which is in turn described by a hierarchy of page category IDs corresponding to various contextual information with different level of granularity [7,8].

The set of all pre-defined page categories is denoted by $\mathbb{C}$ equals to $\{c_1, c_2, ..., c_{|\mathbb{C}|}\}$ where $|\mathbb{C}|$ is the number of categories. For a page visited by user u_j at time-step t_i, its page categories can be shown in the form of array like $[c_1, c_2, c_3, ...]$. For each user $u_j \in U$, we take the history of web pages visited by the user and denote it by $r_{u_j} = \{r_{u_j}^{t_1}, r_{u_j}^{t_2}, ..., r_{u_j}^{t_m}\}$. Because of the variety in the number of websites visited by users, we have $r_{u_j} \in \mathbb{R}^{m \times d}$ where m is the maximum sequence length. Thus, given the historical records of all users as $R = \{r_{u_1}, r_{u_2}, \cdots, r_{u_n}\}$ where $R \in \mathbb{R}^{n \times m \times d}$, $d < |\mathbb{C}|$, our **objective** is using historical user activities as the chronological sequence of requests before an arbitrary time-step t_i to predict (1) the probability that a user may interact with an Ad at t_i by generating a click response, and (2) predict which campaign Ad the user might click.

2.2 LSTM for User Modeling

Recurrent Neural Networks (RNN) is an extension of feed forward networks and has been successfully applied in various sequence data analysis and temporal sequence modeling [9]. While traditional RNN networks are unable to learn all term dependencies in sequences because of vanishing or exploding gradient problem [10], long short-term memory (LSTM) networks were introduced which uses special multiple-gate structured cell to replace hidden layer nodes. Using LSTM cells in these networks has been shown as an efficient way to overcome these problems [10].

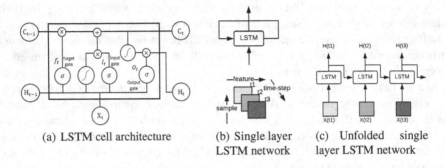

(a) LSTM cell architecture   (b) Single layer LSTM network   (c) Unfolded single layer LSTM network

Fig. 1. LSTM cell and network architecture. (a) shows the detailed view of an LSTM cell, (b) shows a single layer LSTM network with respect to the input, represented in a 3rd-order tensor [sample, feature, time-step], and (c) shows the unfolded structure of the single layer LSTM network in (b). In our research, we stack three LSTM layers to form a deep LSTM network (detailed in the experimental settings).

Two significant challenges in online display advertising to model user response and user interest using deep learning approaches like LSTM networks are that the collection of online user behavior data are (1) in multi-variant categorical form because each page may belong to one or multiple categories, and (2) user sequences of historical data may have different lengths because users' responses and actions vary over time. They result in multi-length sequences, where data points of each time-step may also include variant features. More specifically, in our model, the historical data collected for user modeling contains page category IDs of the pages that a user visited during a short period of time. For a user at a particular time-step, we have an array of category IDs of the page visited by the user discussed in Problem Definition section. Such IDs are represented as $[c_1, c_2, \cdots]$ which are in different lengths. Table 1 shows the sample of input sequential data used for user modeling.

One-Hot-Encoding with Thresholding. To handle multi-length page categories as the features to describe each visited page, we use one-hot-encoding to represent them as sparse binary features. For each user, we have a sequence of visited pages attributed by a couple of page category IDs that correspond to their content. Therefore, each time-step can be shown as binary vector with length equals to the maximum number of categorical variables where 1 indicating the presence of each possible value from the original data. For example, in Table 1, at time-step t_1 the visited page of user u_2 is described by an array of page category IDs as features can be shown as $[0, 1, 1, 0, 0, 0, ..., 0]$. The dimension of vectors for time-step is determined by the number of unique page category IDs in the dataset that in our example it equals to $|\mathbb{C}| = 18$.

Concatenating these vectors generates a matrix with high dimensionality. Therefore, for features like page category IDs with high cardinality, using one-hot-encoding usually leads to extra computational costs. In the past, much research has been done to work with such sparse binary features [11,12]. To address this problem and in order to reduce the dimension of these vectors, we used an alternative to encode more frequent page category IDs based on a threshold based approach. In this case, page category IDs are sorted based on the number of their occurrences. Those with repetitions more than the user-defined threshold will be kept for the next parts.

Table 1. Schema of the data representation. We represent each audience(user) and his/her actions as multi-dimensional temporal sequence. Each row in the table denotes an audience, and $t_1, t_2, \cdots, t_n$ denotes temporal order of the sequence (if $i>j$, then t_i happens after t_j). $c_1, c_2, \cdots, c_m$ denotes the IAB tier-2 page category of the webpage visited by the users. $\langle$**click**$\rangle$ denotes an Ad click event from the audience. Not all sequences results in click events.

			Temporal order of audience response			
User	t_1	t_2	t_3	t_4	...	t_n
u_1	$r_{u_1}^{t_1} = [c_1, c_2, c_3]$	$r_{u_1}^{t_2} = [c_1, c_3]$	$r_{u_1}^{t_3} = [c_1, c_4, c_5, c_6]\langle$click$\rangle$	–	–	–
u_2	$r_{u_2}^{t_1} = [c_2, c_3]$	$r_{u_2}^{t_2} = [c_4]\langle$click$\rangle$	–	–	–	–
u_3	$r_{u_3}^{t_1} = [c_{10}, c_7, c_3, c_{20}]$	$r_{u_3}^{t_2} = [c_1, c_3, c_{15}]$	$r_{u_3}^{t_3} = [c_6, c_{12}, c_{22}, c_{24}, c_1, c_3]$	–	–	–
u_4	$r_{u_4}^{t_1} = [c_8, c_{14}, c_{30}]$	$r_{u_4}^{t_2} = [c_2, c_6]$	$r_{u_4}^{t_3} = [c_{11}, c_{16}, c_{21}]$	$r_{u_4}^{t_4} = [c_4, c_7, c_{11}]\langle$click$\rangle$	–	–
...						

Bucketing and Padding. The variable length of sequences, like samples in Table 1, is another technical challenge. To handle sequences of any length and capture short and long dependencies in input data, padding with constant value (*e.g.* inserting zeros) is a straightforward strategy to make input dimensions fixed. However, applying this approach to train LSTM with wide range of sequence lengths is not only computationally expensive it also adds extra zero values resulting in bias in outcomes and changes input data distribution. Therefore, we propose to combine padding and bucketing to best utilize temporal information in sequences without inserting too many padding symbols.

To combine bucketing and padding, we construct several buckets in training samples, where sequences in each bucket have the same lengths corresponding to the range of sequence length in the dataset. Each sample is assigned to one bucket corresponding to its length. In this case, padding of samples mitigates to inside of buckets being used just for assigned sequences as much as necessary to fit into the bucket. The most important item in the sequence of request page categories is the last item that corresponds to the possible user click response, we use pre-padding approach. It means that each short sample inside buckets with the length lower than bucket size is pre-padded to become a sample with length equal to the maximum length in that bucket.

Following the idea, we designed an ensemble learning method for multi-class classification task. Rather than using the original splitting to generate buckets as the representative subset of samples in input space, we just use the sequence length of buckets to indicate one representation of samples through truncating time-steps. It means that for each representation, all samples in input data are trimmed to the selected sequence length by removing some time-steps from the beginning of sequences. Then in order to obtain classification, we build one LSTM model for each representation. The final result of classification is generated by applying majority voting as the result merger of all models.

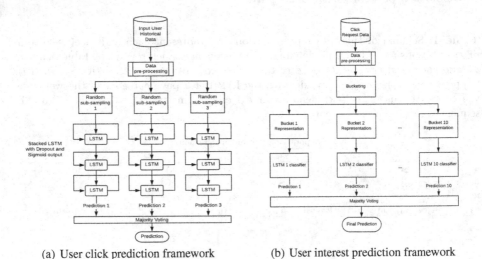

(a) User click prediction framework   (b) User interest prediction framework

Fig. 2. The proposed User click prediction (a) and user interest prediction (b) frameworks. Given user request and response historical data, user click prediction aims to train stacked LSTM classifiers to predict whether a new user is going to click an Ad or not, *i.e.* a binary classification task; and user interest prediction will predict Ad campaign a new user is going to click, *i.e.* a multi-class classification task.

2.3   LSTM Based User Click Prediction Framework

Figure 2(a) briefly describes the structure of our proposed method for user click prediction problem as a binary classification. It includes the stacked LSTM model consisting of three LSTM layers followed by one fully connected layer with sigmoid activation to combine the output of hidden neurons in previous layers to predict click instances. In this case, the loss function is defined as the weighted binary cross entropy which aims to maximize the probability of correct prediction. The weight introduced in this function allows a trade-off between recall and precision in both classes to mitigate the negative effect of the class imbalance problem in our task:

$$L = 1/N \times \sum_{j=1}^{N}(y_i \times -log(p(x_i)) \times w + (1 - y_i) \times -log(1 - p(x_i))) \tag{1}$$

where N is the number of samples in training set. $y_i \in [0,1]$ is target label and $p(x_i) \in [0,1]$ is the network output, which represents the likelihood that how likely the sample x_i has a click response at the end. w is the coefficient which determine the cost of positive error relative to the misclassification error of negative ones.

2.4 LSTM Based User Interest Prediction Framework

Figure 2(b) outlines the model for user interest prediction. It is defined as multi-class classification to classify the number of clicks in 10 different advertising campaigns. The number of buckets are defined uniformly over the range of sequence length in the dataset. Then, for each bucket, one representation of data is generated by trimming all longer samples and pre-padding shorter samples to the selected sequence length. Then prediction is made by following the ensemble learning approach. In this Figure, LSTM block follows the structure mentioned in Fig. 2(a) except the last layer having softmax activation function. In this case, the objective function is similar to Eq. (1) when $w = 1$. It is actually an unweighted categorical cross entropy loss function in which $p(x_i)$ is the output of the network after softmax layer.

3 Experiments

3.1 Benchmark Data

We pulled out data from our industry partner's bidding engine, and prepared two datasets to validate user click and user interest prediction.

Post-View Click Dataset: This dataset is mainly used for validating binary user click prediction. We pulled 5.6 million users' request records from 1 day log events. These anonymous records include a chronological sequence of various request categories which represent user browsing interactions. In this case, there are two types of positive and negative responses from users where success occurs, if a post-view click takes place at end of a chain of visited impressions. Because of rarity of positive responses (click) in digital advertising, this dataset suffered from severe class imbalance problem. Thus, to deal with this issue, we use random down-sampling to get 10:90 positive:negative post sampling class distribution on the dataset.

Multi-Campaign Click Dataset: This dataset includes historical records with positive response in post-view click dataset. The positive response in this case is defined as user clicks on different type of campaigns (we used 10 types of campaigns). This dataset is mainly used for validating multi-class user interest prediction.

One issue we encountered in these datasets is that the sequence lengths are severely skewed where a large proportion of sequences are very short in length even less than 3 time-steps. Our bucketing and padding combined approach, introduced in Sect. 3.2, is specially designed to handle this challenge.

3.2   Experimental Settings and Performance Metrics

We implemented 7 methods for comparisons. All models were implemented through Tensorflow and CUDA to take advantage of using GPU and trained by Adam optimization as a variant of gradient descent. The remaining models are built using Scikit-learn library in Python. For data preprocessing, we convert input sequential data to binary vector by one-hot-encoding and get rid of less frequent categorical campaign IDs. we use a threshold to keep those categories with more 1,000 occurrence in our dataset. To control overfitting problem in neural networks early stopping mechanism is used to stop after 10 subsequent epochs if there is no progress on the validation set. Dropout rate was set at 0.4 for neural networks. For the rest of methods L2 regularization is used in training process. All experiments are evaluated based on 5-fold cross validation.

We use the Area Under Receiver Operating Characteristics Curve (AUC) as the major evaluation metric because it shows the model accuracy of ranking positive cases versus negative ones. We also employ accuracy, F_1-measure, precision, and recall as additional performance metrics.

3.3   Performance Comparison

User Click Prediction Results. As a binary classification task, the performance of proposed method is compared with SVM, Random Forest, Logistic Regression in addition a variant of convolutional neural network (CNN) [13]. Since input data has an extremely imbalanced class distribution with around 5,646,569 no-click user sequences (negative samples) versus 31,144 click user sequences (positive samples), we use random under-sampling and ensemble learning to build the model in Fig. 2(b).

In our proposed method, three dimensional input data($\mathbb{R}^{n \times m \times d}$) are passed into the network where m and d are 70 and 153 corresponding to more frequent sequence lengths and the most frequent number of page category IDs. For remaining methods, the initial 3d input data is projected to the plane of $\mathbb{R}^{n \times d}$ by adding up values in sequence length dimension. Then, each model is trained by minimizing weighted binary cross entropy shown in Eq. (1). By default, we use cost ratio as 5 for positive samples because of the effectiveness seen in our experiments. Table 2 reports the result of prediction using different methods. As our proposed method pays more attention to history of requested pages before

Table 2. User click prediction results (binary classification task)

Method	Precision	Recall	F_1-measure	AUC	Accuracy
NaiveBayes	0.2638	0.2827	0.2729	0.6004	0.8583
Random forest	0.3001	0.2758	0.2875	0.6045	0.8713
Logistic regression	0.3353	0.2995	0.3164	0.6189	0.8782
Linear SVM	0.3534	0.2682	0.3050	0.6086	0.8850
SVM	**0.3815**	0.0333	0.0613	0.5138	0.9039
CNN	0.3492	0.4323	0.3862	0.6742	0.8707
LSTM	0.3111	**0.5196**	**0.3892**	**0.7001**	**0.8466**

click, having higher performance in our proposed method shows the importance of this feature in click prediction.

User Interest Prediction Results. For multi-class user interests prediction task, we compare proposed approach with NaiveBayes, Random Forest (with 100 tree estimators), Logistic Regression and two version of SVM with linear and RBF kernels. The input data are click samples used in previous task for click response prediction. Considering the AUC performance, the overall results in Table 3 illustrate that the proposed method in this classification task outperforms the others. It shows the effective of LSTM networks in detecting the correlation of sequential data and click response.

Table 3. User interest prediction results (multi-class classification task)

Method	Precision	Recall	F_1-measure	AUC	Accuracy
NaiveBayes	0.2486	0.2940	0.2713	0.6708	**0.4016**
Random forest	0.3640	0.3401	0.3697	0.7404	0.3697
Logistic regression	0.3911	0.2425	0.3240	0.7124	0.2233
Linear SVM	0.3858	0.2680	0.3369	0.7113	0.2415
SVM	0.1037	0.0654	0.0691	0.4593	0.0691
LSTM	0.3942	0.2628	0.3312	0.6994	0.2239
LSTM_Bucketing_Padding	**0.4076**	**0.3401**	**0.3835**	**0.7599**	0.3835

4   Conclusion

In this paper, we proposed a new framework for click response and user interest prediction using LSTM based deep learning. By combing padding and bucketing to learn binary user click prediction and multi-class user interest prediction, our method allows sequences to have variable lengths and variable number of dimensions and can maximally leverage temporal information in user sequences for learning. Experiments and comparisons on real-world data collected from industry partner show that our method is able to encode useful latent temporal information to predict user response and interests.

References

1. Zhu, X., Tao, H., Wu, Z., Cao, J., Kalish, K., Kayne, J.: Fraud Prevention in Online Digital Advertising. SCS. Springer, Cham (2017). https://doi.org/10.1007/978-3-319-56793-8
2. IAB, Openrtb api specification version 2.5 (2016). https://www.iab.com/guidelines/real-time-bidding-rtb-project/
3. Liu, H., Zhu, X., Kalish, K., Kayne, J.: Ultr-CTR: Fast page grouping using URL truncation for real-time click through rate estimation. In: Proceedings of IEEE IRI Conference, pp. 444–451 (2017)
4. Ormandi, R., Lu, Q., Yang, H.: Scalable multidimensional hierarchical Bayesian modeling on spark. In: Proceedings of the 4th International Conference on Big Data, Streams and Heterogeneous Source Mining: Algorithms, Sys., Programming Models and Applications, vol. 41, pp. 33–48 (2015)
5. Li, C., Wang, D., Lu, Y., Sandeep, P.: Click-through prediction for advertising in twitter timeline. In: Proceedings of ACM SIG KDD (2015)
6. Guo, H., Tang, R., Ye, Y., Li, Z., He, X.: DeepFM: a factorization-machine based neural network for CTR prediction. In: Proceedings of the International Joint Conference on Artificial Intelligence, pp. 1725–1731 (2017)
7. Chapelle, O., Manavoglu, E., Rosales, R.: Simple and scalable response prediction for display advertising. ACM Trans. Intell. Syst. Tech. **5**(4), 61:1–61:34 (2015)
8. Agarwal, D., Agrawal, R., Khanna, R., Kota, N.: Estimating rates of rare events with multiple hierarchies through scalable log-linear models. In: Proceedings of the KDD Conference (2010)
9. Lipton, Z.C., Berkowitz, J., Elkan, C.: A critical review of recurrent neural networks for sequence learning, CoRR, vol. arXiv:abs/1506.00019 (2015)
10. Zhang, L., Wang, S., Liu, B.: Deep learning for sentiment analysis : a survey, wiley interdiscip. Rev. Data Min. Knowl. Discov. **8**, e1253 (2018)
11. Rendle, S.: Factorization machines, ICDM. In: IEEE, pp. 995–1000 (2010)
12. Qu, Y., et al.: Product-based neural networks for user response prediction. In: IEEE International Conference on Data Mining, pp. 1149–1154 (2016)
13. Yoon, K.: Convolutional neural networks for sentence classification. arXiv:1408.5882 (2014)

Graph Data and Social Networks

In Pursuit of Social Capital: Upgrading Social Circle Through Edge Rewiring

Qian Chen[1], Hongyi Su[1], Jiamou Liu[2(✉)] (iD), Bo Yan[1], Hong Zheng[1],
and He Zhao[1]

[1] Beijing Lab of Intelligent Information Technology,
School of Computer Science and Technology, Beijing Institute of Technology,
Beijing, China
henrysu@bit.edu.cn
[2] School of Computer Science, The University of Auckland, Auckland, New Zealand
jiamou.liu@auckland.ac.nz

Abstract. The paper investigates the dynamics of the social circle of an individual in a social network. Updating social circle affects two kinds social assets: Bonding social capital which affects trusts and social support, and bridging social capital that determines information access. We address a rewiring process that enables the individual to upgrade her social circle. The questions are (1) what strategies would guide the individual to iteratively rewire social ties to gain bridging while maintain bonding social capital, and (2) what structural properties will arise as a result of applying the strategies. For the first problem, we put forward three greedy rewiring strategies based on scoping the network access for the individual. We conduct experiments over four random graph models and five real-world datasets to evaluate these strategies. The results reveal a striking difference between bonding and bridging social capitals, while the community-based strategy is able to achieve a balance between bridging with bonding social capital. For the second problem, we correlate social capital with structural features such as centrality and embeddedness. In this respect, the paper advances understanding to social capital and its connections to network structures.

Keywords: Social network · Social capital · Bonding and bridging · Interpersonal ties · Community

1 Introduction

The quest for a general yet basic model of social circle dynamics has wide implications in social media, public health, organizational management, and many other fields [20,24]. As an individual forges links with others, the social circle of the individual shifts [17]. Such shifts have the potential to significantly change social attributes such as influence, status, and solidarity with social surrounding. These features are important traits of *social capital*, which denotes the investment of an individual to gain the embedded resources within interpersonal

J. Shao et al. (Eds.): APWeb-WAIM 2019, LNCS 11641, pp. 207–222, 2019.
https://doi.org/10.1007/978-3-030-26072-9_15

interactions [16]. By purposefully updating social links, an individual may gain social capital. The question arises as to understand how the updates of social circle affect social capital. This question embodies several challenges.

The first challenge is around a formal definition of social capital. The notion has been the subject of much debate in the last decades due to its multiplex nature [25]. Two seemingly opposing views have been developed [1,5]. The first sees social capital as a byproduct of *network closure*. In this view, social capital lies within solidarity of an individual's ego network and is the cause of many factors such as social norm and trust. This is referred to as *bonding social capital*. The second sees social capital as generated from information brokerage, or *open networks*. In this view, social capital arises from connections to a diverge population thereby improving the accessibility to new information. This notion is referred to as *bridging social capital*. A main goal of this work is to reconcile these two forms of social capital through an egocentric lens.

The second challenge lies with developing a general framework to simulate and intervene social circle dynamics. In general, establishing new social relations upgrades one's social circle. On the other hand, as social relations require efforts and revision to preserve, it is not feasible for an individual to maintain an unbounded number of social relations. Indeed, Dunbar's well-celebrated theory asserts an upper bound of the number of social relations [4]. it is therefore important to take into consideration both linking and severance in our framework.

This paper tackles the two challenges posed above by investigating two research questions. The first question asks how an individual strategically upgrades her social circle and gains social capital. The second question asks how such strategies would affect the structural properties of the network. In particular, this paper makes the following contributions: (1) We propose a new formalism of social capital. The formalism captures both the bonding and bridging roles through network-theoretical notions and is based on well-grounded social capital theory. (2) We define the social circle upgrade problem for a given node v. The goal of v is to maximize its bridging social capital while maintaining bonding social capital through rewiring edges. Due to high computational costs in evaluating the social capital functions, we resort to efficient heuristics that utilize other structural properties of a node in a scope-based greedy framework. Under this framework, we put forward three rewiring strategies. Through experiments over four network models and five real-world networks, we investigate the effects of the strategies to a target node's social capital. Particular attention is placed over networks with a visible community structure. (3) We study social capital in the context of other network metrics. In particular, our experiments verify that the notion of closeness centrality is closely associated with bridging social capital, while embeddedness reflects bonding social capital.

Related Work. Most existing models of link formation either rely on sociological insights such as triadic closures and mutual neighborhoods [11] or game-theoretic notions of stability [12,15]. These models capture general patterns of link formation but fail to reflect strategic decision for an individual. *Network building*, studied in [3,19,26,27], refers to purposely creating edges to bring an agent to a more central position in a network. This work differs by seeing social capital as functional constructs as opposed to structural features, and considering tie rewiring instead of only tie creation. Community structure is a prevalent structural property and amounts to a major theme of social network studies. Many efforts aim to measure the prominence of community structures in a social network [18]. Radicchi et al. in [21] mentions the strength of community structures. Ruan and Zhang compares weak communities with strong communities in [22]. Robustness is also used as a mean to evaluate community structures [13]. It is important for users of social networks to identify appropriate community divisions and community scale choices.

2 Towards a Formalization of Social Capital

Social capital is defined as a form of investment in one's social relations for the benefits embedded within the social network [16]. The notion emerged in 1990s with roots from pioneering works of Granovetter [10] and Coleman [9], and has become one of the most discussed terms in social sciences. The main benefits of social relations include facilitation of information flow, influence and solidarity [1]. Therefore social capital is distinctive from other forms of capital in that:

1. The benefit of social relation comes largely from *network effect*, i.e., it arises due to the structure of the network. This intuition pins social capital firmly in the realm of network theory. As a consequence, the measurement of social capital is done based on the network structural properties.
2. Structural properties of the network, on their own, do not define social capital. Rather, it is the *function of social relations* that determines social capital. Such functions include, e.g., information passing between nodes through social ties, or the delivery of social support through social ties.
3. The benefit of social relation is reflected within ones' local social circles and personal utility. This means that the notion of social capital is necessarily *egocentric*, i.e., the social capital of a node is defined from the perspective of the node itself.

The strong-weak division of interpersonal ties form a key theoretical underpinning of social network theory [10]. Strong ties link those individuals that share large parts of their social circles, while weak ties bridge individuals that sit in disparate parts of the network. Social capital theory distinguishes the benefit brought by these two types of links [1,5]: *Bonding social capital* is associated with benefits such as social norm, trust, and self-identity reinforcement which are largely coming from strong ties, while *bridging social capital* represents the benefits of accessing to new and diverse ideas, power, and ability to influence information flow, which are brought from weak ties.

Despite the consensual intuition above, there has so far been no generally-accepted formalism that defines social capital. A desirable formalism would incorporate tools within the network theory literature, while naturally unifying both bonding and bridging social capital. Existing use of social network in the data analysis and computational social science community either adopts ad-hoc structural metrics that fail to capture the many-facets of social capital [8], or require exogenous and superficial factors such as resources [23]. We thus propose a simple model of social capital to reflect the following considerations: (1) Our formalism presents bonding and bridging social capital as inherently egocentric concepts, i.e., they stem from the local view of an individual. (2) Moreover, we intentionally avoid measuring social capital directly using structural features (e.g. size of network, degree, distance, common neighbors, etc.). The structural features are exogenous variables that influence social capital but not endogenous variables which give social capital its meaning.

Following standard convention, we view a *social network* as an undirected graph $G = (V, E)$, where V is a set of nodes denoting social actors or *agents* in the social network and E is a set of edges on V of the form uv where $u \neq v \in V$ which represent interpersonal ties. If $uv \in E$, we say that u, v are *neighbors* of each other. The *neighborhood* of $v \in V$ is $N(v) = \{u \in V \mid uv \in E\}$. The *distance* between u and v, denoted by $\text{dist}_G(u, v)$, is the length of a shortest path between u and v. We focus on connected graphs and thus $\text{dist}_G(u, v) < \infty$ for any pair $u, v \in V$. From an individualistic perspective, the social circle of a node $v \in V$ contains all ties that v maintains and perceives. This can be captured using the 2-level ego network of v as defined below:

Definition 1. *The 2-level ego network of a node $v \in V$ is the subgraph G_v of G induced by v, v's neighbors, and v's neighbors' neighbors in G, i.e., the nodes of G_v is $V_v = \{u \in V \mid \text{dist}_G(u, v) \leq 2\}$. The node v is called the ego in G_v.*

Bonding Social Capital. Bonding social capital refers to the collective asset enjoyed by members of a group as generated from resources due to strong ties. In this sense, bonding social capital refers to solidarity and cohesion within an agent v's social surrounding G_v. Imagine that each member of G_v holds a certain amount of 'goodwill' in the form of e.g., trust or support. We consider social relations as channels of transfer of such goodwill. In an egocentric perspective, assume the ego v invests a part, say $\alpha \in [0, 1]$ of her total goodwill evenly among her neighbors. Any agent upon receiving this goodwill may either return the goodwill back to v, or evenly distribute this goodwill onto their own neighbors. The bonding social capital is therefore the expected amount of goodwill that is returned to v from its neighbors. More formally, we assign an *expected return score* $\iota_v(u)$ to every node u in G_v and let $\overrightarrow{\iota_v}$ be the vector of expected returns for all these nodes. Let $r_u = 1$ if $u = v$ and $r_u = 0$ otherwise. We have

$$\iota_v(u) := (1 - \alpha)r_u + \alpha(\overrightarrow{\iota_v} \cdot \overrightarrow{E_v(u)} / \deg_v(u)) \tag{1}$$

where $\overrightarrow{E_v(u)}$ is the column vector associated with u in the adjacency matrix E_v of G_v, and $\deg_v(u)$ is the degree of u in G_v. The notion of expected return score can be understood in two ways: The first way views the formulation of $\iota_v(u)$ as capturing an investment: As v shares goodwill with the neighbors and such goodwill circulates within the ego network, $\iota_v(u)$ refers to the amount of goodwill that v's social circle returns to v. This accumulated goodwill is naturally higher with a more cohesive social circle. The second way is to realize that (1) captures the long term probability distribution of a random walk (with return) in the ego network that starts from v. Thus, the notion is essentially *personalized Pagerank*, a standard link prediction score. This means that, for nodes u, v that are not initially connected by an edge, $\iota_v(u)$ has been used as a prediction on the likelihood of tie emergence between two agents u, v in a social network [15]. In our context, since uv is already an edge, it is reasonable to view $\iota_v(u)$ as a prediction on how likely the tie between v and u can be sustained, i.e., tie strength. The bonding social capital $\kappa(v)$ therefore captures *expressive outcome* as defined by Lin in [16] which refers to the preservation of possessed resources.

Definition 2. *Given $G = (V, E)$, the* bonding social capital $\kappa(v)$ *of a node v is $\sum_{u \in V_v} \iota_v(u)$ where ι_v is the expected return score as computed in (1).*

Bridging Social Capital. Bridging social capital represents the position of an individual within the social network as a proprietary advantage. The notion is closely related to structure hole [2] which bridges across disparate groups in the network. Such ties are necessarily weak [7] but they foster information flow. This view is made popular by Granovetter's seminal work and is greatly extended through ideas such as information broker and gate keepers: By building weak ties, an agent gains an advantageous position and improves its *access to diverse information, new ideas, talents, and opportunities*. Hence to measure bridging social capital, one would need to capture how well the position of a node to receive information that is circulating in the network.

Information flow has been studied intensively in social network analysis through diffusion models. In particular, the independent cascade (IC) model has been prevalent in the context of information dissemination and the word-of-mouth effect. Given a network $G = (V, E)$, an IC process starts from a seed set $S \subseteq V$ which are the activated nodes at the beginning. Imagine an information dissemination process, where a message is delivered from activated nodes to others along edges. This means that once a node becomes activated, it has the potential to activate its un-activated neighbors with a certain probability $\beta \in (0, 1]$. Namely, assume w becomes activated at stage i, $u \in N(w)$ is unactivated. Then u is activated through the edge wu at stage $i + 1$ with probability β. The process stops once no more node becomes activated and the total number of activated nodes reveals the influence of S. We measure bridging social capital using the IC model, but instead of focusing on the influence of a node, we aim to capture the ability of a node to receive information. Imagine every agent in the network has the potential to disseminate a piece of information, the bridging social capital is defined as follows.

Definition 3. *Given a network $G = (V, E)$, the* bridging social capital $\lambda(v)$ *of $v \in V$ is the probability that v becomes activated in an IC process that starts from a random node as the singleton seed set.*

The more likely a node is activated by a random IC process, the higher ability its position has in receiving information. This ability naturally is enhanced by building bridges that span diverse regions of the network.

3   Social Circle Upgrade Problem

We now discuss the main problem of the paper which revolves around the dynamics of the social circle of a node in a social network. Fix a network $G = (V, E)$ and a *target node* which is denoted throughout as v. We make the following fundamental assumptions:

1. The first assumption assumes that nodes in G are *agents*, i.e., a social actor who is able to perceive surrounding environments, deliberate, and iteratively make actions. In our context, the agent v aims to improve the social capital through iteratively updating its social circle.
2. The second assumption contrasts between the bridging and the bonding social capital. For an individual in a social network, there is a natural tendency to bond with others to create trust hence to increase the bonding social capital. On the other hand, it is the bridging social capital that has a stronger influence on personal achievements [10]. To do that, an agent would have to go out of her closed social circle and extend beyond to diverse parts of the social network. In this sense, improving bridging social capital would require more efforts and strategic thinking. We therefore focus on the bridging social capital and use it as the main indicator of *performance*. The bonding social capital, on the other hand, will be used as a secondary indicator.
3. A third assumption that we make here refers to the bounded social capability. As a form of resources, social relation requires efforts to maintain. It is well-established that a social actor's capability in maintaining social links is bounded. This means that a node can only keep a fixed number of neighbors. For simplicity, we use the *degree* $\deg(v)$ of v to represent the number of neighbors it may preserve. Thus any new tie created for v must be accompanied by a tie severance of v.

We will define an algorithmic problem by summing up the assumptions above. By an *edge rewiring*, we mean to shift an edge's one endpoint to another node. More formally, for any network $G = (V, E)$ and edge $e = uv$ whose endpoints are in V, we denote using $G \oplus e$ the graph $(V, E \cup \{e\})$, and using $G \ominus e$ the graph $(V, E \setminus \{e\})$.

Definition 4. *Given $G = (V, E)$, an edge rewiring decision for agent $v \in V$ with scope $S(v, G)$ is a pair of nodes (u, w) where $u \in S(v, G)$, $vu \notin E$ and $vw \in E$. The outcome of a rewiring decision (u, w) for v is the graph $G = (G \oplus vu) \ominus vw$.*

Figure 1 illustrates an example of an edge rewiring decision made by an agent v.

While bridging social capital is the main performance metric, bonding social capital also plays an important role in shaping an individual's social circle. Therefore the overall goal is to devise a rewiring strategy that maximally improves bridging social capital while preserving bonding social capital. More formally, the *social circle upgrade problem* is defined as follows: The input to the problem consists of a graph $G = (V, E)$ and a target node $v \in V$. The output of the problem is a sequence of rewiring decisions

$$(u_1, w_1), (u_2, w_2), \ldots, (u_k, w_k)$$

that produces a sequence of graphs $G_0 = G$, $G_1, \ldots, G_k$ where $G_{i+1} := (G \oplus vu_{i+1}) \ominus vw_{i+1}$ so that v has high bridging social capital $\lambda(v)$ in G_k while the bonding social capital $\kappa(v)$ in G_k is not much lower than the $\kappa(v)$ in G. An *edge rewiring strategy* for agent v is a mechanism that directs v to make iterative edge rewiring decisions thereby solving the social circle upgrade problem.

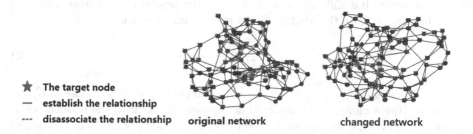

★ **The target node**
— **establish the relationship**
--- **disassociate the relationship**   **original network**   **changed network**

Fig. 1. An example of an edge rewiring decision.

4   Edge Rewiring Strategies

We now study the edge rewiring strategies of a target agent in a social network. One straightforward idea that may be adopted for the social circle upgrade problem is to directly optimize the bridging social capital function $\lambda(v)$ while preserving the bonding social capital function $\kappa(v)$. There are several technical limitations with this approach. Namely, the functions $\lambda(v)$ and $\kappa(v)$ are both defined in terms of a stochastic process and their evaluation requires high computation costs. This is especially the case for $\lambda(v)$ as to measure the effectiveness of v in receiving messages from other nodes, one would need to carry out a large number of repeated trials of information diffusion by simulating the IC model. We therefore present a simple scope-based framework that utilize greedy heuristics. Instead of directly optimizing $\lambda(v)$, we identify some structural properties that are much easier to be evaluated given the network G, and aim to optimize these structural properties. The results will then lead to increase in the social capital of the target node.

The key think that sets the social circle upgrade problem posed above apart from an optimization problem is the need to consider the bonding social capital $\kappa(v)$. Therefore it is not reasonable to directly optimize $\lambda(v)$ across the entire network. We therefore introduce a new dimension of consideration into our heuristics, namely, the scope of network access. More specifically, the *scope* of an agent v refers to a subgraph of G that contains v and defines the perception of v. Moreover, it sets v's ability to establish new links. In other words, v is confined to create links only within its scope. This scope can be determined in terms of proximity to v. When the scope of v is the entire graph G, the agent v has *global view*. On the other extreme, the scope may only contain v and its neighbors thus forming a *local view*. In general, a *scope function* S maps a node

v and a network G (which contains v) with the scope of v in G. In this paper, we consider three possible scopes: the global view, the 2-level ego network G_v, and the scope determined by the community structure of the network. By setting different scopes for the target node, we obtain different edge rewiring strategies that lead to different performances in terms of $\lambda(v)$ and $\kappa(v)$.

We put forward three rewiring strategies for v, all of which uses the same rewiring decision but different scopes. We use the notion of *closeness centrality* $C_G(v)$ to approximate the ability of v to receive information:

$$C_G(v) := \frac{|V| - 1}{\sum_{u \in V} \text{dist}_G(u, v)} \tag{2}$$

A higher value of $C_G(v)$ means that v is closer in general to other nodes in G and thus occupying a more central position in the network. One of the hypotheses of this paper asserts the correlation between $C_G(v)$ and $\lambda(v)$, which will be verified by the experiments. An edge rewiring decision (u, w) is C-*greedy* if the target node v eagerly rewire edges as directed by its closeness centrality in its scope, namely,

1. u belongs to the scope $S(v, G)$ of v.
2. The centrality of v in its own scope, $C_{S(v,G)}(v)$, has the maximum gain through adding the edge vu, as compared with adding any possible edge vu' for u' in S_v.
3. The centrality of v in its own scope, $C_{S(v,G)}(v)$, has the minimum reduction through removing the edge vw, as compared with removing all possible edge vw' for w' adjacent with v.

The rewiring strategies that we put forward generate C-greedy decisions and differ by the scope function $S(v, G)$. See Algorithm 1 for a generic C-greedy strategy.

Algorithm 1. S-greedy Rewiring

INPUT: Network $G = (V, E)$, target node $v \in V$ with scope function S
OUTPUT: An edge rewire decision (u, w).
for $u \in S(v, G)$: $u \neq v$, $uv \notin E$ **do**
    $H_u \leftarrow G \oplus vu$
    Compute $\gamma(u) \leftarrow C_{S(v,H_u)}(v) - C_{S(v,G)}(v)$
$u \leftarrow \arg\max_u \{\gamma(u) \mid u \neq v, uv \notin E\}$.
for $w \in V$: $wv \in E$ **do**
    $F_w \leftarrow G \ominus vw$
    Compute $\gamma'(w) \leftarrow C_{S(v,G)}(v) - C_{S(v,F_w)}(v)$
$w \leftarrow \arg\min_w \{\gamma'(w) \mid wv \in E\}$
return (u, w)

Global Strategy. Here we assume that v has global view to the network, i.e., $S(v, G) = G$. A real-world example of such case is the online platform Weibo, which recommends people to follow the person with higher "follower"-count. The rewiring strategy uses C-greedy decision which we call *global greedy (gg)*.

Local Strategy. Here we assume that v has only access to its local network and thus the scope S_v contains only nodes in the 2-level ego network G_v. This may correspond to a real-world social network site that allows users to see only those who share a mutual friend. We again propose a rewiring strategy with C-greedy decisions, which we call *local greedy* (lg).

Community Strategy. A third strategy is derived based on the assumption that users have access to their communities. The *community structure* divides a network into disjoint subgraphs, i.e., *communities*, that are densely connected internally but sparsely connected externally. For any node $u \in V$, let $C(u)$ denote the community that contains u. Define the *local community network* G_v^C of v as the subgraph of G induced on $C(v) \cup \{u \in V \mid \exists w \in C(v)\, uw \in E\}$, namely, those nodes that are in the same community as v as well as those who are adjacent to $C(v)$. The Louvain algorithm is a community detection algorithm based on modularity [7], which performs well both in efficiency and effectiveness. Our inspiration for this strategy is two-fold. On one hand, social networking sites such as Nextdoor aim to create a social atmosphere within local communities. On the other hand, community structure is a salient feature of social networks. It is thus of interest to bring these real-world and theoretical aspects of communities together. Here, we apply a similar reasoning as gg and lg, but with $S(v, G) = G_v^C$. We refer to this strategy as *local community* (lc).

5   Experiment Setup

We would like to evaluate and compare the global, local and community-based rewiring strategies in the context of social capital. The experiments involve taking a network dataset, selecting a target node v, then applying the rewiring strategy iteratively, and finally evaluating the outcomes.

Network Dataset. As mentioned above, the focus of the paper is on networks with a more-or-less prominent community structure. We use both random network models and real-world datasets. Several network models are able to generate community structures with varying strength [7]. Four well-known community structure network models are used: (1) The first model is Watts and Strogatz small-world network (WS). Small-world networks have not only smaller average distances but also higher clustering coefficients. We selected the WS model in the small-world model. (2) A second small-world model is Newman and Watt's model NW. NW model always generates a connected graph. Despite high clustering coefficient, there is typically no strong community structure exhibited in the graphs generated by the small-world networks, as the communities are largely hidden within the network structure. (3) The planted partition (RP) model builds communities with high internal density explicitly [6]. The communities generated by RP are homogeneous in terms of size and density. (4) The model by Lancichinetti, Fortunato, Radicchi (LFR) also constructs graphs with heterogeneous communities [14] and is also a popular benchmark for various community detection algorithms. For each of the four network models above, we generate four graphs with 100, 200, 500, and 1000 nodes, resp. We use the built-in communities for RP and LFR and the Louvain method to generate the communities for WS and NW. We also conduct experiments over five real-world networks (BlogCatalog, Douban, Youtube, Facebook and Eu-mail), taken from open online sources[1]. The email-Eu data set is a data set with ground-truth community information. For the other four data sets, the community detection algorithm needs to be used for community division. Table 1 lists statistics of the real networks.

Table 1. Real-world datasets: clustering coefficient (Cc), average path length (Apl) number of communities (#com), sizes of the maximum community (MaxC) and minimum community (MinC), and average community size (AvgC).

Network	$|V|$	$|E|$	Cc	Apl	#com	MaxC	MinC	AvgC
Facebook	3927	84210	0.544	6	108	382	2	36
BlogCatalog	4924	12077	0.543	2	28	144	2	17
Youtube	4999	45958	0.273	3	122	130	2	40
Douban	5996	1876	0.0516	4	564	742	2	10
Eu-mail	1005	16506	0.399	3	34	255	1	29

Target Node Selection. The choice of target nodes obviously impacts the result of the edge rewire strategies. We use the following method to determine the target node v in order to estimate results for a generic case. For each network, compute the median closeness centrality of all nodes. we then select 100 target nodes, 50 of which randomly from those nodes whose closeness centrality is below the median, and the other 50 randomly from those whose closeness centrality is above the median. For each target node, we iteratively apply a number of edge rewire decisions according to the three edge rewire strategies, respectively. The number of iterations is calculated as the initial average path length of the graph, which is the average distance among any pairs of distinct nodes in the graph. The intuition for this is that in an ideal situation, this number of edge rewire decisions made would bring a peripheral node (i.e., a node with minimum closeness centrality) to the network center (i.e., gain maximum closeness centrality).

Evaluation Metrics. To evaluate the effectiveness of the proposed strategies in solving the social circle upgrade problem, we measure changes to the bonding and bridging social capital $\kappa(v)$ and $\lambda(v)$. The value of $\kappa(v)$ can be conveniently approximated through an iterative power algorithm. For $\lambda(v)$, we randomly select 1000 nodes, from each of which we run an IC process, then evaluate the proportion of times when v is activated. The outputs of $\kappa(v)$ and $\lambda(v)$ are

---

[1] https://snap.stanford.edu/data/ and http://konect.uni-koblenz.de/.

clearly affected by the parameters α (for $\kappa(v)$) and β (for $\lambda(v)$). It is therefore necessary to calibrate the α and β values.

Shifts in the graph structure changes structural properties such as centrality and degrees; On the other hand, such changes also lead to updated social capital. Our other aim is to discover important correlations that bridge structural properties with social capital using the following evaluation metrics: The first such structural property is the closeness centrality $C_G(v)$ of the target agent v. We are interested in whether $C_G(v)$ increases as it performs a number of greedy edge rewire decision under the constraint on network access. Our two hypotheses include that (i) changes between $C_G(v)$ and $\lambda(v)$ are positively correlated, and (ii) **gg** will have a more significant effect in improving $C_G(v)$.

The second structural property is *embeddedness* which captures the level of engagement of an individual with its social surrounding. The notion has been traditionally defined through mutual neighborhood. We define the *embeddedness score* $\varepsilon(v)$ of the target node by

$$\varepsilon(v) := \frac{\sum_{uv \in E} |\{w \in V \mid wu, wv \in E\}|}{(|V| - 1)(|V| - 2)}. \tag{3}$$

Here again, we have two hypotheses: (1) As $\varepsilon(v)$ is often associated with trust and norm in the literature [7], a correlation exists between $\varepsilon(v)$ and $\kappa(v)$. (2) Since **lg** only connects v to "friends' friends", it will have a positive effect on the embeddedness $\varepsilon(v)$; On the contrary, as **gg** eagerly bridges to far away regions of the network, it will not effectively improve $\varepsilon(v)$.

6   Results and Discussions

Result Set 1. Figures 2 and 3 plot the changes to the four metrics above after applying the three rewiring strategies to the target nodes over the generated networks. Figure 2 plots changes to the closeness centrality and embeddedness while Fig. 3 plots changes to the social capital $\lambda(v)$ and $\kappa(v)$. We fix $\alpha = 0.9$ as it is consistent with existing work on random walk with restart [28], β is tuned to 0.4 to make sure the result fall into an appropriate range. The results show a striking difference between the rewiring strategies' effects on closeness centrality and embeddedness, as well as between the effects on bonding and bridging social capital. In the case of the closeness and bridging social capital, in all cases, **gg** gives the biggest improvement, which is then followed by **lc**, and **lg** comes at the last place. The pattern is reversed for embeddedness and bonding social capital. There, **lg** gives the biggest improvement on embeddedness as well as bonding social capital. **gg** and **lc** improves embeddedness to a much less extent and mostly decrease the value of $\kappa(v)$.

The results validate our hypotheses. There is a close resemblance between the change on $C_G(v)$ and $\lambda(v)$: As a node becomes more central, so will the node be more easy to access information. The **gg** strategy enables v to bridge widely thus gaining in bridging social capital $\lambda(v)$. On the other hand, embeddedness

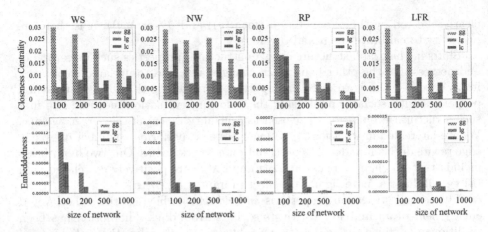

Fig. 2. Change on v's closeness centrality (top) and embeddedness (bottom) in synthetic graphs.

$\varepsilon(v)$ exhibits a correlation with $\kappa(v)$, as the lg strategy tends to improve the most embeddedness while making the least damage to bonding social capital.

At the same time, there are certain noticeable distinctions between the structural features and social capital. While there is a large gap between the changes on $C_G(v)$ due to gg and lc, the gap is very small over the bridging social capital $\lambda(v)$ between these two strategies. While lc improves embeddedness slightly, it generally reduces the bonding social capital $\kappa(v)$, although to a much lesser extent than lg. Therefore the community-based strategy balances the two strategies achieving a very similar improvement in bridging social capital as for gg and a much less drop in bonding social capital than lg.

Increasing the network size generally leads to unfavorable changes to $C_G(v)$, $\varepsilon(v)$ and $\lambda(v)$, although the trend is not apparent for $\kappa(v)$. The results show much consistency across the three network models, although lg improves bridging social capital more in networks with strong community structures (RP and LFR) than in weak community structures such as WS and NW.

Result Set 2. We calibrate the parameters α and β and evaluate $\kappa(v)$ and $\lambda(v)$ over synthetic networks with 1000 nodes. The aim is to see whether consistent results are obtained; See Fig. 4 as α, β vary within $(0, 1)$. $\kappa(v)$ increases as α grows. This is reasonable as α abstractly denotes the level of dyadic interactions and the target node bonds more with neighbors as the interactions increases. Similarly, as β increases, $\lambda(v)$ also increases. This is expected as β controls the easiness of activation in the diffusion model. lc even can outperform gg when β is small, showing the strength of community-based strategies.

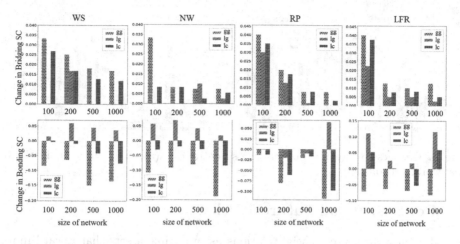

Fig. 3. Change on $\lambda(v)$ (top) and $\kappa(v)$ (bottom) in synthetic graphs.

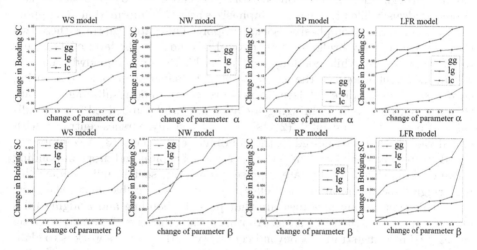

Fig. 4. Calibrate $\alpha \in (0,1)$ for $\kappa(v)$ (above) and $\beta \in [0,1]$ for $\lambda(v)$ (bottom) in synthetic graphs with 1000 nodes.

Result Set 3. We then collect results over the five real-world datasets; see Fig. 5. The changes to all metrics due to the strategies are consistent with those obtained over the synthetic networks. Even though increase to closeness centrality and embeddedness show much fluctuation, the general correlation between them and bridging and bonding social capital, respectively, remain the same. Changes on social capital are also consistent across the five datasets.

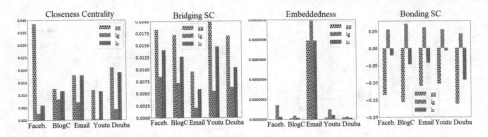

Fig. 5. Changes to target nodes' closeness centrality, $\lambda(v)$, $\varepsilon(v)$, and $\kappa(v)$ in the 5 real-world networks.

7   Conclusion and Future Work

Social capital theory sees social relations as intangible assets that reward individuals through bridging and bonding roles [5]. These two roles are seen as two opposing ideals: while bridging role emphasizes access to information and opportunities, bonding role emphasizes social support. This paper captures these two roles mathematically: bridging social capital of a node is the level of its access to information flow, while bonding social capital is the sum of goodwill that lies within its neighborhood in the ego network. The paper then looks at rewiring strategies assuming different level of network access. The results demonstrate that: (1) centrality reflects bridging social capital while embeddedness an indicator of bonding social capital. (2) As an agent greedily pursues centrality with global view, a drop is witnessed on the bonding social capital. (3) A community-based strategy can balance gains in bridging social capital and drops in bonding social capital, while a local strategy preserves bonding social capital while gaining bridging social capital in a less effective way. It is desirable but at the same time challenging to invent a measure that puts the bonding and bridging roles of social capital on the same scale. Another future challenge is to elaborate on the agents' rewiring strategies beyond centrality. A third future work is to elaborate the the edge rewire decisions to multiple agents and use it to develop an understanding of network evolution.

References

1. Adler, P.S., Kwon, S.W.: Social capital: prospects for a new concept. Acad. Manage. Rev. **27**(1), 17–40 (2002)
2. Burt, R.S.: Structural holes versus network closure as social capital. In: Social capital, pp. 31–56. Routledge (2017)
3. Cai, Y., Zheng, H., Liu, J., Yan, B., Su, H., Liu, Y.: Balancing the pain and gain of hobnobbing: utility-based network building over atributed social networks. In: Proceedings of AAMAS 2018, pp. 193–201 (2018)
4. Carron, P.M., Kaski, K., Dunbar, R.: Calling dunbar's numbers. Soc. Netw. **47**, 151–155 (2016)
5. Coffé, H., Geys, B.: Toward an empirical characterization of bridging and bonding social capital. Nonprofit Volunt. Sect. Q. **36**(1), 121–139 (2007)

6. Condon, A., Karp, R.M.: Algorithms for graph partitioning on the planted partition model. Random Struct. Algorithms **18**(2), 116–140 (2001)
7. Easley, D., Kleinberg, J.: Networks, Crowds, and Markets: Reasoning about a Highly Connected World. Cambridge University Press, Cambridge (2010)
8. Ellison, N.B., Steinfield, C., Lampe, C.: The benefits of facebook "friends:" social capital and college students' use of online social network sites. J. Comput. Mediat. Commun. **12**(4), 1143–1168 (2007)
9. Gamst, F.C.: Foundations of social theory. Anthropol. Work Rev. **12**(3), 19–25 (1991)
10. Granovetter, M.: The strength of weak ties: a network theory revisited. Sociol. Theor. **1**, 201–233 (1983)
11. Huang, H., et al.: Mining triadic closure patterns in social networks. In: Proceedings of WWW 2014, pp. 499–504. ACM (2014)
12. Jackson, M.O., Wolinsky, A.: A strategic model of social and economic networks. In: Dutta, B., Jackson, M.O. (eds.) Networks and Groups. Studies in Economic Design. Springer, Heidelberg (2003). https://doi.org/10.1007/978-3-540-24790-6_3
13. Karrer, B., Levina, E., Newman, M.E.: Robustness of community structure in networks. Phys. Rev. E **77**(4), 046119 (2008)
14. Lancichinetti, A., Fortunato, S., Radicchi, F.: Benchmark graphs for testing community detection algorithms. Phys. Rev. E **78**(4), 046110 (2008)
15. Liben-Nowell, D., Kleinberg, J.: The link-prediction problem for social networks. J. Am. Soc. Inf. Sci. Tech. **58**(7), 1019–1031 (2007)
16. Lin, N.: Building a network theory of social capital. In: Social Capital, pp. 3–28. Routledge (2017)
17. Liu, J., Li, L., Russell, K.: What becomes of the broken hearted?: an agent-based approach to self-evaluation, interpersonal loss, and suicide ideation. In: AAMAS 2017, pp. 436–445 (2017)
18. Liu, J., Wei, Z.: Community detection based on graph dynamical systems with asynchronous runs. In: CANDAR 2014, pp. 463–469. IEEE (2014)
19. Moskvina, A., Liu, J.: How to build your network? a structural analysis. In: Proceedings of IJCAI 2016, pp. 2597–2603 (2016)
20. Moskvina, A., Liu, J.: Integrating networks of equipotent nodes. In: Nguyen, H.T.T., Snasel, V. (eds.) CSoNet 2016. LNCS, vol. 9795, pp. 39–50. Springer, Cham (2016). https://doi.org/10.1007/978-3-319-42345-6_4
21. Radicchi, F., Castellano, C., Cecconi, F., Loreto, V., Parisi, D.: Defining and identifying communities in networks. PNAS **101**(9), 2658–2663 (2004)
22. Ruan, J., Zhang, W.: Identification and evaluation of weak community structures in networks. In: AAAI 2006, pp. 470–475 (2006)
23. Smith, M., Giraud-Carrier, C., Stephens, S.: Measuring and reasoning about social capital: a computational framework. A Computational Framework for Social Capital in Online Communities, p. 27 (2011)
24. Song, Y., Xu, R.: Affective ties that bind: Investigating the affordances of social networking sites for commemoration of traumatic events. Soc. Sci. Comput. Rev. **37**, 333–354 (2018). https://doi.org/10.1177/0894439318770960
25. Villalonga-Olives, E., Kawachi, I.: The measurement of social capital. Gac. Sanit. **29**, 62–64 (2015)
26. Yan, B., Chen, Y., Liu, J.: Dynamic relationship building: exploitation versus exploration on a social network. In: Bouguettaya, A., et al. (eds.) WISE 2017. LNCS, vol. 10569, pp. 75–90. Springer, Cham (2017). https://doi.org/10.1007/978-3-319-68783-4_6

222   Q. Chen et al.

27. Yan, B., Liu, Y., Liu, J., Cai, Y., Su, H., Zheng, H.: From the periphery to the center: information brokerage in an evolving network. In: IJCAI 2018, pp. 3912–3918. AAAI Press (2018)
28. Yin, Z., Gupta, M., Weninger, T., Han, J.: A unified framework for link recommendation using random walks. In: ASONAM 2010, pp. 152–159. IEEE (2010)

Distributed Landmark Selection
for Lower Bound Estimation of Distances
in Large Graphs

Mingdao Li, Peng Peng$^{(\boxtimes)}$, Yang Xu, Hao Xia, and Zheng Qin

Hunan University, Changsha, China
{limingdao,hnu16pp,xy_720,1281877114,zqin}@hnu.edu.cn

Abstract. Given two vertices in a graph, computing their distance is a fundamental operation over graphs. However, classical exact methods for this problem often cannot scale up to the rapidly evolving graphs in recent applications. Many approximate methods have been proposed, including some landmark-based methods that have been shown to have good scalability and estimate the upper bound of the distance in acceptable accuracy. In this paper, we propose a new landmark-based framework based a new measure called *coverage* to more accurately estimate the lower bound of the distance. Although we can prove that selecting the optimal set of landmarks is NP-hard, we propose a heuristic algorithm that can guarantee the approximation ratio. Furthermore, we implement our method through the distributed graph processing systems while considering the characteristic of the distributed graph processing systems. Experiments on large real graphs confirm the superiority of our methods.

1 Introduction

As an increasing number of real applications present their datasets in the form of graphs, studies on graph computations have important applications in many domains. Common examples include online social networks, road networks, biological networks, hyperlink graph. In online social networks, it can be applied to building an intimate network between users and querying the closest social relationships; in road networks, it can be applied to designing the minimum cost line between two locations; in biological networks, it can be applied to exploring the relationship between the various biological components in the food chain; and in the hyperlink graph, the goal is to quickly complete the response process etc. The scale of these graph has continuously increasing in the past few years, in some instances reached trillions of edges, billions of vertices.

The shortest path problem is one basic problem in graph computations. It becomes very significant in graph application with the sustained development and expansion of the graphs. Due to the large size of the graph, computing the exact distance between any two vertices is highly challenging. Therefore, we loosen up this problem into a bound estimate. The estimation of the lower bound of the distance between two vertices is a critical problem that is widely

© Springer Nature Switzerland AG 2019
J. Shao et al. (Eds.): APWeb-WAIM 2019, LNCS 11641, pp. 223–239, 2019.
https://doi.org/10.1007/978-3-030-26072-9_16

used. For example, Zou et al. [20] utilize the lower bounds of distances between vertices to find candidate matches since it is much cheaper to compute. In real life, it can also be applied to cost control, model optimization and so on.

A commonly used method for estimating the bounds of the distance between two vertices is that selecting a subset of nodes as landmarks [5,11,12,16]. However, most of these existing methods are optimized to select landmarks for good upper bounds. They all tend to select central nodes in the graph as landmarks, which perform poorly for estimating lower bounds. This paper proposes an optimized landmark selection method that are tailor made for estimating the lower bounds. Generally speaking, we first propose a new measure, named *coverage*, to evaluate how many pairs of vertices a set of landmarks can provide good distance estimates for them. Then, although selecting the optimal set of vertices with the maximal coverage as the landmarks can be proven to be NP-hard, we propose a heuristic algorithm while guaranteeing the approximation ratio. Extensive experiments confirm that our approach can improve the approximation quality of estimating the distances between pairs of vertices.

Furthermore, as the size of graphs is rapidly increasing, the number of edges has reached the enormous scale and it may cause poor performances of memory access and changing the degree of parallelism during computing process. Therefore, there is a pressing need to utilize the distributed graph processing systems, like Pregel [10], Giraph [15] and GraphX [4], to fix this problem. These distributed graph processing systems are in a vertex-centric computational model that composed of a sequence of supersteps conforming to the Bulk Synchronous Parallel (BSP) model. In this paper, we implement our method through the distributed graph processing systems, which can greatly improve the scalability of our methods. In our implementation, to avoid some redundant computation, we fully utilize the characteristic of the parallel computational model to merge some common computations of shortest path trees.

In summary, our main contributions are summarized as follows:

- We propose a new landmark selection model. We define a new measure, named *coverage*, to evaluate the optimality of landmarks for estimating the lower bounds of distances between pairs of vertices.
- We put forward a landmark-based framework to estimate lower bound of the distance between two vertices. Although we prove that the problem of selecting the optimal set of landmarks is NP-hard, we propose a heuristic strategy to guarantee the approximation ratio and be scalable in graphs.
- We implement our method by using the distributed graph database management system, while fully utilizing the characteristic of the distributed graph processing systems to merge some common computations.
- Finally, we conduct extensive experiment with real graphs including social networks, road networks and knowledge graphs to evaluate our method.

2   Preliminaris

In this section, we review the related concepts through this paper.

2.1   Graph and Path

In this paper, we consider an undirected, unweighted graph $G = (V, E)$, where V is a set of vertices, E is a set of edges. Note that, our approach can be easily generalized to accommodate weighted directed graphs as well.

Given two vertices $s, t \in V$, we define a sequence $p(s, t) = (s, v_1, v_2, \ldots, v_{l-1}, t)$ to denote a path of length between two vertices s and t, where $\{v_1, v_2, \ldots, v_{l-1}\} \subseteq V$ and $\{(s, v_1), (v_1, v_2), \ldots, (v_{l-1}, t)\} \subseteq E$. Then, the distance $d(s, t)$ between vertices s and t is defined as the length of the shortest path between s and t. For example, Fig. 1(a) shows an example graph.

A shortest path tree rooted at vertex v is a spanning tree T of G, such that the distance from v to any other vertex u in T is equal to the distance from v to u in G. Here, we denote the shortest path tree rooted at vertex v as $SPT(v)$. Furthermore, for each vertex v' in $SPT(v)$, its parent in $SPT(v)$ is denoted as $v'.p_v$, and its distance to v is denoted as $v'.d_v$. Here, the parent of v in $SPT(v)$ is v itself. For example, Figs. 1(b) and (c) show $SPT(v_4)$ and $SPT(v_{10})$, two shortest path trees rooted at v_4 and v_{10} for the graph shown in Fig. 1(b) and (c). For v_1, its parent $v_1.p_{v_4}$ in $SPT(v_4)$ is v_2 and $v_1.d_{v_4}$ is 2.

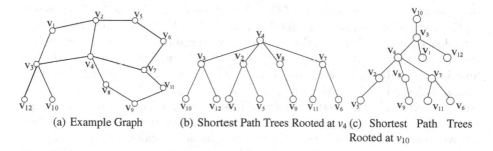

(a) Example Graph   (b) Shortest Path Trees Rooted at v_4 (c) Shortest Path Trees Rooted at v_{10}

Fig. 1. Example graph and the shortest path trees rooted at v_4 and v_{10}

The shortest path distance in a graph is a metric and satisfies the triangle inequality. That is, for any vertex $s, t, v \in V$,

$$|d(s, v) - d(v, t)| \leq d(s, t) \leq d(s, v) + d(v, t)$$

We call the left side of the inequality the lower bound of the distance between s and t, which is denoted as $\tilde{d}(s, t)$. The lower bound becomes an equality if there exists a shortest path $p(v, t)$ which passes through s or a shortest path $p(v, s)$ which passes through t.

2.2   Landmark

As we mentioned above, if we select a vertex v and precompute the distance $d(v, u)$ from this vertex to each other vertex u in the graph, we can get a lower bound approximation $\tilde{d}(s, t)$ between any two vertices s and t.

Furthermore, if we select a set $\{l_1, l_2, \ldots, l_k\}$ of k vertices as *landmarks*, a potentially lower bound approximation can be computed as follows.

$$\tilde{d}(s,t) = \max_{1 \le i \le k} |d(s, l_i) - d(l_i, t)| \tag{1}$$

Most previous studies [5,11,12,16] indicate that the current landmark selection methods cannot select to accurately estimate the lower bounds. In this paper, we propose a novel landmark selection method to improve the approximation ratio of lower bound estimation.

2.3   Computational Model of Distributed Graph Processing System

As Pregel [10] has been proposed to process large graphs, many distributed graph processing systems have been proposed to follow the similar computational model.

In Pregel-like distributed graph processing systems, the computational model is a vertex-centric model that composed of a sequence of supersteps conforming to the Bulk Synchronous Parallel (BSP) model. Given a graph G as the input data, each vertex v is in one of the two states, i.e., active and inactive. The function *superStep* gets the number of the current superstep. In the first superstep, all the vertices are active. The function *voteToHalt*: $V \rightarrow \emptyset$ is called by a vertex v to deactivate itself. The entire computation terminates when all vertices are inactive. Let M be the set of messages. Within a superstep, the user-defined function *compute*(G, M) is executed on each active vertex in parallel. The function *sendMsg*: $V \times V \times M \rightarrow \emptyset$ is to send messages M from one vertex v to another vertex v'. An inactive vertex will be reactivated by incoming messages sent to it. When *compute*(G, M) is invoked on each active vertex v, it (1) first receives messages (i.e., each $m \in M$) sent to v in the previous superstep, (2) gets and/or updates its value, (3) modifies M to generate the set of new messages M', and (4) invokes *sendMsg*(v, v', M') to send M' to the adjacent vertex M'.

3   Coverage

As discussed before, given two vertices s and t, the lower bound as shown in Eq. 1 can become equal to the exact distance when the shortest paths from at least one landmarks to vertex s (or vertex t) pass through vertex t (or vertex s). Here, we call that a vertex v *covers* vertices s and t, if the shortest paths from v to vertex s (or vertex t) pass through vertex t (or vertex s). Obviously, if a vertex v cover vertices s and t, s (or t) is the ancestor of t (or s) in $SPT(v)$. Here, we define a boolean variable $\sigma_{SPT(v)}(s,t)$ to record whether a vertex v covers vertices s and t.

$$\sigma_{SPT(v)}(s,t) = \begin{cases} 1 & \textit{if } v \textit{ covers vertices } s \textit{ and } t; \\ 0 & \textit{otherwise.} \end{cases}$$

Then, given a vertex v in a graph $G = (V, E)$, we define its *covering set* $CS(v, V)$ to maintain the set of all vertices pairs in V that v covers as follows.

$$CS(v, V) = \{(s, t) | \sigma_{SPT(v)}(s, t) = 1\}$$

Furthermore, we define a measure named *coverage* $C(v)$ for vertex v to record the number of vertices pairs that v covers as follows.

$$C(v) = |CS(v, V)|$$

where $|CS(v, V)|$ is the number of vertices pairs in $CS(v, V)$.

Given a vertex v and its shortest path tree $SPT(v)$, the coverage of v is how many pairs of vertices covered by v. For example, for $SPT(v_4)$ and $SPT(v_{10})$ shown in Fig. 1(b) and (c), we can compute out $C(v_4) = 18$ and $C(v_{10}) = 32$. It is obvious that the coverage of v_{10} is much larger than the coverage of v_4, which indicates that v_{10} can cover much more vertices pairs than v_4. Hence, selecting v_{10} as a landmark to estimate the lower bounds of vertices pairs can be more beneficial than selecting v_4.

When we select some landmarks to estimate the distances, an intuitive goal is to select the landmarks that cover as many vertices pairs as possible. In other words, we should select a set of landmarks with the largest coverages. However, as discussed in the following section, selecting the optimal set of k vertices as landmarks is NP-hard, so we propose a heuristic algorithm to select the set of k landmarks while guaranteeing the approximation ratio.

4   Landmark Selection

In real applications, although some vertices have large coverages, the vertices pairs that they covers highly overlap. In other words, these vertices are so close to each other that their shortest path trees highly overlap. Figure 2 shows two example highly overlapped shortest path trees. Vertices v_{10} and v_{12} in Fig. 1(a) are so close to each other that their shortest path trees greatly overlap.

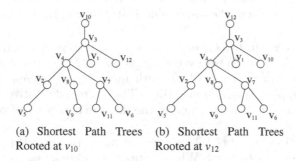

(a) Shortest Path Trees   (b) Shortest Path Trees
Rooted at v_{10} Rooted at v_{12}

Fig. 2. Example highly overlapped shortest path trees rooted at v_{10} and v_{12}

The more two vertices overlap, the fewer vertices pairs they can totally cover. If two vertices cover the similar set of vertices pairs in the graph, it is probably wise to include only one of them. Thus, we select the vertices that less overlap to each other as the landmarks to cover as many vertices pairs as possible.

To this goal, we formally define the *coverage* of a set of k vertices according to the number of vertices pairs that the k vertices cover, in order to measure the benefit of selecting them as landmarks. Given a set of k vertices $L = \{v_1, v_2, \ldots, v_k\}$, the *coverage* of L is defined as follows.

$$C(L) = | \bigcup_{v_i \in L} CS(v_i)| \tag{2}$$

As suggested in Eq. 2, we should select the set of k vertices with the largest coverage as landmarks to cover as many vertices pairs as possible. However, we can prove the problem of selecting the set of k vertices with the largest coverage is NP-hard in the following theorem.

Theorem 1. *The problem of selecting the set of k vertices with the largest coverage as landmarks is NP-hard.*

Proof. Here, we first prove that the function of coverage, $Coverage(L)$ in Eq. 2, is submodular. Here, let $\Delta_{Coverage}(v|L) = C(\{v\} \cup L) - C(L)$ be the discrete derivative of coverage at L with respect to v. In other words, for every $V_1 \subseteq V_2$ and a vertex $v \notin V_2$, we need to prove the following in equation.

$$\Delta_{Coverage}(v|V_1) \geq \Delta_{Coverage}(v|V_2)$$

For vertex v, there are three kinds of vertices pairs in $CS(v, V)$: the set CS_1 of vertices pairs not covered by any vertices in V_2, the set CS_2 of vertices pairs only covered by vertices in $(V_2 - V_1)$, and the set CS_3 of vertices pairs covered by vertices in V_1.

For any vertices pair (s_1, t_1) in CS_1, $\max_{v_i \in V_1}\{\sigma_{SPT(v_i)}(s_1, t_1)\} = \max_{v_i \in V_2}\{\sigma_{SPT(v_i)}(s_1, t_1)\} = 0$. When vertex v is selected, (s_1, t_1) can be covered only by v. Then, both $\max_{v_i \in \{v\} \cup V_1}\{\sigma_{SPT(v_i)}(s_1, t_1)\}$ and $\max_{v_i \in \{v\} \cup V_2}\{\sigma_{SPT(v_i)}(s_1, t_1)\}$ become 1. The discrete derivative of coverage at V_1 with respect to v is the same to the discrete derivative of coverage at V_2 with respect to v.

For any vertices pair (s_3, t_3) in CS_3, $\max_{v_i \in V_1}\{\sigma_{SPT(v_i)}(s_3, t_3)\} = \max_{v_i \in V_2}\{\sigma_{SPT(v_i)}(s_3, t_3)\} = 1$. When vertex v is selected, (s_3, t_3) can be covered by both v and vertices in V_1. Then, both $\max_{v_i \in \{v\} \cup V_1}\{\sigma_{SPT(v_i)}(s_3, t_3)\}$ and $\max_{v_i \in \{v\} \cup V_2}\{\sigma_{SPT(v_i)}(s_3, t_3)\}$ are still 1. The discrete derivative of coverage at V_1 with respect to v is also the same to the discrete derivative of coverage at V_2 with respect to v.

For any vertices pair (s_2, t_2) in CS_2, $\max_{v_i \in V_1}\{\sigma_{SPT(v_i)}(s_2, t_2)\} = 0$ and $\max_{v_i \in V_2}\{\sigma_{SPT(v_i)}(s_2, t_2)\} = 1$. When vertex v is selected, (s_2, t_2) can be covered by both v and vertices in $V_2 - V_1$. Then, $\max_{v_i \in \{v\} \cup V_1}\{\sigma_{SPT(v_i)}(s_2, t_2)\}$ becomes 1 and $\max_{v_i \in \{v\} \cup V_2}\{\sigma_{SPT(v_i)}(s_2, t_2)\}$ is still 1. The discrete derivative of coverage at V_1 with respect to v is larger than the discrete derivative of coverage at V_2 with respect to v.

In conclusion, $\Delta_{Coverage}(v|V_1) \geq \Delta_{Coverage}(v|V_2)$ and the function $C(L)$ is submodular. Since the problem of maximizing submodular functions is NP-hard [1], the problem is NP-hard.   □

4.1 Our Solution

As proved in Theorem 1, selecting the set L of k vertices with the largest coverage as landmarks is NP-complete problem. We propose a greedy algorithm as outlined in Algorithm 1.

In general, for each vertex v, we first compute out its shortest path tree, and traverse its shortest path trees to get its covering set $CS(v, V)$ (Lines 1–5 in Algorithm 1). Then, we iteratively select the vertex v_{max} to maximize the marginal gain of the coverage of selected landmarks until we meet the constraint of the number of the landmarks or cannot find a landmark to increase the coverage (Lines 7–11 in Algorithm 1). Finally, the algorithm outputs L (Line 12 in Algorithm 1).

For example, we assume that we want to select three landmarks in the example graph shown in Fig. 1(a). After we compute the covering sets of all vertices, we can find out that v_{10} and v_{12} have the largest covering sets and their coverages are the same. If we select v_{10} as the first landmark and update all covering sets, the covering set of v_{12} becomes much smaller and v_9 has the largest covering set, so v_9 is selected as the second landmark. Similarly, after we use the covering set of v_9 to update all remaining covering sets, we select v_{11} as the third landmark. Thus, $\{v_{10}, v_9, v_{11}\}$ are the set of three selected landmarks.

Algorithm 1. Landmark Selection

Input: A graph $G = (V, E)$
Output: A set $L \subseteq V$ of k vertices as landmarks
1 **for** *each vertex v in V* **do**
2 | $CS(v, V) \leftarrow \emptyset$;
3 | Compute the shortest path tree $SPT(v)$ rooted at v;
4 | **for** *each ancestor-descendant vertices pair (s, t) in $SPT(v)$* **do**
5 | | $CS(v, V) \leftarrow CS(v, V) \cup \{(s, t)\}$;
6 $L \leftarrow \emptyset$;
7 **for** $i = 1$ *to* k **do**
8 | Find the vertex v_{max} with the largest size of $CS(v, V)$;
9 | $L \leftarrow L \cup \{v\}$;
10 | **for** *each vertex v in V* **do**
11 | | $CS(v, V) \leftarrow CS(v, V) - CS(v_{max}, V)$;
12 Return L;

Theorem 2. *Algorithm 1 obtains a set of k vertices as landmarks of coverage at least $\frac{1}{2}(1 - \frac{1}{e})$ times the value of an optimal solution.*

Proof. In Algorithm 1, since the problem of selecting the set of k vertices with the largest coverage as landmarks is a problem of maximizing a submodular set function subject to a constraint as discussed in Theorem 1, we directly apply the greedy algorithm in [8] to iteratively select the vertex with the largest coverage. [8] proves that the worst-case performance guarantee of the greedy algorithm is $\frac{1}{2}(1 - \frac{1}{e})$, so the coverage of the selected landmarks is at least $\frac{1}{2}(1 - \frac{1}{e})$ of the optimal coverage.   □

In Algorithm 1, we need compute out the shortest path tree of each vertex and enumerate all vertices pairs that it covers. In real applications, the number of vertices is large, so the space and time costs of Algorithm 1 are too large. Hence, we randomly sample a set V' of vertices and use V' to estimate the coverages of all vertices.

Here, given a vertex v and a sample set $V' \subseteq V$, we define its *approximate covering set* $\widetilde{CS}(v, V')$ to maintain the set of all vertices pairs in V' that v covers.

$$\widetilde{CS}(v, V') = \{(s,t)|\sigma_{SPT(v)}(s,t) = 1 \wedge s,t \in V'\}$$

Then, we can define a measure named *approximate coverage* $\widetilde{C}(v)$ for vertex v to record the number of vertices pairs that v covers as follows.

$$\widetilde{C}(v) = |\widetilde{CS}(v, V')|$$

where $|\widetilde{CS}(v, V')|$ is the number of vertices pairs in $\widetilde{CS}(v, V')$.

Based on the above definitions, we propose an approximate landmark selection algorithm in Algorithm 2. In general, we first sample a set V' of vertices (Line 1 in Algorithm 2). Then, we compute out the shortest path trees of these vertices, and traverse these shortest path trees to enumerate the set of covered vertices pairs in V' for each vertex v to compute out $\widetilde{CS}(v, V')$ (Lines 4–10 in Algorithm 2). Finally, we iteratively select the vertex v_{max} to maximize the marginal gain of the approximate coverage of selected landmarks to get the set L of selected landmarks (Lines 12–17 in Algorithm 1).

Algorithm 2. Approximate Landmark Selection

    Input: A graph $G = (V, E)$
    Output: A set $L \subseteq V$ of k vertices as landmarks
1 Sample a set V' of vertices from V;
2 **for** *each vertex v in V* **do**
3 | $\widetilde{CS}(v, V') \leftarrow \emptyset$;
4 **for** *each vertex s in V'* **do**
5 | Compute the shortest path tree $SPT(s)$ rooted at v;
6 | **for** *each vertex v in V* **do**
7 | | **if** $v \in V'$ **then**
8 | | | $\widetilde{CS}(v, V') \leftarrow \widetilde{CS}(v, V') \cup \{(s, v)\}$;
9 | | **if** *vertex t is an ancestor of v in* $SPT(s)$ *and* $t \in V'$ **then**
10 | | | $\widetilde{CS}(v, V') \leftarrow \widetilde{CS}(v, V') \cup \{(s, t)\}$;
11 $L \leftarrow \emptyset$;
12 **for** $i = 1$ *to* k **do**
13 | Find the vertex v_{max} with the largest size of $\widetilde{CS}(v, V')$;
14 | $L \leftarrow L \cup \{v\}$;
15 | **for** *each vertex v in V* **do**
16 | | $\widetilde{CS}(v, V') \leftarrow \widetilde{CS}(v, V') - \widetilde{CS}(v_{max}, V')$;
17 Return L;

---

5 Implementation with Distributed Graph Processing Systems

With the increasing sizes of graphs in real applications, the computational and storage requirements coupled with rapidly growing graphs have stressed the limits of single machine processing. In the presence of large scale graphs, which cannot be stored in a single node, the distributed solutions are required. For the scalability of the above landmark selection algorithm over large graphs, we implement it by using the distributed graph processing systems.

For our landmark selection algorithms, distributed graph processing systems can be utilized to compute the shortest path trees of sample vertices. The straightforward way of utilizing the distributed graph processing systems is to invoke the algorithm developed in the vertex-centric way [10] to compute the shortest path tree of the sample vertices one by one.

However, when multiple shortest path trees of sample vertices are computed, there is room for sharing computation when executing these shortest path trees. For example, the computations of both $SPT(v_4)$ and $SPT(v_{10})$ in Fig. 1 involve the edge between v_3 and v_{12} in the second superstep, so these computations can be merged together.

Based on the above observation, an optimization can be devised to share common computations. When we compute the shortest path trees of the sample vertices, we can merge these computations in one sequence of supersteps. Thus, we propose an algorithm to share the common computations as described in Algorithm 3. Generally speaking, Algorithm 3 is analogous to Algorithm 2. Their main difference is that Algorithm 3 merges the computation of sample vertices' shortest path trees.

Algorithm 3. Landmark Selection over Distributed Graph Processing Systems

---

 Input: A graph $G = (V, E)$
 Output: A set $L \subseteq V$ of k vertices as landmarks
1   Sample a set $V' = \{s_1, s_2, ..., s_{|V'|}\}$ of vertices from V;
2   **for** *each vertex v in V* **do**
3   Initialize two vectors $v.dVec$ and $v.pVec$ of length $|V'|$, and all elements in $v.dVec$ and $v.pVec$ are $+\infty$ and NIL respectively;
4   $compute(G, M)$;
5   **for** *vertex s_i in V'* **do**
6   Construct the shortest path tree $SPT(s_i)$ through all vertices' $pVec[i]$;
7   **for** *each vertex v in V* **do**
8   **if** $v \in V'$ **then**
9      $\widetilde{CS}(v, V') \leftarrow \widetilde{CS}(v, V') \cup \{(s_i, v)\}$;
10   **if** *vertex t is an ancestor of v in $SPT(s_i)$ and $t \in V'$* **then**
11      $\widetilde{CS}(v, V') \leftarrow \widetilde{CS}(v, V') \cup \{(s_i, t)\}$;
12   $L \leftarrow \emptyset$;
13   **for** $i = 1$ *to* k **do**
14   Find the vertex v_{max} with the largest size of $CS(v, V)$;
15    $L \leftarrow L \cup \{v\}$;
16   **for** *each vertex v in V* **do**
17     $CS(v, V) \leftarrow CS(v, V) - CS(v_{max})$;
18   Return L;

In Algorithm 3, we first sample a set $V' = \{s_1, s_2, \ldots, s_{|V'|}\}$ of vertices from V (Line 1 in Algorithm 3). Then, we associate each vertex v with two vectors $v.dVec$ and $v.pVec$ of length $|V'|$, where $v.dVec[i]$ maintains the distance from s_i to v and $v.pVec[i]$ maintains the parent of v in $SPT(s_i)$. Initially, all elements in $v.dVec$ and $v.pVec$ are $+\infty$ and NIL (Lines 2–3 in Algorithm 3). Then, we merge the computations of multiple shortest path trees of sample vertices in one sequence of supersteps in the function $compute(G, M)$ (Line 4 in Algorithm 3).

In Function $compute(G, M)$, in the first superstep, the distance of each sample vertex to itself is initialized as 0, and the parent of each sample vertex in its shortest path tree is itself. Then, at each superstep, there are four steps in $compute(G, M)$ for each vertex v: (1) receives the set M_r of messages from the adjacent vertices, (2) updates its distances and parents of all sample vertices' shortest path trees in $v.dVec$ and $v.pVec$, (3) generate the set M_s of new messages according to the updated distances and parents, and (4) send M_s to the adjacent vertices of v.

Function compute(G,M)

1   **if** $superStep = 1$ **then**
2   **for** $each\ vertex\ s_i\ in\ V'$ **do**
3        $s_i.dVec[i] \leftarrow 0$ and $s_i.pVec[i] \leftarrow s_i$;
4        $M_s \leftarrow \{(s_i.dVec, s_i.pVec)\}$;
5   **for** $each\ adjacent\ vertex\ v'\ in\ v.AdjacentVertices()$ **do**
6          $sendMsg(v', M_s)$;
7   **else**
8   **if** $M_r\ is\ not\ empty$ **then**
9        $M_s \leftarrow \emptyset$;
10   **for** $each\ message\ msg\ in\ M_r$ **do**
11          $(dVec_{min}, pVec_{min}) \leftarrow msg.Value()$;
12   **for** $i = 1\ to\ |V'|$ **do**
13   **if** $dVec_{min}[i] < v.dVec[i]$ **then**
14              $v.dVec[i] \leftarrow dVec_{min}[i]$;
15              $v.pVec[i] \leftarrow msg.sourceVertex$;
16              $M_s \leftarrow \{(v.dVec, p.pVec)\}$;
17   **for** $each\ vertex\ v'\ in\ v.AdjacentVertices()$ **do**
18          $sendMsg(v', M_s)$;
19   **else**
20        $voteToHalt()$;

6   Experiments

To verify the accuracy and efficiency of our methods, we tested our methods with several large-scale datasets.

6.1   Datasets

We test our approach on six real-world datasets, which can be divided into three categories: road networks, social networks and knowledge graphs. In the following, we provide more details and statistics about the datasets.

RoadNet-TX. RoadNet-TX [9] is a dataset describing the road network of Texas. Intersections and endpoints in California are represented by vertices and the roads connecting these intersections or endpoints are represented by edges.

RoadNet-PA. RoadNet-PA [9] is a dataset describing the road network of Pennsylvania. Similarly, intersections and endpoints are vertices and the roads are edges.

CA-AstroPh. CA-AstroPh [9] is a dataset of a collaboration network of Arxiv Astro Physics category. In this dataset, each vertex corresponds to an author. If an author i co-authored a paper with author j, the graph contains an undirected edge between i and j.

Email-EuAll. Email-EuAll [9] is a dataset of an email network of a large European Research Institution. In this dataset, each vertex corresponds to an email address. An edge between two vertices, if the two corresponding email addresses communicate with each at least once.

YAGO. YAGO [14] is a real RDF knowledge graph that is extracted from Wikipedia and integrates its facts with the WordNet thesaurus. It can be represented as a graph, where entities are vertices and relationships between entities are edges.

DBpedia. DBpedia [7] is a crowd-sourced RDF knowledge graph extracted from Wikipedia. Here, we use the high-quality data extracted from Infoboxes in Wikipedia using the strict ontology-based extraction in DBpedia. It can also be represents a graph, where resources are vertices and relationships between resources are edges.

The statistics of these datasets are summarized in Table 1. The statistics include the number of vertices $|V|$, number of edges $|E|$, the average clustering coefficient $\bar{\delta}$, the number of triangles θ, and the longest shortest path $\bar{d}$ (computed on a sample vertex pairs).

Table 1. Statistics of datasets

Dataset	$|V|$	$|E|$	$\bar{\delta}$	θ	$\bar{d}$
RoadNet-TX	1,379,917	1,921,660	0.0470	82,869	1,054
RoadNet-PA	1,088,092	1,541,898	0.0465	67,150	786
CA-AstroPh	18,772	396,160	0.6306	1,351,441	14
Email-EuAll	265,214	420,045	0.0671	267,313	14
YAGO	629,610	1,259,374	0.0851	283,518	20
DBpedia	634,974	1,130,216	0.1032	257,029	23

6.2 Experimental Setup

All experiments are conducted on a cluster with 12 physical nodes in Alibaba Cloud. Each node has four CPUs with 32GB memory and a 100GB disk. We implement our landmark selection method on the distributed graph processing system, GraphX. The version of GraphX that we use is 2.1.0, and all the codes are implemented in Scala 2.11.8.

By default, we sample 400 vertices to select 100 landmarks. We also sample 500 vertex pairs from each graph and use them as the input distance queries to evaluate different methods.

In order to evaluate the performance and cost of methods more intuitively, we use the definition of approximation error [5,11,16] to evaluate the relevant indicators of the methods as follows:

$$|d(s,t) - \tilde{d}(s,t)|/d(s,t) \tag{3}$$

where $d(s,t)$ denotes the shortest distance between two vertices, and $\tilde{d}(s,t)$ denotes the estimated shortest distance.

6.3 Effectiveness Test

Comparison with Different Landmark Selection Methods. We compare our measure coverage with some famous landmark selection measures in terms of approximation error. The competitors include Random, Degree [2], Closeness [2], and Potamias et al. [11]. Our method is denoted as Coverage. Random is a landmark-selection strategy whose idea is to randomly select the vertices as landmarks in a graph; Degree is a landmark-selection strategy whose idea is to select the vertices with the largest degrees in a graph as landmarks; Closeness is a landmark-selection strategy whose idea is to select the vertices with the best closeness centrality as landmarks; Potamias et al. is a landmark-selection strategy to estimate the lower bounds of distances more accurately. Its idea is to select the first landmark randomly, and then select the farthest one as the second landmark. Then, they start the recursion, and at each iteration the vertex with the maximum sum of all the current landmark distances is selected as the next landmark. Table 2 summarizes the approximation error of different strategies across all 6 datasets studied here.

Table 2. Summary of mean approximation errors across datasets, using 100 landmarks (top) and 500 landmarks (bottom)

	Random	Degree	Closeness	Potamias et al.	Coverage
100 landmarks					
RoadNet-TX	0.70	0.62	0.52	0.48	0.39
RoadNet-PA	0.72	0.64	0.50	0.49	0.43
CA-AstroPh	0.47	0.33	0.31	0.33	0.24
Email-EuAll	0.45	0.30	0.27	0.34	0.24
YAGO	0.60	0.55	0.37	0.43	0.33
DBpedia	0.57	0.51	0.45	0.42	0.13
500 landmarks					
RoadNet-TX	0.63	0.52	0.50	0.45	0.27
RoadNet-PA	0.68	0.60	0.48	0.46	0.32
CA-AstroPh	0.47	0.31	0.29	0.28	0.23
Email-EuAll	0.40	0.26	0.25	0.31	0.23
YAGO	0.50	0.51	0.41	0.42	0.22
DBpedia	0.53	0.45	0.43	0.37	0.13

Here, we are using two landmark sizes: 100 and 500 landmarks. By examining Table 2, we can find out that the coverage-based landmark selection are much better than other landmark selections. This is because the selected landmarks based on coverage are more likely to cover the input query vertex pairs.

Varying Number of Selected Landmarks. In this experiment, we study the impact of the number of selected landmarks. We vary the number of selected landmarks from 100 to 500. The results are shown in Fig. 3. We can see that with the increase of landmark number, the approximation errors decrease. In principle, the more landmarks we selected, the lower the approximation errors are. Given two vertices s and t, as the number of landmark increases, the more pairs of vertices can be covered by the landmarks, which means that the probability of $\tilde{d}(s,t) = d(s,t)$ increases. In the case of $\tilde{d}(s,t) \neq d(s,t)$, as the number of landmark increases, the more likely the landmarks are close to a vertex that its shortest path to t passes through s, so the lower the approximation errors are.

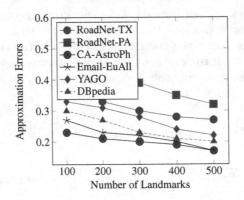

Fig. 3. Evaluating number of selected landmarks

6.4   Efficiency Test

Query Execution Time. In this experiment, we compare the results of our methods, averaged over the vertex pairs, to the average running time of Dijkstra's algorithm. The results are listed in Table 3. Observe that, using the landmark-based estimation algorithm, we can answer distance queries with excellent accuracy within on average 6 milliseconds. Compared with Dijkstra, our landmark-based estimation algorithm can improve the query performance by orders of magnitude.

Table 3. Query execution time (in ms)

	Dijkstra	Landmark-based estimation
RoadNet-TX	147,233	6.6
RoadNet-PA	98,512	6.4
CA-AstroPh	6,552	5.1
Email-EuAll	10,920	5.4
YAGO	35,497	5.9
DBpedia	20,642	6.1

Table 4. Precomputation time (in ms)

	Sampling V' and computing their SPTs	Constructing covering sets	Selecting landmarks	Total time
RoadNet-TX	267,322	7,850,328	4,301,648	12,501,343
RoadNet-PA	185,760	5,904,278	2,886,417	9,002,617
CA-AstroPh	45,474	432,319	273,822	801,420
Email-EuAll	80,783	2,341,788	1,856,974	4,283,178
YAGO	110,382	3,597,748	973,280	4,650,834
DBpedia	107,566	1,857,111	908,153	2,863,457

Precomputation Time. We measure the running time of the stages in landmark selection for all datasets. We divide our landmark selection into three stages: sampling a set V' of vertices and computing their shortest path trees, constructing the approximate covering sets of all vertices, and selecting the optimal set of landmarks. Table 4 provides an overview over the required times. For small datasets like CA-AstroPh, the landmarks can be obtained within 13 min, while the largest dataset (RoadNet-TX) requires about 3.5 h of preprocessing.

7   Related Work

7.1   Landmark-Based Distance Estimation

Landmark-based distance estimation has been widely studied in previous studies [5,11–13,16]. Potamias et al. [11] first evaluates the landmarks-based algorithm for approximate distance estimation in large graphs. The algorithms in [5] extend the above method by storing complete shortest paths to each landmark at each vertex. Qiao et al. first analyze the error bound of the estimated distances [13], and then identify a local landmark close to the specific query vertex [12]. Tretyakov et al. [16] discuss how to maintain the informations of landmarks incrementally under edge insertions and deletions,

7.2   Distributed Graph Processing Systems

As more and more real applications concern large graphs, many distributed graph processing systems with a new computational model for graph processing. The computational model is a vertex-centric model that composed of a sequence of supersteps conforming to the Bulk Synchronous Parallel (BSP) model, while the earliest prototype system is Pregel [10].

After Pregel, many systems are proposed to optimize it. Giraph [15] is the open-source counterpart to Pregel; GraphLab [3] extends the BSP model to an asynchronous data-pulling model; GraphX [4] is Spark's graphs computing API which also implements the Pregel operator; Pregel+ [18] proposes the integration mirroring and message combining as well as a request-respond mechanism; Blogel [17] extends the vertex-centric model to an block-centric model; Quegel [19] extends the vertex-centric model to a query-centric model; TurboGraph++ [6] discusses how to processes large graphs by exploiting external memory without compromising efficiency.

8 Conclusion

In this paper, we describe and evaluate a landmark-based framework to more accurately estimate the lower bounds of distances in large graphs. We propose a new measure called *coverage* to select the landmarks for the lower bound estimation of distances. We propose an algorithm to select the landmarks while guaranteeing the approximation ratio. Moreover, we use the characteristic of the distributed graph processing systems to implement our method. Last, we have conducted extensive experiments to prove the practicability of our algorithms in terms of effectiveness and efficiency.

Acknowledgment. The corresponding author is Peng Peng, and this work was supported by NSFC under grant 61702171, 61772191 and 61472131, Hunan Provincial Natural Science Foundation of China under grant 2018JJ3065, the Fundamental Research Funds for the Central Universities, Science and Technology Key Projects of Hunan Province (Grant No. 2015TP1004, 2016JC2012), and Changsha science and technology project kq1801008.

References

1. Bordeaux, L., Hamadi, Y., Kohli, P.: Tractability: Practical Approaches to Hard Problems. Cambridge University Press, Cambridge (2014)
2. Borgatti, S.P.: Centrality and network flow. Soc. Netw. **27**(1), 55–71 (2005)
3. Gonzalez, J.E., Low, Y., Gu, H., Bickson, D., Guestrin, C.: PowerGraph: distributed graph-parallel computation on natural graphs. In: OSDI, pp. 17–30 (2012)
4. Gonzalez, J.E., Xin, R.S., Dave, A., Crankshaw, D., Franklin, M.J., Stoica, I.: GraphX: graph processing in a distributed dataflow framework. In: OSDI, pp. 599–613 (2014)
5. Gubichev, A., Bedathur, S.J., Seufert, S., Weikum, G.: Fast and accurate estimation of shortest paths in large graphs. In: CIKM, pp. 499–508 (2010)
6. Ko, S., Han, W.: TurboGraph++: a scalable and fast graph analytics system. In: SIGMOD, pp. 395–410 (2018)
7. Lehmann, J., et al.: DBpedia - a large-scale, multilingual knowledge base extracted from Wikipedia. Semant. Web **6**(2), 167–195 (2015)
8. Leskovec, J., Krause, A., Guestrin, C., Faloutsos, C., VanBriesen, J.M., Glance, N.S.: Cost-effective outbreak detection in networks, pp. 420–429 (2007)
9. Leskovec, J., Krevl, A.: SNAP datasets: stanford large network dataset collection, June 2014. http://snap.stanford.edu/data
10. Malewicz, G., et al.: Pregel: a system for large-scale graph processing. In: SIGMOD, pp. 135–146 (2010)

11. Potamias, M., Bonchi, F., Castillo, C., Gionis, A.: Fast shortest path distance estimation in large networks. In: CIKM, pp. 867–876 (2009)
12. Qiao, M., Cheng, H., Chang, L., Yu, J.X.: Approximate shortest distance computing: a query-dependent local landmark scheme. IEEE Trans. Knowl. Data Eng. **26**(1), 55–68 (2014)
13. Qiao, M., Cheng, H., Yu, J.X.: Querying shortest path distance with bounded errors in large graphs. In: Bayard Cushing, J., French, J., Bowers, S. (eds.) SSDBM 2011. LNCS, vol. 6809, pp. 255–273. Springer, Heidelberg (2011). https://doi.org/10.1007/978-3-642-22351-8_16
14. Rebele, T., Suchanek, F., Hoffart, J., Biega, J., Kuzey, E., Weikum, G.: YAGO: a multilingual knowledge base from Wikipedia, wordnet, and geonames. In: Groth, P., et al. (eds.) ISWC 2016. LNCS, vol. 9982, pp. 177–185. Springer, Cham (2016). https://doi.org/10.1007/978-3-319-46547-0_19
15. Sakr, S., Orakzai, F.M., Abdelaziz, I., Khayyat, Z.: Large-Scale Graph Processing Using Apache Giraph. Springer, Cham (2016). https://doi.org/10.1007/978-3-319-47431-1
16. Tretyakov, K., Armas-Cervantes, A., García-Bañuelos, L., Vilo, J., Dumas, M.: Fast fully dynamic landmark-based estimation of shortest path distances in very large graphs. In: CIKM, pp. 1785–1794 (2011)
17. Yan, D., Cheng, J., Lu, Y., Ng, W.: Blogel: a block-centric framework for distributed computation on real-world graphs. PVLDB **7**(14), 1981–1992 (2014)
18. Yan, D., Cheng, J., Lu, Y., Ng, W.: Effective techniques for message reduction and load balancing in distributed graph computation. In: WWW, pp. 1307–1317 (2015)
19. Zhang, Q., Yan, D., Cheng, J.: Quegel: a general-purpose system for querying big graphs. In: SIGMOD, pp. 2189–2192 (2016)
20. Zou, L., Chen, L., Özsu, M.T., Zhao, D.: Answering pattern match queries in large graph databases via graph embedding. VLDB J. **21**(1), 97–120 (2012)

AERIAL: An Efficient Randomized Incentive-Based Influence Maximization Algorithm

Yongyue Sun[1,2], Qingyun Wang[1,2], and Hongyan Li[1,2(✉)]

[1] Key Laboratory of Machine Perception, Ministry of Education,
Peking University, Beijing 100871, China
[2] School of Electronics Engineering and Computer Science,
Peking University, Beijing 100871, China
lihy@cis.pku.edu.cn

Abstract. In social networks, once a user is more willing to influence her neighbors, a larger influence spread will be boosted. Inspired by the economic principle that people respond rationally to incentives, properly incentivizing users will lift their tendencies to influence their neighbors, resulting in a larger influence spread. However, this phenomenon is ignored in traditional IM studies. This paper presents a new diffusion model, IB-IC Model (Incentive-based Independent Cascade Model), to describe this phenomenon, and considers maximizing the influence spread under this model. However, this work faces great challenge under high solution quality and time efficiency. To tackle the problem, we propose AERIAL algorithm with solutions not worse than existing methods in high probability and $O(n^2)$ average running time. We conduct experiments on several real-world networks and demonstrate that our algorithms are effective for solving IM Problem under IB-IC Model.

Keywords: Influence maximization · Incentive · Linear programming

1 Introduction

Nowadays, with the rapid development of Internet, social networks have attracted increasingly interests in various areas, such as viral marketing [1], rumor control [2], politician election [3], recommendation system [4] and so on. Individuals and industries are willing to utilize social networks to enlarge their influence, aiming to make generous profits. One fundamental problem in social network literature is influence maximization (IM) problem. [5] The IM problem asks for k initial users to influence the largest proportion of users in a social network.

In practice, because of the power law distribution of social networks [6], there are a few VIPs having significant impact in online social network. They are so important that they can determine the magnitude of influence eventually to a

© Springer Nature Switzerland AG 2019
J. Shao et al. (Eds.): APWeb-WAIM 2019, LNCS 11641, pp. 240–254, 2019.
https://doi.org/10.1007/978-3-030-26072-9_17

great extend. Thus, if VIPs are more willing to spread their opinions, a larger influence is more likely to boost in a social network. For example, as shown in Fig. 1, a VIP in Sina Weibo wanted to recommend a product eagerly. So she not only sent a microblog, but also placed it on top. As a result, the number of people to be influenced would increase significantly.

Fig. 1. An influential user topped her microblog to boost its influence

Observing the interesting fact described above, and inspired by the economic principle that people respond rationally to incentives, individuals and industries would try to properly incentivize VIPs in social network to boost a larger influence, especially when the budget is limited. For example, as shown in Fig. 2, a drink-maker wants to promote its new product's sales. The company then offers some free samples to a few users in a social network. However, if the company allocates a portion of budget to some influential VIPs, not only offers free samples, but also signs lucrative contract with them, then those VIPs (user A in the figure for example) will be more willing to recommend the new product to their followers, leading to a larger influence eventually, although the budget to favor other users decreases.

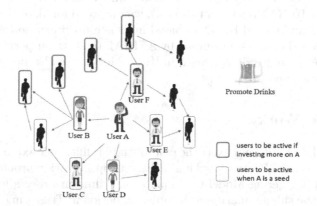

Fig. 2. An example. Suppose user A is a seed node, and it will activate user C, D, E. When we invest more on A, she will make more efforts to persuade B, F, leading to a larger influence.

From the previous example we can conclude that users in social networks respond to the incentives by lifting their tendency to influence their out-neighbors. Intuitively, as users are more willing to spread their opinions, the influence spread will be enlarged. However, existing IM studies ignore the effect of incentives, which may deviate from reality. Motivated by this, we considered solving incentive-based Influence Maximization Problem. However, this work faced two significant challenges as below:

1. We face great challenge when seeking for high solution quality. In general, due to monotonicity and submodularity of the object function, existing IM algorithms always lead to a $(1\text{-}1/e\text{-}\epsilon)$ approximation solution [5,7,8]. However, in this paper, we claim that for incentive-based case, the submodularity of object function no longer holds, meaning that existing algorithms' theoretical guarantees cannot be generalized to incentive-based case. Thus the tradition algorithm framework is invalid for high solution quality, and we are thirsty for new valid approaches.
2. We also face great challenge when seeking for time efficiency. In the traditional case, for a specific user, she is either active or inactive initially. Thus the initial state of the social network is discrete and the solution space is countable. However, in incentive-based case, the initial state of the social network is continuous and the solution space is uncountable. What's worse, for a specific initial state, evaluating the influence spread is #P- hard. Meanwhile, although the traditional algorithm framework is efficient, they fail to obtain high solution quality. These facts block the way for high time efficiency.

To meet these challenges, our contribution can be summarized as follows: We proposed a new diffusion model, Incentive-Based Independent Cascade model (IB-IC Model), to describe the effect of incentives in social networks. Then we proved a series of properties of IB-IC model and claimed the hardness of IM problem under IB-IC Model. On this basis, we proposed an algorithm, AERIAL (An Efficient Randomized Incentive-based influence maximization ALgorithm). We claimed that the solution returned by AERIAL tends to outperform other IM algorithms based on greedy approach in IB-IC Model. We also proved AERIAL incurs quadratic time complexity.

2   Related Work

Domingos and Richardson [1] were the first to study Influence Maximization Problem, while Kempe et al. formulated it as a discrete optimization problem [5]. They also proposed Triggering Model to simulate the diffusion process in social network. Due to the submodularity of the object function in Triggering Model, they then proposed a greedy approximation algorithm. However, there are two significant shortcomings for Kempe's algorithm. The first is huge time cost incurred by Monte-Carlo simulations to estimate the influence of a given seed set. The second is that diffusion models they utilized oversimplified reality. Since then, a variety of subsequent work aimed to overcome the two shortcomings.

For the work aiming to improve the time efficiency, the state-of-art is the reverse influence sampling approach, such as IMM [7] and SSA [8], which returns an (1-1/e-ϵ) approximate solution with linear time complexity. However, these work still only support models based on Trigger Model.

For the work aim to support more realistic scenarios, there are two main categories.

1. Some work extend the semantics of social networks by specifying the meaning of the weight of edges or differring the value of nodes. For example, Babieri et al. [9] differed the weight on a given edge under different topics. Wang et al. [10] redefined the influence spread in a geo-social network and related it with location information. Chen et al. [11] illustrated some nodes in social network may be negative and discussed Negative-aware IM problem.

2. Some work make different assumptions from Triggering Model to support more realistic scenarios. For example, based on the assumption that the attempt to activate neighbors for a newly active user does not limited to the following timestamp, some new models are proposed, such as LAIC Model [12], IC-M model [13], Delayed LT Model, Delayed IC Model [14] and so on. Based on the assumption that latest information are most attractive, some new models are proposed, such as ICND Model [15], TV-IC Model [16], CMDD Model [17], AT Model [18] and so on. Based on the assumption that active users may turn inactive due to neighborhood effect and external effect, Lynn et al. [19] introduced Ising Model in statistical physics to model the social network.

However, to our knowledge, the work supporting users' response to incentives is still in absence. We shall make reasonable assumptions and explore how to deal with incentives to support a more realistic scenarios in this paper.

3   Preliminary

3.1   Incentive-Based Diffusion Model

In Influence Maximization literature, a social network can be modeled as a directed graph $G = \langle V, E \rangle$, where V denotes the set of users and E denotes the set of relationships between users. Each node $v \in V$ has a cost c_v, denoting the expense to persuade the user to adopt the product. As stated above, incentives will improve the initial tendency of users to influence their out-neighbors. To formulate this phenomenon, we introduce a symbol t_v for node v to denote its initial tendency. And we suppose for a specific node, the initial tendency is linear to the expense paid to it. That is, when the expense paid to node v reaches c_v, the initial tendency of v will be fulfilled, namely $t_v = 1$. Also, to set the initial tendency of v to be t_v, the expense paid to v should be $c_v t_v$. Thus we can introduce two vectors, cost vector $\mathbf{c}^T = (c_1, c_2, \cdots, c_n)^T$ and initial tendency vector $\mathbf{t}^T = (t_1, t_2, \cdots, t_n)^T$. Then the total expense in the social network is the inner product of the two vectors, namely $\mathbf{c}^T \mathbf{t}$.

To simulate the impact of incentive on the influence propagation, we are now in a position to construct a more realistic diffusion model. In this paper, we focus on the widely-studied Independent Cascade model. We obey most of the settings in IC model and make a few extensions to properly describe the impact of incentive. The resulting diffusion model is called Incentive-Based Independent Cascade model (IB-IC model).

Obeying the fundamental settings in IC model, the diffusion process in IB-IC model performs as below:

1. At timestamp 0, only the nodes with initial tendency greater than 0 are interpreted as active. These nodes are called seed nodes, and the other nodes are called non-seed nodes.
2. Suppose a node u is activated at timestamp τ. Then at timestamp $\tau + 1$, u will try to activate its inactive out-neighbors.
3. Once a node become active, it won't turn back to inactive state in the following process. The process terminates when no more nodes can be activated.

The differences between IB-IC model and traditional IC model are specified as below:

1. The influence spread in IB-IC model is not the total number of nodes activated in the diffusion process. Otherwise, consider a marketing strategy similar to advertising, each node in the social network is allocated with a tiny initial tendency. Then the number of seed nodes will be the size of the social network, resulting to a ridiculous influence spread. So when calculating the influence spread in IB-IC model, the contribution of a seed node itself should be treated as its initial tendency rather than 1. Then the influence spread is the sum of total initial tendencies of seed nodes and the number of non-seed nodes activated in the process.
2. The success probability for a node to activate its out-neighbor depends on two aspects, both the tendency of source node to activate its out-neighbor and the tendency of sink node to be convinced by its in-neighbor. Namely, the probability u activate v is denoted by $Pr(u \rightarrow v) = IT(u)CT(v)$, where $IT()$ denotes the influence tendency and $CT()$ denotes the convinced tendency. In IB-IC model, if a non-seed node is activated at timestamp τ, then we interpret its tendency to activate its out-neighbors is fulfilled, and the success probability only depends on the out neighbor's convinced tendency.

In other words, in IB-IC model, only seed nodes behave differently from traditional IC model. Their contribution to influence spread should be carefully considered. And the success probabilities for them to activate their out-neighbors differ from the traditional case.

3.2   Influence Maximization Under IB-IC Model

To formulate IM problem under IB-IC model, firstly we should formulate the influence spread as discussed above. For the diffusion process as above, we introduce a indicative function ϕ to indicate a node's contribution to the influence spread. Suppose the tendency vector is $\mathbf{t}$ initially. For a seed node v, $\phi(\mathbf{t}, v) = t_v$. For a non-seed node v, if it is active eventually, $\phi(\mathbf{t}, v) = 1$, otherwise $\phi(\mathbf{t}, v) = 0$. Then the influence spread denoted by $I(\mathbf{t})$ reads $I(\mathbf{t}) = \sum_{v \in V} \phi(\mathbf{t}, v)$. Once given the structure of the graph and the state of either "live" or "block" for every edges, $I(\mathbf{t})$ is a function only depending on initial tendency vector $\mathbf{t}$.

What we care about is maximizing the function $I(\mathbf{t})$ with limited budget. However, note that $I(\mathbf{t})$ is actually a random variable, the quantity we wish to optimize in practice is the expectation of $I(\mathbf{t})$, namely $\mathbf{E}(I(\mathbf{t}))$. Thus we can formulate IM Problem under IB-IC model as follows:

$$\mathbf{t}^* = \arg \min_{\mathbf{t}} \mathbf{E}(I(\mathbf{t}))$$
$$s.t. \quad \mathbf{c}^T \mathbf{t} \leq k \tag{1}$$

Note that the object is to maximize the expectation of influence spread, while the constraint implies the total expense doesn't exceed the budget.

The hardness of IM problem under IB-IC model lies in the hardness of evaluating the influence spread, which can be stated as the following theorem:

Theorem 1. *Under Incentive-Based Independent Cascade model, given a specific tendency vector $\mathbf{t}$, computing the influence spread $I(\mathbf{t})$ is #P-hard.*

Proof. Given a specific initial tendency vector $\mathbf{t}$ to value its influence spread, we can reduce the problem into valuing influence spread under traditional IC model in polynomial time as follows. Let S denote the set containing all seed nodes with initial tendency greater than 0. For $u \in S$, $(u, v) \in E$, we compute the success probability for u to activate v and allocate it to the edge (u, v). Then consider the diffusion process under IC model we have $\mathbf{E}(I(\mathbf{t})) = \mathbf{E}(I(S)) - |S| + \|\mathbf{t}\|_1$. Thus the theorem comes true from the #P-hardness of computing influence spread under IC model.

3.3   Properties of IB-IC Model

In the previous subsection we claimed the hardness of solving IM problem under IB-IC model, which means seeking for exact solution is unrealistic. Settle for second best, we turn to chase for an approximate solution efficiently. Before that, we summarize some important properties of CIT Model to help us design a corresponding algorithm.

Firstly we can show the continuity of the object function.

Theorem 2. *The objection function $\mathbf{E}(I(\mathbf{t}))$ of IM problem under IB-IC Model depends continuously on the initial tendency vector $\mathbf{t}$.*

Proof. Suppose there are two distinct initial tendency vector $\mathbf{t}^{(1)}$ and $\mathbf{t}^{(2)}$. If we have showed that $\lim_{\mathbf{t}^{(2)} \to \mathbf{t}^{(1)}} \mathbf{E}(I(\mathbf{t}^{(2)})) = \mathbf{E}(I(\mathbf{t}^{(1)}))$, then we can claim $\mathbf{E}(I(\mathbf{t}))$ depends continuously on $\mathbf{t}$.

Without losing generality, suppose $\mathbf{t}^{(1)}$ and $\mathbf{t}^{(2)}$ are only different on the v-th component. Namely, only the tendency of node v differs. Also, suppose $t_v^{(2)} > t_v^{(1)}$ and study the process of $t_v^{(2)}$ approaching $t_v^{(1)}$.

The key of the proof is to split the node v. Namely, we construct another node v' as a copy of v as follows. Define $V' = V \cup \{v'\}$. Then define an edge set $E^{(new)}$. If there exists u satisfying $(u, v) \in E$, then $(u, v') \in E^{(new)}$. If there exists u satisfying $(v, u) \in E$, then $(v', u) \in E^{(new)}$. Define $E' = E \cup E^{(new)}$. The result graph is denoted by $G' = \langle V', E' \rangle$. Set $t_{v'}^{(1)'} = \delta \to 0^+$, $t_v^{(2)'} = t_v^{(1)'} = t_v^{(1)}$, $t_{v'}^{(2)'} = t_v^{(2)} - t_v^{(1)}$, then v, v' are both seed nodes in G'.

Now consider an arbitrary node u satisfying $(u, v) \in E$. Due to the nonlinearity, when $t_{v'}^{(2)'} \neq 0$ $(t_v^{(2)} \neq t_v^{(1)})$, the active probability of node u in G and G' shouldn't be same. However, when $t_{v'}^{(2)'} \to 0$, the active probability of node u in G' will converge to that in G. Thus if $t_{v'}^{(2)'} \to 0^+$, $\mathbf{E}(I_G(\mathbf{t}^{(2)})) = \mathbf{E}(I_{G'}(\mathbf{t}^{(2)'}))$. Meanwhile, it is obvious that $\mathbf{E}(I_G(\mathbf{t}^{(1)})) = \mathbf{E}(I_{G'}(\mathbf{t}^{(1)'}))$.

Note that in G', when $t_{v'}^{(2)'} \to 0^+$, the contribution of v' to the influence spread will converge to 0, leading to $\lim_{t^{(2)'} \to t^{(1)'}} \mathbf{E}((I_{G'}(\mathbf{t}^{(2)'}))) = \mathbf{E}(I_{G'}(\mathbf{t}^{(1)'}))$. Thus $\lim_{\mathbf{t}^{(2)} \to \mathbf{t}^{(1)}} \mathbf{E}(I_G(\mathbf{t}^{(2)})) = \mathbf{E}(I_G(\mathbf{t}^{(1)}))$, which shows that the objection function depends continuously on the initial tendency vector $\mathbf{t}$.

Then we can show the monotonicity of the object function based on the previous theorem.

Theorem 3. *The objection function $\mathbf{E}(I(\mathbf{t}))$ of IM problem under IB-IC Model is monotonous.*

Proof. The key is to prove the partial derivatives of $\mathbf{E}(I(\mathbf{t}))$ at $\mathbf{t}$ are positive. We can construct an initial tendency vector in its neighborhood and apply the analysis similar to the previous theorem. The influence spread can be viewed as the sum of the contribution from both the initial tendency vector $\mathbf{t}$ and its differential $d\mathbf{t}$, where $d\mathbf{t}$'s contribution is positive. So the partial derivatives of $\mathbf{E}(I(\mathbf{t}))$ at $\mathbf{t}$ are positive, leading to the theorem.

Based on the properties above, we claim that the object function is not submodular.

Remark 1. The objection function $\mathbf{E}(I(\mathbf{t}))$ of IM problem under IB-IC model is not submodular.

Recall that the submodularity property can only describe functions with discrete independent variables, while the objection function of IM problem under IB-IC Model depends continuously on its continuous independent variable, then the remark automatically comes true. This fact reflects that traditional greedy algorithms can not be directly applied to IM problem under IB-IC Model.

4 Proposed Solution

In this section we proposed AERIAL (An Efficient Randomized Incentive-based influence maximization ALgorithm) to solve the problem efficiently. Firstly we generalized the definition of reverse reachable sets (RR sets) [20] which are applied broadly in IM literature. Then we reexpressed the problem by sampling random RR sets and rewrote the problem into a linear programming form. Finally we solved the problem and showed the theoretical guarantee.

4.1 Generalization of Reverse Reachable Sets

Firstly we revisit the definition of RR sets in traditional IM literature.

Definition 1 *(Reverse Reachable Set)* [20]. *A reverse reachable set for node v in graph G is generated by reversing all the edges in the graph and determining each edge's connectivity, then finding out all nodes which can reach v.*

In traditional IM literature, the significance of RR Sets lies in that the probability that a diffusion process from seed set S can activate node v equals the probability that S overlaps an RR set for v [20]. Based on this fact one can develop an unbiased estimation of $\mathbf{E}(I(S))$ by randomly sampling RR sets and finding out the proportion of RR sets which are covered by the seed set S. The estimation can be formulated as follows:

Suppose x_i denotes whether seed set S covers an RR set R_i or not. If S overlaps R_i, then $x_i = 1$, otherwise $x_i = 0$. Then $\mathbf{E}(I(S)) = \frac{n}{\theta} \sum_{i=1}^{\theta} x_i$, where θ is the number of RR sets sampled.

Now we begin to generalize a similar conclusion for RR set under IB-IC model. Consider a seed node u with its initial tendency t_u is greater than zero, and it is contained in an sample of RR set for node v, namely R_v. Now we only care about the possibility that node u's impact finally transmits to node v in this sampled graph, namely $Pr(u \rightsquigarrow v)$. Note that the possibility node u activates its out-neighbor w is $Pr(u \rightarrow w) = IT(u)CT(w)$. So different from the traditional case that $Pr(u \rightsquigarrow v) = 1$, although u is in R_v, whether u's impact can finally transmit to v depends on $IT(u)$, namely t_u, and $Pr(u \rightsquigarrow v) = t_u$ in this sampled graph.

Let set S contain all seed nodes. We say, S is the set induced by initial tendency vector $\mathbf{t}$. Now we get ready to show the following theorem.

Theorem 4. *Let x_i denotes the probability that set S induced by initial tendency vector $\mathbf{t}$ covers R_i, then $x_i = 1 - \prod_{u \in R_i \cap S}(1 - t_u)$*

Proof. If node u is in the RR set R_v and the set S induced by $\mathbf{t}$, then the probability that u's impact cannot transmit to v finally is $1 - Pr(u \rightsquigarrow v) = 1 - t_u$. Now consider the probability that node v is not active finally. Because the events that different nodes' impact cannot transmit to v are independent, so the probability of v is not active finally is equal to the product of these events' probabilities, namely $\prod_{u \in R_v \cap S}(1 - t_u)$. So $x_v = 1 - \prod_{u \in R_v \cap S}(1 - t_u)$.

Based on Theorem 4, one can efficiently estimate $\mathbf{E}(I(\mathbf{t}))$ when given the initial tendency vector $\mathbf{t}$. However, due to the non-linear form of x_i, this theorem indicates that optimizing $\mathbf{E}(I(\mathbf{t}))$ as a function of $\mathbf{t}$ is no longer as simple as the traditional case, which blocks the way for us to solve IM problem efficiently. We shall pave the way in the next subsection.

4.2 Optimal Tendency Vector Solution

Note that the expression of x_i is very similar to the configuration integral in evaluating the partition function of real gases [21]. So we apply a approximation similar to Mayer's cluster expansion. We rewrite $1 - x_i$ in a Taylor series form and preserve the first two terms, assuming higher order terms are not large:

$$
\begin{aligned}
1 - x_i &= \prod_{u \in R_i \cap S} (1 - t_u) \\
&= 1 - \sum_{u \in R_i \cap S} t_u + \sum_{u,w \in R_i \cap S \& u \neq w} t_u t_w - \cdots \\
&\approx 1 - \sum_{u \in R_i \cap S} t_u
\end{aligned}
\tag{2}
$$

Equation 2 implies $x_i \approx \sum_{u \in R_i \cap S} t_u$, which is only valid when $\sum_{u \in R_i \cap S} t_u$ is not large. If $\sum_{u \in R_i \cap S} t_u$ is large enough, then the following up terms in expansion in Eq. 2 can not be omitted. Intuitively, when $\sum_{u \in R_i \cap S} t_u$ is large, x_i will be near 1. For example, if there exists a node u in S and R_i simultaneously with $t_u = 1$, then the case degenerate to the traditional case, where x_i absolutely equals to 1.

Based on the observation above, we introduce an threshold $\gamma < 1$ to monitor whether $\sum_{u \in R_i \cap S} t_u$ is large enough. If $\sum_{u \in R_i \cap S} t_u$ exceeds γ, to minimize the bias as much as possible we set $x_i = \frac{1+\gamma}{2}$. In this paper we set $\gamma = 1$, and showed under this setting we tend to obtain a high quality solution in the following subsection.

Under this setting, we use $\frac{n}{\theta} \sum_{i=1}^{\theta} x_i$ to estimate the influence caused by an initial tendency vector t, where θ is the sampling size. Thus our goal is to maximize $\sum_{i=1}^{\theta} x_i$. Note that x_i equals to either $\sum_{u \in R_i \cap S} t_u$ or $\frac{1+\gamma}{2}$ (which equals to 1 when $\gamma = 1$). Also the total cost should not exceed the given budget k, namely $c^T t \leq k$, so we can express our goal and constraint as below:

$$
\begin{aligned}
Maximize \quad & \sum_{i=1}^{\theta} x_i \\
s.t. \quad & \sum_{j \in R_i} t_j \geq x_i \quad i = 1, 2, \ldots, \theta \\
& \sum_{j=1}^{\theta} c_j t_j \leq k \\
& 0 \leq x_i \leq 1 \quad i = 1, 2, \ldots, \theta \\
& 0 \leq t_j \leq 1 \quad j = 1, 2, \ldots, n
\end{aligned}
\tag{3}
$$

It can be seen clearly that this expression is actually a linear programming problem, we can solve it in polynomial time using interior point method. In this paper we apply simplex method [25] which performs excellently in reality with $O(n^2)$ average time complexity. To invoke simplex algorithm we should rewrite the form in (3) into standard form [22]. Note that there are $\theta + n$ variables, namely $x_1, x_2, \cdots, x_\theta, t_1, t_2, \cdots, t_n$. For convenience, we use the symbol $x_{i+\theta}$ to denote t_i, then we can use a vector $\mathbf{x}$ to denote all these variables. Then the standard form reads:

$$Maximize \quad \mathbf{h}^T \mathbf{x}$$
$$s.t. \quad A\mathbf{x} \le \mathbf{b} \tag{4}$$
$$\mathbf{x} \ge 0$$

Note that $\mathbf{h}^T = (1, 1, \cdots, 1, 0, 0, \cdots, 0)^T$ is a $(\theta + n)$-dim vector with its first θ elements equal to 1 and others equal to 0. $\mathbf{b}$ is a $(n + 2\theta + 1)$-dim vector with its first θ elements equal to 0, and last $(n + \theta)$ elements equal to 1, while the rest element in the middle is k. The $(n + 2\theta + 1) \times (\theta + n)$matrix A is constructed as follows:

1. For the first θ rows, $A_{ii} = 1$, and if node j is in the i-th RR set, $A_{i,j+\theta} = -1$.
2. For the $(\theta + 1)$-th row, $A_{\theta+1,\theta+j} = c_j$, namely the activate cost for node j.
3. For the last $(n + \theta)$-th row, it is a unit matrix, namely $A_{\theta+1+i,i} = 1$.
4. The other elements not mentioned above are set to be zero.

One can easily verify the form in (4) is the standard form for (3) if $A, \mathbf{b}, \mathbf{h}$ are constructed as above.

4.3   Theoretical Guarantee

In this subsection we will show guarantees on both solution quality and time efficiency. Firstly we claim that if $\frac{n}{\theta} \sum_{i=1}^{\theta} x_i$ is a good indicator of the influence spread, then the solution returned by AERIAL tend to be state-of-the-art.

Remark 2. If $\frac{n}{\theta} \sum_{i=1}^{\theta} x_i$ is a good indicator of the influence spread, then the solution returned by AERIAL will be no worse than that of other RR Set based traditional IM algorithms when the RR sets sampled are the same.

The core of the remark lies in expressing the goal of RR Set based traditional IM algorithm into integer programming form.

Note that in the node selection phase in these algorithms, such as IMM, the aim is to select k seeds that can cover the maximum reverse reachable sets. We can rewrite the aim as follows:

$$Maximize \quad \sum_{i=1}^{\theta} x_i$$
$$s.t. \quad \sum_{j \in R_i} t_j \ge x_i \quad i = 1, 2, \ldots, \theta$$
$$\sum_{j=1}^{\theta} c_j t_j \le k \tag{5}$$
$$x_i = 0 \quad or \quad 1 \quad i = 1, 2, \ldots, \theta$$
$$t_j = 0 \quad or \quad 1 \quad j = 1, 2, \ldots, n$$

Thus these algorithms solving IM problem under traditional models actually solve an integer programming problem as (5), while the linear programming in (3) is its relaxation form.

Because the solution to linear programming is always no worse than the corresponding integer programming [23], the remark comes true automatically.

Secondly we consider about the time complexity of AERIAL. To bound the time complexity we need the lower bound of the sampling size θ to ensure the solution returned by AERIAL is near the actual value of $\mathbf{E}(I(t))$. Note that x_i is no longer an independent Poisson test, which blocks our efforts to apply corresponding bound estimation techniques such as Chernoff Bound [24] or the martingale approach [7] similar to IMM. However, inspired by Remark 2, we can firstly solve IM problem under traditional IC Model by methods like IMM or SSA. Then we use the RR sets generated just now to run the linear programming in (3). Finally, we can obtain the solution. We express these steps above as the AERIAL algorithm below:

Algorithm 1. AERIAL

Input: Social Network G, budget k
Output: Initial tendency vector t^*

1 Choose proper value for parameters ϵ, l
2 $\mathscr{R} =$ IMM.Sampling (G, k, ϵ, l)
3 Find out A, b, h in the standard form
4 $\mathbf{x} =$ simplex(A, b, h)
5 for $i \leftarrow 1$ to n do
6 $\quad\lfloor\; t_i^* \leftarrow x_{i+|\mathscr{R}|}$

7 return t^*

The following theorem shows the time complexity of AERIAL:

Theorem 5. *Algorithm AERIAL runs in $O(n^2)$ time in average.*

Proof. The first phase of AERIAL applies IMM algorithm to generate considerable RR sets, with its time complexity is $O((k + l)n \log n \cdot \epsilon^{-2}/OPT)$, which is near linear to the input. The second phase of AERIAL applies simplex method to solve the linear programming problem, with its average time complexity is $O((n + \theta)^2)$, which can be expressed as $O(n^2)$ because $n > \theta$. Combine the two results, we finally get the result that the time complexity of AERIAL is $O(n^2)$.

5 Experiments

We conduct experiments on two real-world social networks to test the performance of our algorithms. All the algorithms are implemented with C/C++ and run in Windows 10 environment. The CPU is Intel Core i7-6700HQ, the main frequency is 2.60 GHz, and the memory is 16G.

5.1   Data Description

The datasets we invoked in this paper are presented in the Table 1 below.

Table 1. Description of datasets

Name	n	m	Type	Corresponding site
NetHEPT	15233	62774	Directed	arxiv.org
Brightkite	58228	214078	Directed	brightkite.com

NetHEPT. NetHEPT is an academic collaboration network from the "High Energy Physics Theory" section of arXiv from 1991 to 2003, where nodes represent the authors and each edge represents one paper co-authored by two nodes. There are 15,233 nodes and 62774 directed edges in the NetHEPT dataset.

Brightkite. Brightkite was once a location-based social networking service provider where users shared their locations by checking-in. The friendship network was collected using their public API, and consists of 58,228 nodes and 214,078 directed edges.

5.2   Baselines

In this experiment, we implement AERIAL in a heuristics way to speed up the algorithm. Literature [26] shows that the top-k influential nodes are mainly selected from those with large degrees. For example, on the NetHEPT and Flickr datasets, the top-50 influential nodes are selected from the top 1.79% and 0.84% nodes in degree. Inspired by this fact, we consider only incentivizing top 5% nodes in degree. Thus we can shrink the numbers of variables greatly and speed up the optimization process while the solution quality suffers little loss.

Besides AERIAL, we test four other algorithms in the experiment. We denote them as IMM-full, IMM-part, SSA-full, SSA-part.

IMM-full. We apply IMM directly to IM Problem under IB-IC Model. That is, if a node is chosen to be a seed node, then its initial influence tendency is fulfilled, namely equals to 1. IMM-full is corresponding to the situation that industries invest mainly on influential users and make their tendency to influence others be as large as possible.

IMM-part. We apply IMM to IM Problem under IB-IC Model in a manner described as following. If a node is chosen to be a seed node, then its initial influence tendency is set to be 0.5. IMM-part is corresponding to the situation that industries make users' tendency to influence others be at a normal level.

SSA-full. Similarly we apply SSA directly to IM Problem under CIT Model and set the tendency of seed nodes to be 1.

SSA-part. We apply SSA to IM Problem under CIT Model and set the tendency of seed nodes to be 0.5.

5.3   Experiment Results

In our experiments, for every solution returned by testing algorithms, we apply 10000 times Monte-Carlo simulations to obtain its expected influence spread. We also set k to be 5, 10, 15, 20, 25 respectively.

Firstly we compare influence spread results of the five algorithms. The results on NetHEPT are presented in Table 2, while the results on Brightkite are presented in Table 3.

Table 2. The performance of influence spread on NetHEPT

Algorithm	$k = 5$	$k = 10$	$k = 15$	$k = 20$	$k = 25$
AERIAL	**41.8**	**71.1**	**104.8**	**113.7**	**152.1**
IMM-full	22.1	42.4	57.3	73.9	80.3
IMM-part	21.9	38.4	49.4	65.5	78.0
SSA-full	21.9	42.2	57.4	73.3	80.0
SSA-part	21.6	38.0	49.2	65.3	77.8

Table 3. The performance of influence spread on Brightkite

Algorithm	$k = 5$	$k = 10$	$k = 15$	$k = 20$	$k = 25$
AERIAL	**27.1**	**75.3**	**140.5**	**198.0**	**384.9**
IMM-full	10.5	35.1	78.3	121.3	303.7
IMM-part	17.1	52.1	79.6	108.1	163.0
SSA-full	10.2	39.2	77.2	125.4	298.8
SSA-part	19.1	49.5	78.4	110.2	170.0

As we can see, AERIAL outperforms other algorithms in both NetEHPT and Brightkite dataset. This result reveals the fact that differently investing on different users has significant effect. Also we observe that the difference between setting selected users' tendency to be 1 and to be 0.5 is not significant, which implies in viral marketing scenarios, to boost more influence, industries should select some seeds with normal tendency as well as some seeds with high tendency.

We also compare running time of these algorithms. The results are presented in Tables 4 and 5.

As we can see, AERIAL's running time consumed is higher than the other four baselines, but is still acceptable. SSA outperforms IMM as claimed in [8] because its number of RR sets sampled is much smaller. The reason why AERIAL consumes more time is because it utilizes IMM to generate sufficient numbers of RR sets, and the linear programming phase involves large sparse matrix A, which slows the algorithm. How to reduce time consumption in this phase will be included in future work.

Table 4. The running time on NetHEPT, in seconds

Algorithm	$k = 5$	$k = 10$	$k = 15$	$k = 20$	$k = 25$
AERIAL	93.7	81.2	77.2	95.4	97.0
IMM-full	15.8	25.7	26.0	31.9	34.7
IMM-part	16.4	22.5	27.7	35.5	37.7
SSA-full	**7.8**	**10.2**	**14.5**	**20.7**	**22.9**
SSA-part	8.5	12.1	15.0	22.6	25.6

Table 5. The running time on Brightkite, in seconds

Algorithm	$k = 5$	$k = 10$	$k = 15$	$k = 20$	$k = 25$
AERIAL	231.2	290.0	319.7	346.6	488.9
IMM-full	58.3	58.0	75.9	84.7	135.7
IMM-part	55.6	60.9	68.1	95.9	118.7
SSA-full	**20.1**	**24.0**	**27.6**	**30.3**	**35.9**
SSA-part	22.7	24.5	28.8	34.6	37.4

6 Conclusion

In this paper, we put forward a new diffusion model closer to reality, IB-IC Model, which describes how people response to incentives. We formulated IM Problem under IB-IC Model and claimed its hardness. We also showed the object function is continuous, monotonous but not submodular. To tackle the problem we generalized the sense of RR sets and developed an randomized approach to estimate the influence of a given initial tendency vector. Then we rewrote the problem into a linear programming form and invoked simplex algorithm. We also showed the guarantee on the solution quality and time efficiency. We conducted experiments on real datasets to verify our conclusions.

References

1. Domingos, P., Richardson, M.: Mining the network value of customers. In: KDD, pp. 57–66 (2001)
2. Budak, C., Agrawal, D., El Abbadi, A.: Limiting the spread of misinformation in social networks. In: WWW, pp. 665–674 (2011)
3. Mwanza, S., Suleman, H.: Measuring network structure metrics as a proxy for socio-political activity in social media. In: ICDM Workshops, pp. 878–883 (2017)
4. Li, Y., Bao, Z., Li, G., Tan, K.: Real time personalized search on social networks. In: ICDE, pp. 639–650 (2015)
5. Kempe, D., Kleinberg, J.M., Tardos, E.: Maximizing the spread of influence through a social network. In: KDD, pp. 137–146 (2003)
6. Barabasi, A., Albert, R.: Emergence of scaling in random networks. Science **286**, 509–512 (1999)

7. Tang, Y., Shi, Y., Xiao, X.: Influence maximization in near-linear time: a martin-gale approach. In: SIGMOD, pp. 1539–1544 (2015)
8. Nguyen, H., Thai, M., Dinh, T.: Stop-and-stare: optimal sampling algorithms for viral marketing in billion-scale networks. In: SIGMOD, pp. 695–710 (2016)
9. Barbieri, N., Bonchi, F., Manco, G.: Topic-aware social influence propagation models. In: ICDM, pp. 81–90 (2012)
10. Wang, X., Zhang, Y., Zhang, W., Lin, X.: Efficient distance-aware influence maximization in geo-social networks. IEEE Trans. Knowl. Data Eng. **29**(3), 599–612 (2017)
11. Chen, Y., Li, H., Qu, Q.: Negative-aware influence maximization on social networks. In: AAAI, pp. 8063–8064 (2018)
12. Liu, B., Cong, G., Xu, D., Zeng, Y.: Time constrained influence maximization in social networks. In: ICDM, pp. 439–448 (2012)
13. Chen, W., Lu, W., Zhang, N.: Time-critical influence maximization in social networks with time-delayed diffusion process. In: AAAI (2012)
14. Mohammadi, A., Saraee, M., Mirzaei, A.: Time-sensitive influence maximization in social networks. J. Inf. Sci. **41**(6), 765–778 (2015)
15. Feng, S., Chen, X., Cong, G., Zeng, Y., Chee, Y., Xiang, Y.: Influence maximization with novelty decay in social networks. In: AAAI, pp. 37–43 (2014)
16. Ohsaka, N., Yamaguchi, Y., Kakimura, N., Kawarabayashi, K.: Maximizing time-decaying influence in social networks. In: Frasconi, P., Landwehr, N., Manco, G., Vreeken, J. (eds.) ECML PKDD 2016. LNCS (LNAI), vol. 9851, pp. 132–147. Springer, Cham (2016). https://doi.org/10.1007/978-3-319-46128-1_9
17. Zhang, Z., Wu, H., Yue, K., Li, J., Liu, W.: Influence maximization for cascade model with diffusion decay in social networks. In: Che, W., et al. (eds.) ICYCSEE 2016. CCIS, vol. 623, pp. 418–427. Springer, Singapore (2016). https://doi.org/10.1007/978-981-10-2053-7_37
18. Sun, H., Cheng, R., Xiao, X., Yan, J., Zheng, Y., Qian, Y.: Maximizing social influence for the awareness threshold model. In: Pei, J., Manolopoulos, Y., Sadiq, S., Li, J. (eds.) DASFAA 2018. LNCS, vol. 10827, pp. 491–510. Springer, Cham (2018). https://doi.org/10.1007/978-3-319-91452-7_32
19. Lynn, C., Lee, D.: Maximizing activity in ising networks via the TAP approximation. In: AAAI, pp. 679–686 (2018)
20. Borgs, C., Brautbar, M., Chayes, J., Lucier, B.: Maximizing social influence in nearly optimal time. In: SODA, pp. 946–957 (2014)
21. Greiner, W., Neise, L., Stoocker, H.: Thermodynamics and Statistical Mechanics, 1st edn. Springer, New York (1995). https://doi.org/10.1007/978-1-4612-0827-3
22. Cormen, T., Leiserson, C., Rivest, R., Stein, C.: Introduction to Algorithms, 3rd edn. MIT Press, Cambridge (2001)
23. Vazirani, V.: Approximation Algorithms, 1st edn. Springer, New York (2002). https://doi.org/10.1007/978-3-662-04565-7
24. Motwani, R., Raghavan, P.: Randomized Algorithms, 1st edn. Cambridge University Press, Cambridge (1995)
25. Dantzig, G.: Linear Programming and Extensions, 1st edn. Princeton University Press, Princeton (1963)
26. Liu, X., Liao, X., Li, S., Lin, B.: Towards efficient influence maximization for evolving social networks. In: Li, F., Shim, K., Zheng, K., Liu, G. (eds.) APWeb 2016. LNCS, vol. 9931, pp. 232–244. Springer, Cham (2016). https://doi.org/10.1007/978-3-319-45814-4_19

Cider: Highly Efficient Processing of Densely Overlapped Communities in Big Graphs

Yadi Chen[1,2], Wen Bai[1,2], Runyuan Chen[1,2], Di Wu[1,2(✉)], Guoqiao Ye[1,2], and Zhichuan Huang[1,2]

[1] School of Data and Computer Science,
Sun Yat-sen University, Guangzhou 510006, China
{chenyd28,baiw6,chruny,yegq3}@mail2.sysu.edu.cn,
wudi27@mail.sysu.edu.cn, zhihu1@umbc.edu
[2] Guangdong Key Laboratory of Big Data Analysis and Processing,
Guangzhou 510006, China

Abstract. As one of the most fundamental operations in graph analytics, community detection is to find groups of vertices that are more densely connected internally than with the rest of the graph. However, the detection of densely overlapped communities in big graphs is extremely challenging due to high time complexity. In this paper, we propose an effective and efficient graph algorithm called *Cider* to detect densely overlapped communities in big graphs. The intuition behind our algorithm is to exploit inherent properties of densely overlapped communities, and expand the community by minimizing its conductance. To make *Cider* more efficient, we extend the algorithm to expand the community more quickly by merging vertices in batches. We explicitly derive the time complexity of our algorithm and conclude that it can be implemented in near-linear time. Besides, we also implement a parallelized version of *Cider* to further improve its performance. Experimental results on real datasets show that our algorithms outperform existing approaches in terms of *Flake Out Degree Fraction* (FODF) and F_1 *Score*.

Keywords: Community detection · Densely overlapped communities · Large scale graphs

1 Introduction

Graph is a kind of abstract data structure that can naturally represent structural relationships among objects, such as friendships in social networks, links between webpages, etc. These graphs often contain hundreds of millions of vertices and edges. In terms of big graph processing, *community detection* is one of the most fundamental operations, which aims to discover a special subgraph structure containing a set of vertices with more internal connections than with the rest of

J. Shao et al. (Eds.): APWeb-WAIM 2019, LNCS 11641, pp. 255–263, 2019.
https://doi.org/10.1007/978-3-030-26072-9_18

the network. It can be applied in many scenarios such as friend recommendation in social networks and credit evaluation based on social circle.

To detect communities in a graph, *modularity* [2] and *conductance* [3] are two widely adopted evaluation metrics to quantify how "well-knit" the subgraph is. Unfortunately, the problem of searching for the optimal values of modularity (or conductance) is NP-hard. In some cases, communities are heavily overlapped and even nested, which further increases the difficulty of community detection. For example, it is possible that individuals may belong to multiple different communities simultaneously according to their hobbies and relationships (e.g., football club community, high-school classmate community). The problem of overlapped communities has received significant attention in recent years. However, traditional algorithms partition a graph such that each vertex belongs to exactly one community while the vertex may participate in multiple communities.

In this paper, we propose an effective and efficient community detection algorithms called *Cider* to detect densely overlapped communities in big graphs. To tackle the challenges incurred by large graph size and overlapped communities, we first exploit the inherent properties of overlapped communities and redefine the conductance metric. To decrease the computational cost, we propose an improved version, called *bCider*, which can expand the community more quickly by merging vertices in batches. Additionally, we also provide a parallel version of *Cider* to improve the efficiency of our algorithms. Experiments on real graph datasets show that *Cider* can achieve significant improvements in terms of detection accuracy and execution efficiency.

The main contributions of our paper can be summarized as follows:

(1) We identify a few important properties of densely overlapped communities, and propose efficient graph algorithms called *Cider* and its improved version *bCider* to detect densely overlapped communities in big graphs based on the conductance metric.
(2) We derive the time complexity of our proposed algorithms and also implement a parallel version of *Cider* to further improve its performance. The worst time complexity of our algorithm increases linearly with the size of the graph.
(3) With the known ground-truth, experimental results on real graph datasets show that our algorithms outperform existing approaches in terms of *Flake Out Degree Fraction* (FODF) and F_1 *Score*.

The rest of this paper is organized as follows. Section 2 reviews related work on community detection. Section 3 describes our theoretical findings and proposes our community detection algorithms for big graphs. Section 4 evaluates performance of our proposed algorithms. Finally, Sect. 5 concludes the paper and discusses the future work.

2   Related Work

During past decades, researchers had proposed many approaches for community detection. In this section, we review different definitions of communities, non-overlapped and overlapped community detection algorithms.

The problem of community detection has been extensively studied in the past years. The definition of community is not consistent in different papers. Yang and Leskovec [6] assessed 13 commonly used community definitions based on the ground-truth communities, and concluded that *conductance* and Triad participation ratio are the best criteria to identify ground-truth communities.

Existing algorithms can be generally divided into two categories: one is for non-overlapped communities, the other is for overlapped communities. The graphs themselves can also be divided into two types: sparse graphs and dense graphs. Compared with previous work, our algorithm is able to detect densely overlapped (or even nested) communities without knowing the number of communities. In addition, instead of removing edges or vertices once a community has been detected, our algorithm will not change the graph structure during the execution process.

3   Design of Cider Algorithm

We first introduce our rationale of community detection in big graphs, then provide the details of *Cider* and its improved version *bCider*. Finally, we implement a parallelized version of *Cider* to further improve the performance.

3.1   Rationale

We first introduce the notations used later. In this paper, we consider the problem of community detection in the unweighted and undirected big graph. Denoted the graph as $G = (V, E)$, where V and E are vertex set and edge set of G respectively. For two arbitrary vertices $u, v \in V$, we use w_{uv} to denote the weight of edge $(u, v) \in E$. As we consider unweighted graphs, $w_{uv} = 1$.

For a given community, let S be the set of vertices in the community, and $N(S) = \{v \in V \setminus S : \exists u \in S \wedge \{u, v\} \in E\}$ be the neighbor set of vertices in S. We continue to define the internal weight of edges in S, which is denoted by $m_S = \sum_{u,v \in S} w_{uv}$, and the external weight of cut edges of S, which is denoted by $c_S = \sum_{u \in S, v \in N(S)} w_{uv}$. As the graph is unweighted, m_S is simply the number of edges in S and c_S is the number of cut edges of S. Given S and a vertex $u \in N(S)$, denote $O(u, S) = \sum_{v \notin S} w_{uv}$ as the outer weights of u and $I(u, S) = \sum_{v \in S} w_{uv}$ as the inner weights of u.

Next, we proceed to introduce the definition of community. For a simple graph $G(V, E)$, a vertex set $S \subset V$ is a community if it satisfies $\phi(S \cup \{u\}) > \phi(S)$. Formally, the conductance $\phi(S)$ of a community S is defined as: $\phi(S) = \frac{c_S}{2 \cdot m_S + c_S}$.

When determining whether a given vertex set is a community or not, the traditional definition is not well suitable for the practical implementation. To assess the distance between S and a vertex u in $N(S)$, we define the degree of alienation between u and S as $DoA(u, S) = \frac{O(u,S) - I(u,S)}{I(u,S) + O(u,S)}$.

Definition 1. *A vertex set $S \subseteq V$ is a community such that $DoA(u, S) > \phi(S)$ for all $u \in N(S)$.*

Given a community S, it can increase by expanding its neighbor vertices in $N(S)$. If a vertex $u \in N(S)$ satisfies $DoA(u, S) \leq \phi(S)$, we should insert u into S. $\forall u \in N(S)$, if $DoA(u, S) > \phi(S)$ holds, then S stop expanding. To reduce the computational cost, if the minimal value of $DoA(u, S)$, denoted by $DoA(u, S)^*$, satisfies $\phi(S) < DoA(u, S)^*$ for every $u \in N(S)$, we stop expanding S.

3.2 Basic Algorithms

Algorithm 1. Cider: Community Detection Algorithm in Big Graphs
---

Input: $G(V, E)$: an undirected graph,
   δ : a threshold value.
Output: $\mathcal{C}$: a set of communities.

1 $\mathcal{C} \leftarrow \emptyset$;
2 **for** $u \in V$ **do**
3      $k \leftarrow 0$; $S \leftarrow \{u\}$; $\phi(S) \leftarrow 1$;
4   Let Q be an empty queue; Q.enQueue(w) for $w \in N(S)$;
5   **while** $!Q.empty()$ **do**
6          $v \leftarrow Q$.deQueue();
7   **if** $DoA(v, S) > \phi(S)$ **then**
8              $\mathcal{C} \leftarrow \mathcal{C} \cup S$; $k = k + 1$;
9   break the loop if $k \geq \delta$;
10         $S = S \cup \{v\}$, update $\phi(S)$;
11   Q.enQueue(w) if $w \notin N(S)$ for $w \in N(\{v\})$;
12   sort $w \in Q$ by $DoA(w, S)$ in an increasing order;

13 **return** $\mathcal{C}$;

The basic idea of *Cider* (see Algorithm 1) is to choose a vertex as the seed of a community and then expand it until its conductance cannot be further reduced. Based on Definition 1, the size of S is increased if $DoA(u, S) \leq \phi(S)$ for an arbitrary $u \in N(S)$. To overcome the problem of local optima, we introduce a mistake threshold δ, which allows that incorrect expansions can occur at most δ times. To reduce the computational cost, we compare $DoA(u, S)^*$ with the $\phi(S)$ for an arbitrary S and $u \in N(S)$. If the conditions in Definition 1 hold, the algorithm will stop expanding.

Let n be the size of $|V|$ and k_i be the size of the i-th community, the main time-consuming part in the expansion process of the ith seed community is to maintain the queue, which needs $O(k_i \cdot log k_i)$. Since every vertex can be chosen as the community seed, the worst time complexity of our algorithm is

$O(\sum_i^n k_i \cdot \log k_i) \approx O(n \cdot max(k_i \cdot \log k_i))$. If the size of a community is much smaller than the number of vertices in the graph, the execution time of our algorithm increases almost linearly with the graph size.

As the operation of $DoA(u, S)$ needs to sort all $u \in N(S)$, the time complexity is at least $O(|N(S)|log|N(S)|)$, which is too time-consuming. In this case, we should decrease the number of vertices to be sorted in $N(S)$. Therefore, only vertices that can decrease $\phi(S)$ will be taken into account. Anyway, there still exist some vertices whose influence cannot be inferred. For these vertices, Definition 1 is used for processing.

Theorem 1. $\forall u \in N(S)$, $\phi(S) > \phi(S \cup \{u\})$ if $O(u, S) \le I(u, S)$.

Proof. Let $S' = S \cup \{u\}$, we have

$$\phi(S') = \frac{c_{S'}}{2 \cdot m_{S'} + c_{S'}} = \frac{c_S + O(u, S) - I(u, S)}{2 \cdot m_S + c_S + O(u, S) + I(u, S)}$$

$$< \frac{c_S}{2 \cdot m_S + c_S} + \frac{O(u, S) - I(u, S)}{2 \cdot m_S + c_S} <= \frac{c_S}{2 \cdot m_S + c_S} = \phi(S)$$

Procedure. Block-based update community(BUC)

Input: S: a community waited to expand, Q_1: the vertices in $N(S)$.
Output: S: a updated community, Q_2: the vertices in $N(S)$ need to be sorted.

1 Let Q_2 be an empty queue;
2 **while** $!Q_1.isEmpty()$ **do**
3 | $v \leftarrow Q_1.deQueue()$;
4 | continue the loop if $v \in S$ if $O(v, S) \le I(v, S)$ **then**
5 | | $S \leftarrow S \cup \{v\}$;
6 | | $Q_1.enQueue(w)$ if $w \notin N(S)$ for $w \in N(\{v\})$);
7 | **else**
8 | | $Q_2.enQueue(v)$;

By Theorem 1, we can observe that a vertex $u \in N(S)$ can be directly inserted into S if $O(u, S) \le I(u, S)$. Therefore, it is unnecessary to put these vertices into Q and thus we decrease the sorting cost. According to the above result, we propose an extended algorithm called $bCider$ (as illustrated in Algorithm 2). The major difference from $Cider$ is that $bCider$ updates S by a set of vertices (also called a block), instead of a single vertex. The update process is shown in Procedure BUC. Compared with $Cider$, $bCider$ further decreases sorting cost.

Since $Cider$ will not change the graph structure during the execution, we implement a parallelized version of $Cider$ algorithm called $pCider$ to further boost the performance of community detection in big graphs. In the $pCider$ algorithm, several seed communities are selected in the beginning, and the expansion of these communities can be conducted simultaneously so that the execution

---

Algorithm 2. *bCider*: Block-based Community Detection Algorithm

---

Input: $G(V, E)$: a simple undirected graph, δ: the mistake threshold
Output: $\mathcal{C}$: a set of communities

1   $\mathcal{C} \leftarrow \emptyset$;
2   **for** $u \in V$ **do**
3     $k \leftarrow 0$; $S \leftarrow \{u\}$; $\phi(S) \leftarrow 1$;
4   let Q_1 be an empty queue; Q_1.enQueue(w) for $w \in N(S)$;
5     $S, Q_2 \leftarrow \text{BUC}(S, Q_1)$;
6   **while** $!Q_2.empty()$ **do**
7       $v \leftarrow Q_2$.deQueue();
8   **if** $DoA(v, S) > \phi(S)$ **then**
9         $\mathcal{C} \leftarrow \mathcal{C} \cup S$; $\delta = \delta + 1$;
10   break the loop if $k \geq \delta$;
11   Q_1.enQueue(w) if $w \notin N(S)$ for $w \in N(\{v\})$;
12       $S, Q_2 \leftarrow \text{BUC}(S, Q_1)$;
13   sort $w \in Q_2$ by $DoA(w, S)$ in an increasing order;

14 **return** $\mathcal{C}$;

---

efficiency is highly improved. Indeed, *pCider* can be regarded as a parallel framework, so it can also compatible with *bCider*. It is not complicated to implement *pCider* so we do not provide the details.

4   Performance Evaluation

In this section, we evaluate the performance of our proposed algorithms using the community benchmark datasets provided by SNAP[1], which includes the ground-truth communities. To the best of our knowledge, these are the largest datasets with densely overlapped communities, as reported in Table 1.

Table 1. Statistics of three networks and execution time

Name	Nodes	Edges	Communities	pCider	pbCider
Amazon	334, 863	925, 872	75, 149	55 s	**15 s**
DBLP	317, 080	1, 049, 866	13, 477	70 s	**27 s**
Youtube	1, 134, 890	2, 987, 624	8, 385	9, 391 s	**1, 715 s**

4.1   Dataset Description

Amazon: This graph depicts a network of Amazon products collected by crawling Amazon website. Each vertex represents a product and each edge represents

---

[1] http://snap.stanford.edu.

two products frequently co-purchased. The ground-truth community is defined according to each product category by Amazon.

DBLP: This graph represents the collaboration relationship among researchers in computer science, which is provided by the DBLP computer science bibliography. The vertices are authors and the ground-truth community is defined according to the publication venue.

Youtube: This graph represents a social network of video sharing in Youtube, which is the largest connected component in Youtube social network. The ground-truth community is defined according to the groups created by users.

4.2 Baseline and Metrics

In our paper, we focus on the detection of densely overlapped communities. We compare with a few state-of-the-art algorithms proposed recently, such as I^2C [4], Bigclam [5] and Coda [7] on all datasets mentioned in Sect. 4.1.

In our paper, we evaluate different algorithms with various metrics. Firstly, we compare the *Flake Out Degree Fraction*(FODF) [1] among different methods, which is used to evaluate the community tightness according to the topological structure without knowing the number of ground-truth communities. Secondly, we compare F_1 *score*, which is used to evaluate the community detection accuracy with the ground-truth.

4.3 Experiment Results

The computer used in the experiments has: 12 CPUs Intel Xeon E5-2430 at 2.5 GHz, 64 GB of RAM and CentOS Linux with kernel 2.6.32-696.10.2.el6.x86_64. In our paper, all the baselines are executed using their default parameters. Since I^2C and Coda run more than 7 day on YouTube dataset, we cannot obtain their Flake ODF and F_1 *Score*.

Due to different definitions in baseline algorithms, we cannot compare their runtime directly. Instead, we just evaluate the execution efficiency of our algorithms. Table 1 shows us the execution time of *pCider* and *pbCider*(parallel *bCider*). We can find that *pbCider* can significantly decrease the execution time than *pCider*, especially for the large dataset.

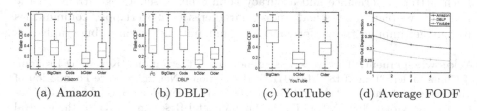

(a) Amazon   (b) DBLP   (c) YouTube   (d) Average FODF

Fig. 1. Flake ODF of different algorithms

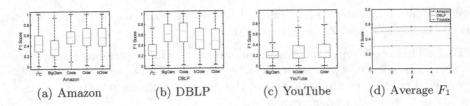

(a) Amazon   (b) DBLP   (c) YouTube   (d) Average F_1

Fig. 2. F_1 *Score* of different algorithms

Figure 1 depicts the FODF distribution of communities based on different algorithms. We can observe that the communities detected by *Cider* and *bCider* can keep the value of FODF in a relatively low level on all datasets which means the communities are relatively dense. Figure 2 depicts the F_1 *Score* distribution of the communities based on different algorithms and shows that *Cider* and *bCider* have higher F_1 *Score* which means more accurate community detection. Though in Fig. 2(b) our algorithms have average performance, they perform better than I^2C which is also based on the conductance. In our proposed algorithms, there is only one tunable parameter δ. Figure 1(d) depicts the average FODF and Fig. 2(d) depicts the average F_1 *Score* under various δ. From two figures, we observe that a higher δ leads to the lower average FODF and the higher average F_1 *Score*. According to these results, we can conclude that a higher δ leads to the more accurate detection of communities.

From the above results, we observe that our algorithms can improve the quality of community detection. And *bCider* obtains better performances than *Cider* in both time cost and accuracy metrics.

5   Conclusion and Future Work

In this work, we proposed an efficient graph algorithm called *Cider* for the detection of densely overlapped communities in big graphs. By exploiting the inherent properties of densely overlapped communities, the time complexity of our proposed algorithm increases almost linearly with the size of the graph. We also implemented a parallelized version of *Cider* called *pCider* to further improve the detection efficiency. We conducted a series of experiments on real datasets with ground-truth information. The results show that our algorithm can obtain much better performance and accuracy than other baseline algorithms. In the next step, we plan to investigate more characteristics of overlapped communities and further improve the performance of community detection in big graphs.

Acknowledgement. This work was supported by the National Key R&D Program of China under Grant 2018YFB0204100, the National Natural Science Foundation of China under Grant 61572538 and Grant 61802451, Guangdong Special Support Program under Grant 2017TX04X148, the Fundamental Research Funds for the Central Universities under Grant 18LGPY61.

References

1. Flake, G.W., Lawrence, S., Giles, C.L.: Efficient identification of web communities. In: Proceedings of the ACM SIGKDD International Conference on Knowledge Discovery and Data Mining, pp. 150–160 (2000)
2. Girvan, M., Newman, M.E.J.: Community structure in social and biological networks. Proc. Natl. Acad. Sci. **99**(12), 7821–7826 (2002)
3. Kannan, R., Vempala, S., Vetta, A.: On clusterings: good, bad and spectral. J. ACM **51**(3), 497–515 (2004)
4. Lu, Z., Wen, Y., Cao, G.: Community detection in weighted networks: algorithms and applications. In: IEEE International Conference on Pervasive Computing and Communications, pp. 179–184 (2013)
5. Yang, J., Leskovec, J.: Overlapping community detection at scale: a nonnegative matrix factorization approach. In: Proceedings of ACM International Conference on Web Search and Data Mining, pp. 587–596 (2013)
6. Yang, J., Leskovec, J.: Defining and evaluating network communities based on ground-truth. Knowl. Inf. Syst. **42**(1), 181–213 (2015)
7. Yang, J., McAuley, J., Leskovec, J.: Detecting cohesive and 2-mode communities indirected and undirected networks. In: Proceedings of the ACM International Conference on Web Search and Data Mining, pp. 323–332 (2014)

Time Optimal Profit Maximization in a Social Network

Yong Liu$^{(\boxtimes)}$, Zitu Liu, Shengnan Xie, and Xiaokun Li

HeiLongJiang University, Harbin, China
{liuyong001,lixiaokun}@hlju.edu.cn

Abstract. Influence maximization aims to seek k nodes from a social network such that the expected number of activated nodes by these k nodes is maximized. However, influence maximization is different from profit maximization for a real marketing campaign. We observe that when promotion time increases, the number of activated nodes tends to be stable eventually. In this paper, we first use real action log to propose a novel influence power allocation model with time span called **IPA-T**, and then present time optimal profit maximization problem called **TOPM** based on IPA-T. To address this problem, we propose an effective approximation algorithm called **Profit-Max**. Experimental results on real datasets verify the effectiveness and efficiency of Profit-Max.

Keywords: Influence maximization · Profit maximization ·
Time optimal

1 Introduction

In recent years, many large-scale social networks, such as Facebook, Twitter, Friendster, Microblog are becoming more and more popular. New ideas and information can be spread to a large number of users in a short time through social networks.

Based on the ideas above, Kempe et al. [1] proposed an influence maximization problem. Given a social network, we aim to select k influential users called a set of seeds that maximizes the activated number of users under a given propagation model. The literatures [2–6] have studied the influence maximization problem extensively and proposed a variety of solutions. However, the existing works usually assume that the influence spread is equal to the product profit and do not consider the cost of product promotion. In fact, as the promotional time increases, the promotional cost will increase continuously but the influence spread of the seed set will increase more and more slowly. The following experiments verify the view. We use the algorithm in literature [7] to select top-10 influential nodes on a real dataset, and then select 7 time spans with different length and observe the actual influence spread of the seed set w.r.t different time span. The experimental result is shown in Table 1, where the unit length of T is a day, and *spread* represents the number of activated nodes.

© Springer Nature Switzerland AG 2019
J. Shao et al. (Eds.): APWeb-WAIM 2019, LNCS 11641, pp. 264–272, 2019.
https://doi.org/10.1007/978-3-030-26072-9_19

As shown in Table 1, the number of activated users increases rapidly at the initial stage of propagation. But as the propagation time increases, the growth rate of activated users becomes slow, and the number of activated users will become stable eventually. Moreover, we also discover an interesting phenomenon. When the length of propagation time span is different, the most influential seed set is also different. Accordingly, in the area of vital marketing, it is important to determine the optimal length of propagation time span and the corresponding seed set.

In this paper, We use real action log to propose a new propagation model with time span which is called **IPA-T** propagation model, and then propose a new research problem, called time optimal profit maximization problem **TOPM**. We prove that the problem is NP-hard. In order to solve **TOPM** problem, we design an effective approximation algorithm **Profit-Max**. Experimental results on several real data sets show that **Profit-Max** can solve **TOPM** problem effectively and efficiently.

Table 1. The *spread* vs different time span T in Digg77461 dataset

T (days)	10	20	30	40	50	60	70
Spread (nodes)	165	240	282	295	301	305	308

2 Related Work

Kempe et al. [1] first proposed the influence maximization problem and presented a greedy algorithm with a $(1 - 1/e)$ approximation ratio. Chen et al. [2] proposed a heuristic algorithm Degree discount, which is two orders of magnitude faster than the greedy algorithm. Chen et al. proposed a heuristic algorithm called PMIA under IC model [3] and a heuristic algorithm called LDAG under LT model [4] to estimate the influence spread. Chen et al. [5] and Liu et al. [6] extended IC propagation model and proposed time constraint influence maximization problem. Because the affected people may not buy the product, Wei et al. [8] proposed profit maximization problem under LT-V model. Bhagat et al. [9] proposed product adoption maximization problem based on LT-C model and divided the state of nodes into three categories to further simulate the propagation of information. Li et al. [10] proposed location-aware influence maximization problem, which seek a seed set that maximizes the expected influence spread in certain region. Li et al. [11] proposed the targeted influence maximization problem, which seek a seed set that maximizes the expected influence over users who are relevant to a given advertisement.

The above methods assume that the influence probabilities on edges are obtained in advance, which cannot simulate the spread of influence in the real world accurately. To this end, Saito et al. [12] and Goyal et al. [13] used the observed data to predict the influence probabilities on edges. Goyal et al. [7]

proposed a data-based algorithm to solve the influence maximization problem. This algorithm directly leveraged available action logs to learn the process of influence propagation, proposed a CD model and developed an approximation algorithm. However, the CD model did not consider the time factor, and thus obtain the same seed set for different time span.

3 Problem

3.1 IPA-T Model

A social network is modeled as a directed graph $G = (V, E)$, where V represents the users and E represents the relationships between users. The promotional time is denoted as a time span T. Each record in action log L is a tuple (u, a, t), which indicates user u performs action a at time t. A represents the universe of actions. For $\forall u \in V, \forall a \in A$, $t(u, a)$ denotes the time that user u performs action a. We assume that each user performs an action at most once. If user u performs action a before user v, then $t(u, a) < t(v, a)$. If user u does not perform action a, its execution time is infinite, i.e. $t(u, a) = +\infty$. If $0 \le t(v, a) - t(u, a) \le T$ and $(u, v) \in E$, then we believe that action a is propagated from node u to node v in the time span T. In this case, we call that v is influenced by u, v needs to allocate an influence power to u and ancestors of u as well.

The node set that influence node v directly in time span T w.r.t action a is denoted as $N_{in}(v, a, T) = \{u | (u, v) \in E, 0 \le t(v, a) - t(u, a) \le T\}$. Thus, if node v performs action a, then v assigns direct influence power to $\forall u \in N_{in}(v, a, T)$, which is denoted as $\alpha_{v,u}(a, T)$, and satisfies that $\sum_{u \in N_{in}(v,a,T)} \alpha_{v,u}(a, T) \le 1$. For the sake of convenience, we set $\alpha_{v,u}(a, T) = 1/N_{in}(v, a, T)$, i.e., we give the equal value to each neighbor of node v. At the same time, node v assigns indirect influence power to each neighbor w of node u iteratively. If the time interval between node v and ancestor w is no more than time span T, i.e., $0 \le t(v, a) - t(w, a) \le T$, then v assigns indirect influence power to w. For action a, the total influence power that node v assigns to its ancestor w in time span T is $\Gamma_{v,w}(a, T) = \sum_{u \in N_{in}(v,a,T)} \alpha_{v,u}(a, T) * \Gamma_{u,w}(a, T)$, where the base of the recursion is $\Gamma_{u,u}(a, T) = 1$. Similarly, for action a, the total influence power that node v assigns to a node set S in time span T is $\Gamma_{v,S}(a, T) = \sum_{u \in N_{in}(v,a,T)} \alpha_{v,u}(a, T) * \Gamma_{u,S}(a, T)$. If $u \in S$, then $\Gamma_{u,S}(a, T) = 1$. Finally, we aggregate all actions and compute the total influence.

In time span T, the total influence power that node v assigns to node w is defined as follows. $\Gamma_{v,w}(T) = \sum_{a \in P_v} \Gamma_{v,w}(a, T)/|P_v|$, where $|P_v|$ represents the number of actions that node v has performed. Similarly, for any set $S \in V$, in time span T, the total influence power that node v assigns to set S is defined as follows. $\Gamma_{v,S}(T) = \sum_{a \in P_v} \Gamma_{v,S}(a, T)/|P_v|$. The propagation model with the time constraints above is denoted as IPA-T model. Thus, in time span T, the influence spread of set S is $\delta_{IPA-T}(S, T) = \sum_{v \in V} \Gamma_{v,S}(T)$ under IPA-T model.

3.2   Problem Definition

Time Optimal Profit Maximization (TOPM): Given a directed graph $G = (V, E)$, an action log L, the unit *price* of promotion product, the unit time *cost* of promotion, and a positive integer k, we aim to discover an optimal time span T, and a seed set S with size k. By targeting S, the net profit of products $price * \delta_{IPA-T}(S, T) - cost * T$ is maximized under IPA-T model, where $price * \delta_{IPA-T}(S, T)$ stands for the total sales in time span T, $cost * T$ represents the total cost of promotion in time span T.

TOPM is NP-hard because influence maximization is the special case of TOPM.

4   Profit-Maximization Algorithm

We firstly divide the entire time into M time units, and each time unit may be a day, a week or a month. For each increasing length of time, we use algorithm **Initialization** to initialize the influence allocation list $Inflink$, then call algorithm **Greedy-CELF** to find current seed set S_temp w.r.t. current time span and compute the $profit$ of S_temp. The pseudo-code of **Profit-Max** algorithm is shown in Algorithm 1. We can show that the approximation ratio of **Profit-Max** is $1 - 1/e - \beta/e$, where β is the ratio between cost and profit in optimal solution.

Algorithm 1. Profit-Max

Input: social network $G = (V, E)$, an action log L, the *price* of promotion product, the unit time *cost* of promotion, and a positive integer k
Output: the optimal time span T and corresponding seed set S

1: $max\_profit = 0$;
2: divide the entire time into M time units and let t be the length of unit time;
3: **for** $i = 1$ to M **do**
4:     $Inflink =$ **Initialization**$(G, L, i * t)$;
5:     $S\_temp =$ **Greedy-CELF**$(Inflink, k)$;
6:     $profit = price * \delta_{IPA-T}(Stemp, i * t) - cost * i * t$;
7:     **if** $profit > maxp\_profit$ **then**
8:         $maxp\_profit = profit$; $S = S\_temp$; $T = i * t$;
9: Return S and T;

4.1   Scan Action Log and Initialize $Inflink$

Algorithm **Initialization** scans the action log L sequentially. For each action $a \in A$, we initializes the influence allocation list $Inflink[a][u]$ for each node $u \in V$. $Inflink[a][u]$ stores the influence power of node u assigned to other nodes. *Active* stores the activated nodes in the current action. $P[u]$ denotes the number of actions that user u has performed, $father[u]$ stores the parents of

Algorithm 2. Initialization

Input: $G = (V, E)$, the action log L, time span T
Output:the influence power allocation list $Inflink[a][u]$ for each $a \in A, u \in V$

  1: **for** each $u \in V$ **do**
  2:    create in-degree adjacency list $in\_edge[u]$;
  3: **for** each action $a \in A$ **do**
  4:    $active \leftarrow \emptyset$;
  5:    **for** each (u, a, t) **do**
  6:      $father[u] \leftarrow \emptyset; P[u] + +$;
  7:      **for** each $w \in (in\_edge[u] \bigcap active)$ **do**
  8:        **if** $(0 < time(u) - time(w) < T)$ **then**
  9:          $father[u] \leftarrow father[u] \bigcup \{w\}$;
10:      **for** each $x \in father[u]$ **do**
11:        compute the direct influence power $\alpha_{u,x}$,add x to $Inflink[a][u]$;
12:        **for** each $v \in Inflink[a][x]$ **do**
13:          **if** $(0 < time(u) - time(v) < T)$ **then**
14:            add v to $Inflink[a][u]$;
15:    $active \leftarrow active \bigcup \{u\}$;
16: Return $Inflink$;

---

user u, and $time[u]$ denotes the time that user u performs the current action. We create the in-degree adjacency list $inedge[u]$ for each $u \in V$. We scans each tuple (u, a, t) sequentially. For each node $w \in in\_edge[u]$, if $w \in active$ and the execution time of u minus the execution time of w is no more than length T, then we add w to the influence allocation list $Inflink[a][u]$ and allocate influence power to w. Then we allocates influence powers to the parents of w iteratively. The pseudo-code of **Initialization** is shown in Algorithm 2.

4.2 CELF Optimization to Find Influential Nodes

When the time span T is fixed, IPA-T model has the property of submodular, thus we can use the greedy algorithm with CELF optimization to select seed set. The algorithm **Greedy-CELF** uses algorithm **Margin-Compute** to obtain margin gain $u.margin$ for each $u \in V$, then updates the number of iteration of u and inserts u into the priority queue Q. Q is kept in descending order according to the value of $margin$. **Greedy-CELF** greedily choose a node u with the maximal margin gain each time and put u into seed set S according to the following procedure, until seed set S contains k nodes.

In each iteration, if the iteration number of u is the current iteration number, then we add u to seed set S, call algorithm **Influence-Update** to update the influence power allocation list of u and the influence power given to set S by u; else use algorithm **Margin-Compute** to calculate $u.margin$, renew the iteration number of u and insert u into the priority queue Q. The pseudo-code of **Greedy-CELF** is shown in Algorithm 3.

Algorithm 3. Greedy-CELF

Input: $Inflink[a][u]$ for each $a \in A, u \in V$, a positive number k.
Output: seed set S

1: $SA \leftarrow \emptyset, S \leftarrow \emptyset, Q \leftarrow \emptyset$;
2: **for** each $u \in V$ **do**
3:    $u.margin = $**Margin-Compute**$(Inflink, SA, u)$;
4:    update the iteration number of node u and inserts u into the priority queue Q
5: **while** ($|S| < k$) **do**
6:    $u = pop(Q)$;
7:    **if** (the iteration number of node u is the current iteration) **then**
8:       $S \leftarrow S \cup \{u\}$;
9:       **Influence-Update**$(Inflink, SA, u)$;
10:   **else**
11:      $u.margin = $**Margin-Compute**$(Inflink, SA, u)$;
12:      update the iteration number of node u and inserts u into the priority queue Q
13: Return S;

4.3 Aggregating Actions and Updating the Influence Power Allocation List

The algorithm **Margin-Compute** calculates the margin gain when node x is added to the current seed set. we firstly aggregate the influence power of node x on action a and store this influence power into $gain[a]$. $SA[a][x]$ represents the influence power which is assigned to the current seed set S by node x on action a. Thus the margin gain of adding x into seed set S for action a is $gain[a] * (1 - SA[a][x])$. The pseudo-code of **Margin-Compute** is shown in Algorithm 4.

After node x is added into seed set S, the algorithm **Influence-Update** updates $Inflink$ and the influence power of assigned to seed set S by other nodes. The pseudo-code of **Influence-Update** is shown in Algorithm 5.

Algorithm 4. Margin-Compute

Input: $Inflink[a][u]$ for each $a \in A$ and $u \in V$, SA, node x.
Output: $x.margin$

1: $margin = 0$;
2: **for** each action a **do**
3:    $gain[a] = 0$;
4:    **for** each $u \in V$ **do**
5:       **if** ($x \in Inflink[a][u]$) **then**
6:          $gain[a] = gain[a] + \Gamma_{u,x}(a)/P[u]$;
7:       $margin = margin + gain[a] * (1 - SA[a][x])$;
8: Return $margin$;

Algorithm 5. Influence-Update

Input: $Inflink[a][u]$ for each $a \in A$ and $u \in V$, SA, node x.
Output: $Inflink$ and SA

1: **for** each action a **do**
2:    **for** each $u \in V$ **do**
3:       **if** $(x \in Inflink[a][u])$ **then**
4:          **for** each $v \in Inflink[a][x]$ **do**
5:             delete the influence power assigned to v through node x;
6:       update $SA[a][u]$;

5   Experimental Results and Analysis

5.1   Experiment Setup

We use two network datasets to evaluate the performance of our proposed algorithm. The first dataset is the music social network site Last.fm. We extract 2100 nodes, 25435 edges and the corresponding action log. This log contains 1000 actions and 21,646 records. A record (u, a, t) means that user u comments singer a at time t. We denote the first dataset as Last21646. The second dataset is the social news review site Digg. We extract 5000 nodes, 395,513 edges and the corresponding action log. The action log contains 500 actions and 77,461 records. The second dataset is denoted as Digg77461.

5.2   Comparing the Effects of Seed Size k and Time Span T

In this subsection, we study the effects of different seed size k and different time span T on *spread* and *profit*. As shown in Figs. 1 and 3, when k is fixed, the influence *spread* increases with the growth of time span T until it becomes stable. As shown in Figs. 2 and 4, when k is fixed, with the increase of time span T, *profit* begins to increase, after reaching a certain point, *profit* begins to decline. Thus, in order to maximize profit, it is important to choose the best promotional time span T.

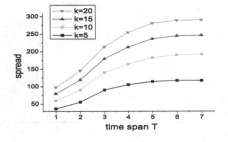
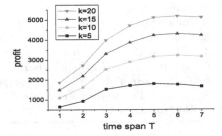

Fig. 1. *spread* vs T in Last21646 dataset

Fig. 2. *profit* vs T in Last21646 dataset, $price = 20$, $cost = 100$

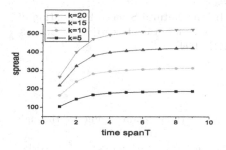

Fig. 3. *spread* vs T in Digg77461 dataset

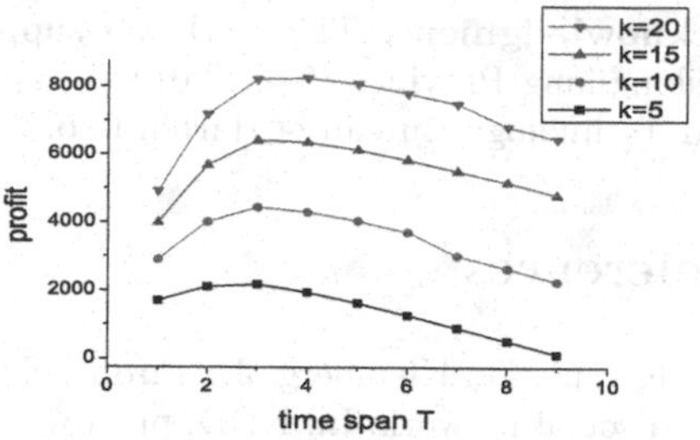

Fig. 4. *profit* vs T in Digg77461 dataset, *price* = 20, *cost* = 400

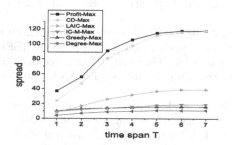

Fig. 5. *spread* vs T in Last21646 dataset with $k = 10$

5.3   Comparing Profit-Max with Different Algorithms

In this subsection, we compare Profit-Max with the following algorithms in terms of the influence *spread* and running time. (1) Degree-Max: select seed set based on node's degree; (2) Greedy-Max: use Monte Carlo simulations to estimate the influence spread; (3) CD-Max: use greedy strategy to select seed set under CD model [13]; (4) LAIC-Max: select seed set under LAIC model [5] according to the length of time; (5) IC-M-Max: select seed set under IC-M model [6]. We use the six algorithms described above to choose seed set with $k = 10$ and compute the influence *spread* under different length of time span on Last21646 dataset. Experimental results are shown in Fig. 5. Profit-Max can achieve the maximal spread under different time span T.

6   Conclusion

In this paper, we first propose an novel influence power allocation model with time span IPA-T. According to the model, we propose the time optimal profit maximization problem called TOPM. We prove that TOPM is NP-hard. We design an effective approximation algorithm Profit-Max for TOPM. Extensive experiments demonstrate the effectiveness and efficiency of Profit-Max.

Acknowledgment. This work was supported by the Natural Science Foundation of Heilongjiang Province (No. F201430), and the Innovation Talents Project of Science and Technology Bureau of Harbin (No. 2017RAQXJ094).

References

1. Kempe, D., Kleinberg, J., Tardos, E.: Maximizing the spread of influence through a social network. In: KDD, pp. 137–146. ACM, New York (2003)
2. Chen, W., Wang, Y., Yang, S.: Efficient influence maximization in social networks. In: KDD, pp. 199–208. ACM, New York (2009)
3. Chen, W., Wang. C., Wang, Y.: Scalable influence maximization for prevalent viral marketing in large-scale social networks. In: KDD, pp. 1029–1038. ACM, New York (2010)
4. Chen, W., Yuan, Y., Zhang, L.: Scalable influence maximization in social networks under the linear threshold model. In: ICDM, pp. 88–97. IEEE, Piscataway (2010)
5. Chen, W., Lu, W., Zhang, N.: Time-critical influence maximization in social networks with time-delayed diffusion process. In: AAAI, pp. 592–598. AAAI, Menlo Park (2012)
6. Liu, B., Cong, G., Xu, D., et al.: Time constrained influence maximization in social networks. In: ICDM, pp. 439–448. IEEE, Piscataway (2012)
7. Goyal, A., Bonchi, F., Lakshmanan, L.: A data-based approach to social influence maximization. Proc. VLDB Endow. 5(1), 73–84 (2012)
8. Wei, L., Lakshmanan, L.: Profit Maximization over Social Networks. In: ICDM, pp. 479–488. IEEE, Piscataway (2012)
9. Bhagat, S., Goyal, A, Lakshmanan, L.: Maximizing product adoption in social networks. In: WSDM, pp. 603–612. ACM, New York (2012)
10. Li, G., Chen, S., Feng, J., et al.: Efficient location-aware influence maximization. In: SIGMOD, pp. 87–98. ACM, New York (2014)
11. Li, Y., Zhang, D., Tan, K.: Realtime targeted influence maximization for online advertisements. Proc. VLDB Endow. 8(10), 1070–1081 (2015)
12. Saito, K, Kimura, M., Nakano, R.: Prediction of information diffusion probabilities for independent cascade model. In: KES, pp. 67–75 (2008)
13. Goyal, A., Bonchi, F., Lakshmanan, L.: Learning influence probabilities in social networks. In: WSDM, pp. 241–250. ACM, New York (2010)

Iterative Hypergraph Computation Based on Hyperedge-Connected Graphs

Kaiqiang Yu, Yu Gu$^{(\boxtimes)}$, Shuo Yao, Zhen Song, and Ge Yu

Northeastern University, Shenyang, China
yukaiqiang1994@163.com, guyu@mail.neu.edu.cn, 1194109464@qq.com,
songzhen_neu@163.com, yuge@mail.neu.edu.cn

Abstract. A hypergraph allows a hyperedge to connect arbitrary number of vertices, which can be used to capture the complex and high-order relationships. By analyzing the iterative processing on bipartite graphs, a method of converting the original hypergraph into a hyperedge-connected graph and corresponding iterative processing method are proposed. Then, the iterative processing solution based on hyperedge-connected graphs is combined with Push-based and Pull-based message acquisition mechanisms. On top of the distributed graph processing system HybridGraph, a hypergraph iterative processing framework HyraphD is implemented. Finally, extensive experiments are conducted on several real-world datasets and hypergraph learning algorithms. Experimental results confirm the efficiency and the scalability of HyraphD.

Keywords: Hypergraph · Iterative processing · Graph computing · Distributed computing

1 Introduction

Big graph analysis has been becoming increasingly significant in a broad spectrum of applications, such as knowledge graphs [6]. As a generalization of the graph model, hypergraphs are recognized as a powerful tool to capture complex and high-order relationships among data objects. Specifically, different from a common graph, each hyperedge in a hypergraph can contain any number of vertices. Also, extensive hypergraph analysis and learning algorithms have been proposed and applied to various tasks such as web link analysis [1], recommender systems [8], wireless communication [3], medical diagnosis [7] and data mining [5].

Existing hypergraph iterative processing methods are majorly developed based on the star-expansion (SE) model, which converts the hypergraph into a bipartite graph. Based on the effective design of the message exchange, the vertex-centric distributed graph computation framework such as Giraph [2] and GraphX [4] can be adapted to process the converted bipartite graphs by alternately evaluating hyperedges and vertexes. Although the SE-based method can guarantee the result correctness, it still faces serious efficiency issues, because the scale of the bipartite graph may be considerably inflated by several orders of

© Springer Nature Switzerland AG 2019
J. Shao et al. (Eds.): APWeb-WAIM 2019, LNCS 11641, pp. 273–280, 2019.
https://doi.org/10.1007/978-3-030-26072-9_20

magnitude and a large amount of communications are incurred among vertexes and hyperedges.

Specifically, we observe that in many hypergraph applications, the number of the hyperedges and vertexes are skewed. Intuitively, if we can directly pass messages among hyperedges and correspondingly update the values of the hyperedges, the communication and computation costs will be significantly reduced. In this paper, we explore a new I/O efficient processing scheme by constructing the hyperedge-connected graph. The major contributions are concluded as follows.

- We propose to directly exchange data among hyperedges to solve the hypergraph skewed distribution problem for the first time.
- We explore the effective construction method of the hyperedge-connected graph, and the auxiliary message generation schemes to guarantee the correctness for hypergraph analysis tasks.
- We further discuss the I/O efficient iterative computation optimization mechanism based on HybridGraph [9] to alleviate the size inflation problem.

The rest of this paper is organized as follows. Section 2 presents the computation model based on hyperedge-connected graphs, and analyzes the vertex information generation strategies for PageRank. Section 3 discusses the I/O efficient implementation techniques and Sect. 4 reports experimental results. Finally, the paper is concluded in Sect. 5.

2   Hypergraph Iterative Computation Optimization Based on Hyperedge-Connected Graphs

2.1   Construction of Hyperedge-Connected Graph

In star-expansion, we update vertices and hyperedges in two adjacent super-steps, respectively. The delivery path of messages is hyperedge-vertex-hyperedge. The role of vertices in this process can be seen as a pivot, so it is possible to send messages straightforwardly from one hyperedge to another, thus reducing the number of iterations and advancing the convergence of algorithms. While a hyperedge is generating messages, the information about the relevant vertices is necessary to ensure the correctness of computation. We use $\Gamma(v)/\Gamma(h)$ to represent a set of hyperedges/vertices incident to vertex v/hyperedge h, and $N(v)/N(h)$ to represent the neighbor vertex/hyperedge set of v/h.

In order to enable direct communication among hyperedges, the original hypergraph requires to be transformed. In this paper, we call this converted graph as **Hyperedge-Connected Graph**. Specifically, if two hyperedges have vertices in common, then an edge is constructed between them. For each hyperedge h_i, its degree and the vertices assigned to it need to be stored. In addition, the h_i needs to store the information about common vertices between h_i and its neighbor hyperedge h_j. The vertex information is used for iteratively updating the hyperedges and vertices in the last iteration. The original hypergraph

$G = (V, H)$ is transformed into a hyperedge-connected graph $G^* = (V^*, H^*)$, where $V^* = H, H^* = \{\{h_i, h_j\} : u \in h_i, u \in h_j, h_i \in H, h_j \in H\}$, as shown in Fig. 1.

In this paper, we call the iterative processing method over hyperedge-connected graphs as HC. Assuming that there are a total of N supersteps, the computation process of HC is concluded as follows. In the first superstep, each hyperedge uses the saved vertex information to compute the initial hyperedge value, then generates and sends a message to the neighbor hyperedges. From the second to the $(N-1)$th supersteps, each hyperedge utilizes the received messages to update its own value, and then generates corresponding messages for the neighbor hyperedges. The content of these messages includes hyperedge value, hyperedge information and information about common vertices between the current hyperedges and their neighbors. In the Nth superstep, the vertices in each hyperedge are updated once with the messages from their incident hyperedges.

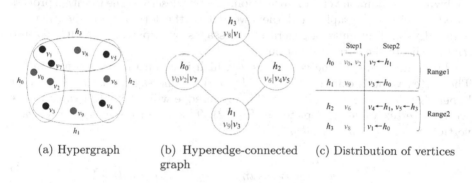

(a) Hypergraph   (b) Hyperedge-connected   (c) Distribution of vertices
graph

Fig. 1. A hypergraph and the converted hyperedge-connected graph

2.2   Distribution of Vertices

In the last superstep of HC, the vertices need to be updated, so it is necessary to guarantee the workload balance. In order to make all the vertices be evenly distributed in the hyperedge and enable the vertices to obtain messages from incident hyperedges, we take the following two steps to assign each vertex v_i to some associated hyperedge h_i, where $h_i \in \Gamma(v_i)$.

First, we look for all the vertices that exist only in one hyperedge. We traverse all the hyperedges and denote the hyperedge currently being processed as $curH$. If a vertex v_i in $\Gamma(curH)$ satisfies $v_i.deg = 1$, it has only one incident hyperedge, indicating that v_i appears only in $curH$. Such vertex v_i needs to be stored by $curH$.

Next, we traverse all the vertices, and denote the vertex currently being processed as $curV$. If $curV$ exists only in one hyperedge, this vertex should be ignored. Otherwise for each hyperedge that is incident to $curV$, firstly we find the *partition* with fewest vertices. Then we find the hyperedge with the smallest number of vertices in the *partition*, denoted as *minH*. We take *MinH* to be the hyperedge that holds *curV*. After all the vertices get assigned, the vertices processed in step 1 only need to receive messages from the hyperedges that keep them, and the vertices processed in step 2 need messages also from the neighbor hyperedges.

After Range partitioning [2], all the hyperedges are divided into two *partitions*. Figure 1 illustrates this with an example that hyperedge h_0 and h_1 are divided to Range1, while h_2 and h_3 are divided to Range2 similarly. Figure 1(c) accounts for the distribution of vertices.

2.3 Obtaining Common Vertex Information

In different hypergraph analysis tasks, the updating processes of hyperedges and vertices are quite different. In this section we take *PageRank* as an example to discuss the information that needs to be stored in each hyperedge. When analyzing the common vertex information, we first describes the iterative process based on bipartite graphs, and then we derive the formulas of updating the hyperedges and formulas of generating messages by hyperedges in HC. We use *itr* to indicate the current iteration.

The initial value of the vertices is d, which is a constant between 0 and 1. The vertex value is computed as Eq. (1), and the message sent to the relevant hyperedges is $v.msg = v.val/v.deg$. The hyperedge will set its own value to the received vertex value, as depicted in Eq. (2). The message sent to the relevant vertices is $h.msg = h.val/h.deg$.

$$v.val = \begin{cases} d, itr = 1 \\ d + (1-d) \times sum(h.msg), h \in \Gamma(v), itr > 1 \end{cases} \tag{1}$$

$$h.val = sum(v.msg), v \in \Gamma(h) \tag{2}$$

Afterwards we can get the computation formula of the hyperedges in HC as Eq. (3), where $v \in \Gamma(h), p \in N(h), q \in \Gamma(h) \cap \Gamma(p)$. $\sum \dfrac{1}{v.deg}$ is the sum of the reciprocals of all vertices in h. Given $h.vInfo = \sum \dfrac{1}{v.deg}$, and *msg* represents the messages h received from its neighbor hyperedges before the last superstep, where $msg = \dfrac{p.val}{p.deg} \times \sum \dfrac{1}{q.deg}$, then Eq. (3) can be transformed into Eq. (4).

$$h.val = d \times \sum \frac{1}{v.deg} + (1-d) \times (\frac{h.val}{h.deg} \times \sum \frac{1}{v.deg} + \sum(\frac{p.val}{p.deg} \times \sum \frac{1}{q.deg})) \tag{3}$$

$$h.val = d \times h.vInfo + (1-d) \times (\frac{h.val}{h.deg} \times h.vInfo + \sum msg) \tag{4}$$

$$h.msg = \begin{cases} \dfrac{h.val}{h.deg} \times \sum \dfrac{1}{q.deg}, itr < last \\ \dfrac{h.val}{h.deg}, itr = last \end{cases} \tag{5}$$

In HC, the message sent by hyperedge h to hyperedge p is as shown in Eq. (5), where $p \in N(h), q \in \Gamma(h) \cap \Gamma(p)$. In the last superstep, the updating formula of vertices is Eq. (1) where $itr > 1$. Each hyperedge h needs to keep three parts of information: (1) $\sum \dfrac{1}{v.deg}$, where $v.deg$ represents the degree of each vertex v in the hyperedge h; (2) $\sum \dfrac{1}{q.deg}$, where $q.deg$ represents the degree of each vertex q that hyperedge h have in common with each of its neighbors p; (3) the degree of hyperedge h.

3 I/O Efficient Implementation Techniques

3.1 Data Storage on Disk

On top of the distributed graph processing system HybridGraph [9], a hypergraph iterative framework HyraphD is implemented. Adjacency lists are used to store HC. The storage format of HC is represented by a quintuple $<h_i, deg, info,$ $\Psi(h_i), N(h_i)>$. Where h_i is the ID of the current hyperedge, deg is the degree of h_i, and $info$ is the vertex information to be saved by h_i. $\Psi(h_i)$ is the vertices assigned to h_i, and $N(h_i)$ is a neighbor hyperedge set of h_i. Meanwhile the message format in HC is represented by a triple $<srcId, dstId, msgValue>$. Where $srcId$ represents the ID of the hyperedge that generates messages, $dstId$ represents the ID of the neighbor hyperedge, and $msgValue$ is the message value.

Hypergraph data and message data are stored on disk. We modify the *GraphRecord* class in HybridGraph to make it correspond to the hypergraph data, and we modify the *MsgRecord* class to make it correspond to the message data. With regard to adjacency lists, in *GraphRecord* we implement the serialization and deserialization for each part in the above quintuple, and in *MsgRecord* we implement those for each part in the above triple. After loading the graph data from HDFS in HybridGraph, the vertices in the graphs will be organized into N Vblocks and the edges will be organized into a data structure Eblock. Vblock and Eblock in the original HybridGraph system are modified to make them meet the condition of iterative processing for hypergraph data. In HC the information about hyperedges is stored in Vblock, including hyperedge ID, hyperedge value, degree of the hyperedges, and additional vertex information. The correspondence between hyperedges and the neighbors is stored in Eblock.

3.2 Combination of Push or Pull

Combination of HC and Push. The hypergraph is converted into HC and iteratively processed with a message acquisition mechanism based on Push. This method is called as HC-Push in this paper. Before the last superstep, after the hyperedge updates its own value, it actively sends a message to the neighbor hyperedges. The vertices in each hyperedge get updated in the last superstep.

Combination of HC and Pull. The hypergraph is converted into HC and iteratively processed with a message acquisition mechanism based on Pull. This method is called the HC-Pull in this paper. Before each superstep gets started, the hyperedge in each Vblock actively acquires messages from the neighbor hyperedges.

4   Experimental Settings and Results Analysis

4.1   Experimental Settings

The experiments is executed in a cluster of local servers composed of 13 nodes. Each node has 4 cores running at 3.3 GHz with 16 GB memory and 931.5 GB disk. We use gigabit ethernet switch in the experiments.

HyraphD is dependent on the underlying distributed graph processing system HybridGraph [9] which is built on top of Hadoop-0.20.2. HyraphD is implemented in Java, and the runtime environment is JDK 1.8. We use CentOS Linux release 7.3.1611 (Core) as the operating system.

Dataset. In the experiments, two real datasets are used, including: (1) *com-Amazon*: a product information dataset with 317,194 products attributed to 75,149 types (https://snap.stanford.edu/data/com-Amazon.html); (2) *wiki-topcats*: a knowledge base with 1,791,498 articles and 17360 categories (https://snap.stanford.edu/data/wiki-topcats.html).

In order to study the iterative processing techniques for disk-resident hypergraph data and message data, the datasets are inflated by twenty-fold or fifty-fold. The suffix infers the replicating times in this section (e.g. AM20 is the dataset obtained by replicating the com-Amazon dataset 20 times).

Baseline Method. As the baseline solution, we implement star-expansion (SE) on HybridGraph and combine SE with Push-based and Pull-based mechanisms, which are denoted by SE-Push and SE-Pull.

4.2   Experimental Results Analysis

We use five nodes (including one master node and four slave nodes) in the default performance evaluation experiment, and up to 13 nodes for the scalability tests. The experiments run on four hypergraph learning algorithms, including PageRank, Random Walks, Connected Components and Single Source Shortest Hyperedge Path (SSSHP) as shown in Fig. 2. We only analyze the experimental results on PageRank due to the space limitation.

For SE and HC, we execute both for 10 iterations. The result is illustrated in Fig. 2 (a). On PageRank, HC-Pull outperforms the other three methods, and delivers up to 140% speed-up on AM20 and AM50; on WK20 and WK50, for HC-Push and HC-Pull, 45%–80% time costs are reduced compared to SE-Push and SE-Pull, respectively. We explain this by the items participating in the computation in HC being fewer than that in SE. In SE, hyperedges and vertices separately perform 10 iterations, and in HC, only 10 hyperedge iterations and 1 vertex updating are in need, that is, there is a reduction in the synchronization

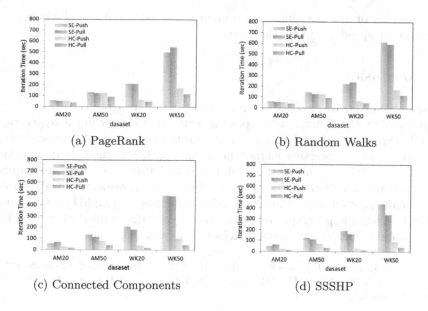

(a) PageRank

(b) Random Walks

(c) Connected Components

(d) SSSHP

Fig. 2. Iterative computation time

overhead after each superstep. In addition, there are fewer outgoing edges in HC, so the messages generated during the iterative processing of HC diminishes. Due to the above three reasons, the iterative processing of HC achieves superior efficiency, and since HybridGraph itself optimizes the Pull algorithm for disk-resident data, HC-Pull performs better than HC-Push.

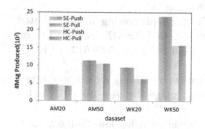

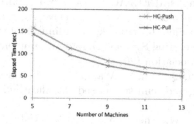

Fig. 3. Number of generated messages   **Fig. 4.** Scalability

The total number of messages generated in each iteration is demonstrated in Fig. 3, and HC obviously performs better than SE. Finally on dataset AM50, we evaluate the scalability of HyraphD via extensive experiments, and the results are illustrated in Fig. 4. We use 5, 7, 9, 11 and 13 machines respectively for experiments. It can be seen that as the number of machines increases, the processing time of the hypergraph tasks gets degraded. Nevertheless, an increment in the number of machines increases communication overhead, so the increment in computing speed gradually slows down.

5   Conclusion

Based on the extensive analysis the characteristics of hypergraph data and hypergraph learning algorithms, we propose a method of converting a hypergraph into a hyperedge-connected graph, and design the corresponding storage structure and iterative processing method in this paper. Given the hypergraph data residing on disk, we combine the Push-based and Pull-based message acquisition mechanisms with the iterative processing method, respectively. We implement the distributed hypergraph processing framework HyraphD that can support disk-resident hypergraph computation on hyperedge-connected graphs. The experimental results reveal that HC achieves superior efficiency.

Acknowledgements. This work is supported by the National Nature Science Foundation of China (61872070, U1811261) and the Fundamental Research Funds or the Central Universities (N171605001).

References

1. Berlt, K., de Moura, E.S., da Costa Carvalho, A.L., et al.: A hypergraph model for computing page reputation on web collections. In: da Silva, A.S. (ed.) XXII Simpósio Brasileiro de Banco de Dados, pp. 35–49. SBC (2007)
2. Ching, A., Edunov, S., Kabiljo, M., et al.: One trillion edges: graph processing at facebook-scale. PVLDB **8**(12), 1804–1815 (2015)
3. Gao, J., Zhao, Q., Ren, W., et al.: Dynamic shortest path algorithms for hypergraphs. IEEE/ACM Trans. Netw. **23**(6), 1805–1817 (2015)
4. Gonzalez, J.E., Xin, R.S., Dave, A., et al.: GraphX: graph processing in a distributed dataflow framework. In: Flinn, J., Levy, H. (eds.) 11th USENIX Symposium on Operating Systems Design and Implementation, OSDI 2014, pp. 599–613. USENIX Association (2014)
5. Liu, Q., Sun, Y., Wang, C., et al.: Elastic net hypergraph learning for image clustering and semi-supervised classification. IEEE Trans. Image Process. **26**(1), 452–463 (2017)
6. Shi, W., Zheng, W., Yu, J.X., et al.: Keyphrase extraction using knowledge graphs. Data Sci. Eng. **2**(4), 275–288 (2017)
7. Somu, N., Raman, M.R.G., Kannan, K., et al.: Hypergraph based feature selection technique for medical diagnosis. J. Med. Syst. **40**(11), 239:1–239:16 (2016)
8. Tan, S., Bu, J., Chen, C., et al.: Using rich social media information for music recommendation via hypergraph model. TOMCCAP **7**(2), 22 (2011)
9. Wang, Z., Gu, Y., Bao, Y., et al.: Hybrid pulling/pushing for I/O-efficient distributed and iterative graph computing. In: Özcan, F., Koutrika, G., Madden, S. (eds.) Proceedings of the 2016 International Conference on Management of Data, SIGMOD Conference 2016, pp. 479–494. ACM (2016)

How to Reach: Discovering Multi-resolution Paths on Large Scale Networks

Zhaokun Zhang[1], Ning Yang[1(✉)], and Philip S. Yu[2]

[1] School of Computer Science, Sichuan University, Chengdu, China
zzk@stu.scu.edu.cn, yangning@scu.edu.cn
[2] Department of Computer Science, University of Illinois at Chicago, Chicago, USA
psyu@uic.edu

Abstract. Reachability query is a fundamental problem which has been studied extensively due to its important role in various application domains, such as social networks, communication networks, biological networks, etc. However, few existing works pay attention to the problem of how two vertices are connected. In this paper, we investigate the problem of discovering paths of multiple resolutions connecting two vertices in large scale networks. We propose a new structure, called Muti-Resolution Path (MRP), to describe how two vertices are connected at different resolution levels. To facilitate the building of MRPs on a network of large scale, we propose a new search structure, called Hierarchical Compressed Network (HCN), which can represent a network at multiple resolution levels and can be built offline. At last, extensive experiments conducted on real-world datasets verify the effectiveness and efficiency of the proposed approach.

Keywords: Reachability query · Multi-resolution path · Hierarchical compressed network

1 Introduction

In this paper, we investigate the problem of discovering paths of multiple resolutions connecting two vertices in a given network, where there are two challenges we have to overcome. One challenge is how to represent the paths of multiple resolutions, and the other one is how to build them efficiently. To address these challenges, we propose a new structure referred to as Multi-Resolution Path (MRP) to represent the resulting paths connecting given two vertices at different resolution levels. To build MRP efficiently, we propose a new search structure, Hierarchical Compressed Network (HCN), to represent a network at multiple resolution levels, which can be built offline from the given original network, and lower the resolution level, smaller the scale. Due to HCN, an MRP at some resolution level for two given target vertices can be built efficiently by a

© Springer Nature Switzerland AG 2019
J. Shao et al. (Eds.): APWeb-WAIM 2019, LNCS 11641, pp. 281–288, 2019.
https://doi.org/10.1007/978-3-030-26072-9_21

quick search over the HCN at the same resolution level. The main contributions of this paper can be summarized as follows:

(1) As an extension of reachability query, we present the problem of discovering Multi-Resolution Path (MRP) connecting two given vertices. To our best knowledge, this is the first research on it.
(2) We propose a new search structure, Hierarchical Compressed Network (HCN), which is built offline and due to which the MRP can be built efficiently and the traditional reachability query can be speeded up.
(3) The extensive experiments conducted on real-world datasets verify the effectiveness and efficiency of the proposed HCN, MRP, and their building algorithms.

The rest of this paper is organized as follows. The concepts of HCN and MRP are defined in Sect. 2. The details of HCN construction are described in Sect. 3. The details of MRPD are described in Sect. 4. We analyze the experimental results in Sect. 5. At last, we review the related work in Sect. 6 and conclude in Sect. 7.

2   HCN and MRP

We denote an undirected network as $G = (V, E)$, where V represents the set of vertices and E represents the set of edges of G. An edge $e \in E$ is a pair (u, v), where $u, v \in V$. The number of vertices and edges in G are denoted as $|V|$ and $|E|$, respectively. $N(u)$ denotes the direct neighbors of u, i.e., $N(u) = \{v|(v, u) \in E\} \cup \{v|(u, v) \in E\}$. Throughout this paper, we suppose there are totally r resolution levels considered, and larger the number i, $1 \leq i \leq r$, higher the resolution of the i-th level.

Definition 1 (Hierarchical Compressed Network (HCN)). *Given an original network $G = (V, E)$, its HCN at the i-th resolution level is an undirected network $G^i = (V^i, E^i)$, where V^i represents the set of vertices at the i-th resolution level, and E^i represents the set of edges at the i-th resolution level. A vertex $v \in V^i$ represents a connected sub-network (a cluster of vertices) G^{i+1}_v of G^{i+1}. There exists an edge $(u, v) \in E^i$ if in G^{i+1}, there exists at least one edge connecting the two sub-networks G^{i+1}_u and G^{i+1}_v. The HCN at r-th resolution level (the highest resolution level) is the original network itself, i.e., $G^r = G$.*

Definition 2 (Vertex Covering). *A vertex u of G^i covers the vertices of the V^{i+1}_u and the vertices covered by V^{i+1}_u, where V^{i+1}_u is the vertex set of the sub-network G^{i+1}_u.*

The structure of HCN is illustrated in the right block of Fig. 1, where $r = 3$.

Definition 3 (Multi-Resolution Path (MRP)). *For two vertices u and v in a given original network G, its MRP at the i-th resolution level, denoted as $\boldsymbol{p}^i_{u,v}$, is a sequence of vertices of its HCN at the i-th resolution level G^i, i.e., $\boldsymbol{p}^i_{u,v} = (z_1, z_2, ..., z_q)$, where q is the length, z_1 covers u, z_q covers v, and $(z_j, z_{j+1}) \in E^i$ for $1 \leq j < q$.*

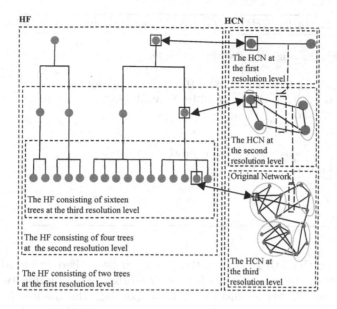

Fig. 1. Illustration of HCN and HF

Problem Statement: Given an original network $G = (V, E)$ and the number of the resolution levels r, the goal is answering whether there exists MRP connecting two given vertices $u \in V$ and $v \in V$, and if it does, building $p_{u,v}^i$ for every resolution level i, $1 \le i \le r$.

3 Building Hierarchical Compressed Network

Definition 4 *(Hierarchical Forest (HF)).* *Given an original network $G = (V, E)$ and its HCNs $G^i = (V^i, E^i)$, $1 \le i \le r$, the HF at the i-th resolution level, denoted as F^i, is defined as a set of trees each of which corresponds to a vertex of G^i, i.e., $F^i = \{t_v | v \in V^i\}$, where t_v is a merge of a subset of trees in F^{i+1}. The HF at the r-th resolution level (i.e., the highest resolution level) F^r is formed by the trees each of which consists of only one vertex of the original network G.*

According to Definition 4, one tree t_v of the HF at the i-th resolution level F^i records the detail information of the sub-network (cluster) G_v^{i+1} of the HCN at the $(i + 1)$-th resolution level G^{i+1}. The structure of HF is illustrated in the left block of Fig. 1.

3.1 Building One HCN

A vertex of an HCN at the i-th resolution level represents a cluster of the vertices of the HCN at the $(i + 1)$-th resolution level, and there exists an edge between

---

Algorithm 1. *BuildAllHCNs(G)* : Building all HCNs at all resolution levels

---
Input:
$\quad$ $G = (V, E)$: the original network;
Output:
$\quad$ $\mathcal{F} = \{F^1, F^2, ..., F^r\}$: the set of all HFs at all resolution levels,
$\quad$ $\mathcal{G} = \{G^1, G^2, ..., G^r\}$: the set of all HCNs at all resolution levels;
1: Initialize F^r and G^r with G;
2: $i = r - 1$;
3: **repeat**
4: $\quad$ $G^i, F^i = BuildOneHCN(G^{i+1}, F^{i+1})$;
5: $\quad$ $\mathcal{F} = \mathcal{F} \bigcup \{F^i\}$;
6: $\quad$ $\mathcal{G} = \mathcal{G} \bigcup \{G^i\}$;
7: $\quad$ $i = i - 1$;
8: **until** i=0

---

two vertices at the i-th level if at the $(i+1)$-th level there exists at least one edge connecting any two vertices belonging to the corresponding two clusters, respectively. we pursue a clustering that maximizes the modularity which is defined as follow:

$$Q = \frac{1}{2m} \sum_{u,v} [A_{u,v} - \frac{k_u k_v}{2m}] \delta(C(u), C(v)), \tag{1}$$

where $A_{u,v}$ is the weight of the edge between vertex u and vertex v, $k_u = \sum_v A_{u,v}$ is the sum of the weights of the edges attached to vertex u, $C(u)$ returns the number of the cluster to which vertex u is assigned, and the δ-function $\delta(C(u), C(v))$ is 1 if $C(u) = C(v)$, 0 otherwise, and $m = \sum_{uv} A_{u,v}$.

3.2 Building All HCNs

Building all HCNs and all HFs for a given original network $G = (V, E)$ is relatively straightforward. As shown in Algorithm 1, the HCNs at different resolution levels are built in a bottom-up fashion, i.e., from highest resolution level to lowest resolution level. At first, the HF and HCN at the highest resolution level are initialized with G (Line 1), and then the HCNs and HFs at different resolution levels are built iteratively by invoking the procedure of building one HCN.

4 Building Multi-resolution Path

4.1 Building Top-k Shortest Paths

According to the definition of MRP (Definition 3), there might be more than one MRP in an HCN at a resolution level for two given vertices in an original network. In this paper, we pursue the top-k shortest MRPs where k is a positive integer given in advance, as shortest path is really popular in real-world applications [2]. However, any other criteria can be taken into account for the building of

Algorithm 2. $BuildMRP(G^i, F^i, u, v, k)$: Discovering the top-k shortest MRP at the i-th resolution level

Input:
    $G^i = (V^i, E^i)$: the HCN at the i-th resolution level,
    F^i: the HF at the i-th resolution level,
    u and v: two vertices for which we want to build MRP,
    k: the number of shortest MRPs;
Output:
    $\mathcal{P} = \{p_1, p_2, ..., p_k\}$: the set of the top-$k$ shortest MRP at the i-th resolution level;
1: **for** each tree t in F^i **do**
2:    **if** t contains u **then**
3:       $t_{u'} = t$;
4:    **end if**
5:    **if** t contains v **then**
6:       $t_{v'} = t$;
7:    **end if**
8: **end for**
9: u' = the vertex $t_{u'}$ corresponds to in G^i;
10: v' = the vertex $t_{v'}$ corresponds to in G^i;
11: $\mathcal{P} = KSP(G^i, u', v', k)$;

MRPs. Building top-k shortest paths is a well studied problem and a few of methods have been proposed for it. Here we utilize the classical K-Shortest-Path (KSP) algorithm proposed by Yen [11] for our purpose. KSP takes an undirected network G, two vertices u and v in G, and a positive integer k as the inputs, and returns the set of the top-k shortest paths connecting u and v.

4.2 Building MRP

Given two vertices, u and v, Algorithm 2 gives the whole procedures of the top-k shortest MRPs discovery on HCN at the i-th resolution level G^i. We first obtain the vertices u' and v' that u and v correspond to respectively on G^i (Lines 1–10). Then we can obtain the top-k shortest path connecting u' and v' on G^i by invoking KSP (Line 11). One can note that if two vertices are connected in the original network, Algorithm 2 can always find out their MRPs at different resolution levels, because for any two connected vertices u and v in the original network, their corresponding vertices at different resolution levels of the HCN are also connected.

5 Experiments

The experiments are conducted on a Spark cluster consisting of 3 PCs where each PC equipped with a 2.7 GHz INTEL CPU of 4 cores and 32 GB RAM, and all the algorithms are implemented in JAVA.

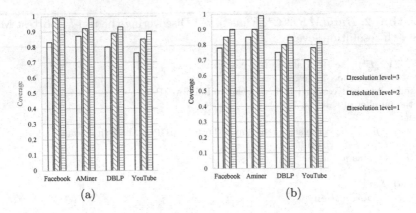

Fig. 2. The coverage when (a) $k = 3$ and (b) $k = 5$

5.1   Experimental Setting

Datasets

- **Facebook** dataset consists of $4,039$ vertices and $88,234$ edges, where a vertex represents a user and an undirected edge represents the friendship between two users.
- **DBLP** dataset is a collaboration network consisting of $317,080$ vertices and $1,049,866$ edges, where the vertices represent authors, and there is an edge between two authors if they co-author at least one paper.
- **AMiner**[1] dataset is also a collaboration network, from which we randomly select $400,000$ papers, whose co-authorship information refers to $114,432$ vertices and $192,497$ edges.
- **YouTube**[2] dataset consists of $1,134,890$ vertices and $2,987,627$ edges, where a vertex represents a user and an undirected edge represents a friendship between two users.

Metrics. As we have mentioned before, we evaluate the quality of an MRP in terms of its ability to summarize the detailed paths. For this purpose, we propose a metric called *Coverage*, of which the details are presented in Subsect. 5.2.

5.2   Verification of MRP Building

It is reasonable to evaluate the quality of MRPs discovered by Algorithm 2 in terms of their ability to summarize the shortest paths in the original network. Here we define a new metric, **coverage**, to measure the summarizing ability of an MRP. At first, we define the concept of path covering as follow:

---

[1] https://cn.aminer.org/billboard/aminernetwork.
[2] http://snap.stanford.edu/data/com-Youtube.html.

Definition 5 *(Path Covering). An MRP p covers another MRP p', if each vertex of p' is covered by one of the vertices of p.*

Let $p = (z_1, z_2, ..., z_q)$ be an MRP, and let V_1 and V_q be the set of vertices in the original network that are covered by z_1 and z_q, respectively. Then the **coverage** c of p is defined as

$$c(p) = \frac{|S_p|}{|S_t|}, \tag{2}$$

where S_p is the set of shortest paths in the original network that are covered by p, and S_t is the set of all the shortest paths in the original network that connect the vertices in V_1 and the vertices in V_q. The more shortest paths are covered by p, the larger the value of $c(p)$ is, which indicates that p has a good ability to summarize the shortest paths in the original network that connect the vertices in V_1 and those in V_q.

We generate 100 pairs of connected vertices for each dataset and build the top-k shortest MRPs at $r = 3$ resolution levels for each dataset. We can see from Fig. 2(a), when $k = 3$, the coverages on all datasets are larger than 70%, and on each dataset, with the resolution level decreases, the coverage increases, which is because the vertices on HCN at lower resolution level cover more vertices of the original network. Figure 2(b) shows the coverages when $k = 5$. We can see the coverages on all datasets are better than those when $k = 3$. Especially on Facebook, the coverage becomes close to 1 on the HCN at the second resolution level. For DBLP and YouTube the coverages are larger than 0.9. Figure 2 validates the ability of MRPs generated by Algorithm 2 to summarize the underlying paths accurately.

6 Related Work

The existing methods for reachability query roughly fall into three categories, Label-Only, Label+G, and random approaches. The Label-Only approach is to answer reachability queries using the labels only, where the labels are computed by compressing TC (Transitive Closure) offline [1,4,5,9]. The Label+G approach makes use of the labels computed as much as possible, but it may need to traverse the network G [3,7–9,12,13]. The main idea behind Label+G is to reduce the time to construct an effective index, since the computing cost to compress TC can be very high. In addition to the deterministic algorithms, recently researchers also propose to introduce randomness into the reachability query. For example, IP [10] can generate a probabilistic answer with the probability guarantee based on k-min-wise independent permutation. Su et al. [6] propose a Bloom-filter labeling, which further improves the performance of IP using every bit for pruning. However, in contrast with our work, all the existing methods only answer whether a vertex can reach another and can not answer how to reach.

7   Conclusions

In this paper, we present and address the problem of path discovery in large scale networks. We propose a new structure MRP (Muti-Resolution Path) to represent the paths that connect two vertices at different resolution levels. To build MRP efficiently, we propose a new search structure HCN (Hierarchical Compressed Network), which can represent an original network at multiple resolution levels. At last, the results of the extensive experiments conducted on real-world datasets verify the effectiveness and efficiency of HCN, MRP, and their building algorithms.

References

1. Agrawal, R., Borgida, A., Jagadish, H.V.: Efficient management of transitive relationships in large data and knowledge bases, vol. 18. ACM (1989)
2. Akiba, T., Hayashi, T., Nori, N., Iwata, Y., Yoshida, Y.: Efficient top-k shortest-path distance queries on large networks by pruned landmark labeling. In: Proceedings of the 29th AAAI Conference on Artificial Intelligence, pp. 2–8 (2015)
3. Chen, L., Gupta, A., Kurul, M.E.: Stack-based algorithms for pattern matching on DAGS. In: Proceedings of the 31st International Conference on Very Large Data Bases, pp. 493–504. VLDB Endowment (2005)
4. Cohen, E., Halperin, E., Kaplan, H., Zwick, U.: Reachability and distance queries via 2-hop labels. SIAM J. Comput. 32(5), 1338–1355 (2003)
5. Jagadish, H.: A compression technique to materialize transitive closure. ACM Trans. Database Syst. (TODS) 15(4), 558–598 (1990)
6. Su, J., Zhu, Q., Wei, H., Yu, J.X.: Reachability querying: can it be even faster? IEEE Trans. Knowl. Data Eng. 29(3), 683–697 (2017)
7. Trißl, S., Leser, U.: Fast and practical indexing and querying of very large graphs. In: Proceedings of the 2007 ACM SIGMOD International Conference on Management of Data, pp. 845–856. ACM (2007)
8. Veloso, R.R., Cerf, L., Meira Jr, W., Zaki, M.J.: Reachability queries in very large graphs: a fast refined online search approach. In: EDBT, pp. 511–522 (2014)
9. Wang, H., He, H., Yang, J., Yu, P.S., Yu, J.X.: Dual labeling: answering graph reachability queries in constant time. In: Proceedings of the 22nd International Conference on Data Engineering, ICDE 2006, p. 75. IEEE (2006)
10. Wei, H., Yu, J.X., Lu, C., Jin, R.: Reachability querying: an independent permutation labeling approach. Proc. VLDB Endow. 7(12), 1191–1202 (2014)
11. Yen, J.Y.: Finding the k shortest loopless paths in a network. Manag. Sci. 17(11), 712–716 (1971)
12. Yildirim, H., Chaoji, V., Zaki, M.J.: Grail: scalable reachability index for large graphs. Proc. VLDB Endow. 3(1–2), 276–284 (2010)
13. Yıldırım, H., Chaoji, V., Zaki, M.J.: Grail: a scalable index for reachability queries in very large graphs. VLDB J. Int. J. Very Large Data Bases 21(4), 509–534 (2012)

Analysis and Management to Hash-Based Graph and Rank

Yangtao Wang[1], Yu Liu[1], Yifei Liu[1], Ke Zhou[1(✉)], Yujuan Yang[1],
Jiangfeng Zeng[1], Xiaodong Xu[1], and Zhili Xiao[2]

[1] Huazhong University of Science and Technology, Wuhan, China
{ytwbruce,liu_yu,yifeiliu,k.zhou,gracee,
jfzeng,xiaodong-xu}@hust.edu.cn
[2] Tencent Inc., Shenzhen, China
tomxiao@tencent.com

Abstract. We study the problem of how to calculate the importance score for each node in a graph where data are denoted as hash codes. Previous work has shown how to acquire scores in a directed graph. However, never has a scheme analyzed and managed the graph whose nodes consist of hash codes. We extend the past methods and design the undirected hash-based graph and rank algorithm. In addition, we present addition and deletion strategies on our graph and rank.

Firstly, we give a mathematical proof and ensure that our algorithm will converge for obtaining the ultimate scores. Secondly, we present our hash based rank algorithm. Moreover, the results of given examples illustrate the rationality of our proposed algorithm. Finally, we demonstrate how to manage our hash-based graph and rank so as to fast calculate new scores in the updated graph after adding and deleting nodes.

Keywords: Analysis · Management · Hash-based · Graph · Rank

1 Introduction

Using graph structure [1,3] to construct data correlation and manage data is a popular method for data experts to mine data and extract knowledge. Calculating the global importance rank for each data contained in a graph has always been an important research topic in data analysis and information retrieval domain [13,14]. Graph-based algorithms have achieved great success in this aspect. Especially, as one of the most important graph-based algorithms, PageRank [11] has been widely applied and extended. However, one of the difficulties is to define the correlation between nodes on a graph. Several previous researches have explored this issue. PageRank [11] considers out-degree of related nodes as impact factor for data rank. [12] applies random walk to ranking community images for searching. [7] introduces the concept of probability to improve the RegEx in PageRank. However, above graph-based rank algorithms all focus on in-degree and out-degree, neglecting the weight on edges, resulting that they are

© Springer Nature Switzerland AG 2019
J. Shao et al. (Eds.): APWeb-WAIM 2019, LNCS 11641, pp. 289–296, 2019.
https://doi.org/10.1007/978-3-030-26072-9_22

not competent for quantization with weighted graphs. TexRank [10] and SentenceRank [5] take the weights on edges into consideration, both of which apply PageRank to improving their respective algorithms, but none of them provide detailed proof of convergence.

Above researchers generally use feature vectors represented by floating point numbers to measure the correlation between nodes while dealing with weighted graphs. However, vast resource cost caused by feature vectors makes it not applicable to utilize this method with the increase of scale of data. Especially for analysis of large-scale image data, the dimension and complexity of image features lead to greater complexity. For example, [2,4] extracted image features through content perception, built the graph using Euclidean or Cosine distance, and further acquired recommended result using improved PageRank algorithm. However, feature vectors will inevitably lead to huge storage overhead. At the same time, the metrics just like Euclidean distance will also bring unacceptable time cost. Moreover, along with massive data quantity expansion recent years, it becomes more and more computationally complex to obtain the correlation between nodes which denote high-dimensional floating point numbers. On the other side, hash techniques are often used in storage and retrieval fields. For example, Hua et al. [6] map data to hash code using Locality-Sensitive Hashing (LSH). Taking advantage of hash in retrieval, they can fast perform some operations like query. Besides, due to the easy "XOR" operation, it will be simple and convenient to measure the correlation between two objects denoted as hash codes.

In this paper, we combine hash with graph structure to establish a semantic information management theory paradigm, which can not only serve for big data analysis but also enrich the operations for graph database. Assuming that we have obtained corresponding hash codes, we can build a undirected weighted hash-based graph by leveraging the Hamming distance [8,15] between nodes. Our defined hash-based graph is a kind of graph of which the node value is a hash value and edge value is the Hamming distance between nodes. Based on this, we design a hash-based rank algorithm which can effectively compute the importance of each node. We will give a complete mathematical proof and analysis of our algorithm. In addition, in order to reduce computational overhead as much as possible when graph changes, we also provide a series of graph data management operations such as addition and deletion.

2 Convergence Analysis

Matrix A and B are both $n \times n$ square matrix and each column sum of them is 1. i and j are positive integers. Generally, we respectively denote A and B as

$$A = \begin{bmatrix} a_{11} & a_{12} & \cdots & a_{1n} \\ a_{21} & a_{22} & \cdots & a_{2n} \\ \cdots & \cdots & \cdots & \cdots \\ a_{n1} & a_{n2} & \cdots & a_{nn} \end{bmatrix} \quad B = \begin{bmatrix} b_{11} & b_{12} & \cdots & b_{1n} \\ b_{21} & b_{22} & \cdots & b_{2n} \\ \cdots & \cdots & \cdots & \cdots \\ b_{n1} & b_{n2} & \cdots & b_{nn} \end{bmatrix} \quad (1)$$

where $\forall\, i \in [1,\, n]$ satisfies $\sum_{j=1}^{n} a_{ji} = 1$ and $\sum_{j=1}^{n} b_{ji} = 1$.

2.1 Matrix Product Convergence

Given matrix $C = AB$, we denote C as $\begin{bmatrix} c_{11} & c_{12} & \cdots & c_{1n} \\ c_{21} & c_{22} & \cdots & c_{2n} \\ \cdots & \cdots & \cdots & \cdots \\ c_{n1} & c_{n2} & \cdots & c_{nn} \end{bmatrix}$. We find that each

column sum of C also satisfies: $\forall\, i \in [1, n]$,

$$\sum_{j=1}^{n} c_{ji} = \sum_{j=1}^{n} a_{1j}b_{ji} + \sum_{j=1}^{n} a_{2j}b_{ji} + \cdots + \sum_{j=1}^{n} a_{nj}b_{ji}$$
$$= b_{1i}\sum_{j=1}^{n} a_{j1} + b_{2i}\sum_{j=1}^{n} a_{j2} + \cdots + b_{ni}\sum_{j=1}^{n} a_{jn} \qquad (2)$$
$$= b_{1i} + b_{2i} + \cdots + b_{ni} = 1$$

2.2 Vector Convergence

R is a column vector whose column sum is r. We denote R as $R = [r_1\ r_2\ \cdots\ r_n]^T$, where $\sum_{j=1}^{n} r_j = r$. Given $R' = AR = [r'_1\ r'_2\ \cdots\ r'_n]^T$, we find that the column sum of R' satisfies:

$$\sum_{j=1}^{n} r'_j = \sum_{j=1}^{n} a_{1j}r_j + \sum_{j=1}^{n} a_{2j}r_j + \cdots + \sum_{j=1}^{n} a_{nj}r_j$$
$$= r_1\sum_{j=1}^{n} a_{j1} + r_2\sum_{j=1}^{n} a_{j2} + \cdots + r_n\sum_{j=1}^{n} a_{jn} \qquad (3)$$
$$= r_1 + r_2 + \cdots + r_n = r$$

Also, for $\forall\, k \in Z^+$, we can conclude that each column sum of $A^k R$ is also 1. Consequently, $A^k R$ will never diverge as k becomes larger if A^k converges.

3 Hash-Based Graph

Given a graph G consisting of n nodes, each node is denoted as a l-bits hash code. N_* denotes the $*$-th node of Graph G and $H(N_*)$ denote the hash code of N_*. We define XOR operation as $\oplus$ and threshold $\Omega \in [1, l] \cap Z^+$. Different from the work of the predecessors [9], we stipulate that two nodes are connected only if the Hamming distance between them does not exceed threshold Ω. Therefore, the Hamming distance weight on undirected edge between N_i and N_j is defined as:

$$d_{ij} = \begin{cases} H(N_i) \oplus H(N_j) & i \neq j, H(N_i) \oplus H(N_j) \leq \Omega, \\ NULL & otherwise. \end{cases} \qquad (4)$$

As a result, our hash-based graph has been established.

4   Hash-Based Rank

In this Section, we demonstrate our designed hash-based rank algorithm on our weighted undirected hash-based graph. Our goal is to calculate the importance score of each node. For $\forall\ i \in [1, n]$, T_i is defined as the set including orders of all nodes connected with N_i, where $T_i \subset [1, n]$.

As defined in Sect. 3, l denotes the length of hash code and d_{ij} denotes weight of the edge between N_i and N_j. We denote $R(N_*)$ as the importance score of N_*. Referring to PageRank, we also intend to calculate the ultimate $R(N_*)$ by means of iteration. Draw impact factor $I(N_{ij})$ for N_j to N_i which measures how N_j contributes to N_i, where $I(N_{ij})$ is defined as:

$$I(N_{ij}) = \begin{cases} \dfrac{l - d_{ij}}{\sum\limits_{t \in T_j} l - d_{tj}} R(N_j) & \exists d_{ij}, \\ 0 & otherwise. \end{cases} \tag{5}$$

Theoretically, we design Eq. (5) according to two principals. Firstly, the less d_{ij} is, the greater influence N_j contributes to N_i is. Meanwhile, the longer hash code (l) is, the more compact the similarity presented by d_{ij} is. Secondly, PageRank considers all (unweighted) edges as the same, but we extend it to be applied to different weights on edges. Specially, when all weights on edges are the same, our hash-based rank algorithm will turn into undirected PageRank. Consequently, $R(N_i)$ should be equal to the sum of the impact factors of all nodes connected to N_i, where N_i is expressed as: $R(N_i) = \sum\limits_{j=1, j \neq i}^{n} I(N_{ij})$.

Let f_{ij} represent the coefficient of $R(N_j)$ in $I(N_{ij})$, where

$$f_{ij} = \begin{cases} \dfrac{l - d_{ij}}{\sum\limits_{t \in T_j} l - d_{tj}} & \exists d_{ij}, \\ 0 & otherwise. \end{cases} \tag{6}$$

We define coefficient matrix D as $\begin{bmatrix} 0 & f_{12} & \cdots & f_{1n} \\ f_{21} & 0 & \cdots & f_{2n} \\ \cdots & \cdots & \cdots & \cdots \\ f_{n1} & f_{n2} & \cdots & 0 \end{bmatrix}$ and calculate each column sum of D according to Eq. (6), we take the jth column as

$$\begin{aligned} & f_{1j} + f_{2j} + \cdots + f_{nj} \\ &= \frac{l - d_{1j}}{\sum\limits_{t \in T_j} l - d_{tj}} + \frac{l - d_{2j}}{\sum\limits_{t \in T_j} l - d_{tj}} + \cdots + \frac{l - d_{nj}}{\sum\limits_{t \in T_j} l - d_{tj}} \\ &= \sum\limits_{t \in T_j} \frac{l - d_{tj}}{\sum\limits_{t \in T_j} l - d_{tj}} = 1 \end{aligned} \tag{7}$$

Usually, the initial value is set as $R^0 = [R^0(N_1) \; R^0(N_2) \; \cdots \; R^0(N_n)]^T = [1 \; 1 \; \cdots \; 1 \;]^T$. We draw iteration formula as

$$R^{k+1} = DR^k \tag{8}$$

where $R^k = [R^k(N_1) \; R^k(N_2) \; \cdots \; R^k(N_n) \;]^T$, and k is the number of iteration rounds. According to Eq. (3), vector R^k will converge when k becomes larger.

We define the termination condition as $R^{k+1}(N_m) - R^k(N_m) \leq \varepsilon$, where $m \in [1, n]$. Meanwhile, ε is set to a small constant (say 0.0001).

Thus, our designed hash-based rank algorithm converges. Then we will illustrate the result of the algorithm. As shown in Fig. 1, we use a graph G_1 with 10 nodes to verify our algorithm. Each node is a 48-bits hash code. We set termination condition $\varepsilon = 1.0E{-}8$, threshold $\Omega = 24$ (see Eq. (4)) and $[R^0(N_1) \; R^0(N_2) \; \cdots \; R^0(N_{10})]^T = [1 \; 1 \; \cdots \; 1]^T$. The score and rank for each node are displayed in Table 1.

Table 1. Score and rank in graph G_1.

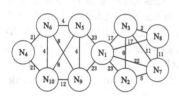

Fig. 1. Example graph G_1 with 10 nodes.

Node	Hash code	Score	Rank
N_1	FFFFFFFFFFFF	1.14788732	1
N_2	FFFFFF800000	1.05633802	6
N_3	FFFFFFFE0000	1.09859154	2
N_4	000000000000	0.38028169	10
N_5	C000007FFFFF	1.0774648	5
N_6	0000001FFFFF	1.09154931	3
N_7	FBFF7F8000E0	1.00704224	9
N_8	FFFFFF7E0080	1.08450703	4
N_9	C0003079FFFF	1.02112677	8
N_{10}	0300001FFE7F	1.03521128	7

As shown in graph G_1, each node is influenced by both of the edges and weights. If a node owns more edges with lower weights, it will obtain higher score and rank. For example, N_1 owns the most connections, so it acquires the highest score and rank. N_3 has the same number of connections as N_9, but the weights on edges connected with N_3 are lower than that of N_9. Thus, N_3 owns a higher rank than N_9. N_4 obtains the lowest score and rank because of fewest connections with high weights. The result in Table 1 is deemed reasonable.

5   Management to Hash-Based Graph and Rank

In this Section, we demonstrate how to manage our hash-based graph and rank algorithm when adding or deleting nodes.

Actually, if we intend to calculate the score and rank for each node in a updated graph, we have to obtain the corresponding updated coefficient matrix D. In Sect. 4, without isolated nodes, the graph consists of n nodes and the coefficient matrix D has been calculated according to Eq. (6). In the next part of this Section, we mainly introduce how to perform minimal change to coefficient matrix D when adding and deleting a node in the graph.

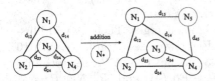

Fig. 2. Example graph G_2 by addition operation.

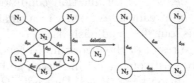

Fig. 3. Example graph G_3 by deletion operation.

5.1   Addition

Generally, when a new node is added to graph G, this node will be marked as N_{n+1} by default if it is connected with one of the n nodes. (As shown in Fig. 2, N_* is added to graph G_2 which contains 4 nodes. We directly mark N_* as N_5 because N_* is connected with N_1 and N_4.) And T_{n+1} (defined in Sect. 4) denotes the set including orders of all nodes connected with N_{n+1} where $T_{n+1} \subseteq [1, n]$. Then we analyze how matrix D will change and calculate the scores and ranks for $n + 1$ nodes.

Algorithm 1. Calculate D_{n+1} and scores for $n + 1$ nodes when adding a node.

1: Calculate $d_{i(n+1)}$ and set T_{n+1} for $\forall\, i \in \{1, 2, \cdots, n\}$.
2: Judge whether T_{n+1} is a empty set and calculate the following D_{n+1}.
3: Directly Calculate the ith column elements of D_{n+1} based on D_n and update T_i for $\forall\, i \in \{1, 2, \cdots, n\} \cap T_{n+1}$.
4: Directly Calculate the ith column elements of D_{n+1} based on D_n for $\forall\, i \in \{1, 2, \cdots, n\} \backslash T_{n+1}$.
5: Calculate the $(n + 1)$th column elements of D_{n+1} according to Equation (6).
6: Calculate scores for $n + 1$ nodes according to Equation (8) using D_{n+1}.

For convenience, we denote the $n \times n$ matrix D as D_n. When adding a node, we need to calculate D_{n+1} based on D_n. As shown in Algorithm 1, we describe the steps that calculate matrix D_{n+1} and scores for $n + 1$ nodes.

5.2   Deletion

Similarly, if we delete a node from graph G, how can we fast adjust the matrix D_n and calculate score for each node in the new graph? For example, as shown in Fig. 3, N_2 is deleted from G_3 which contains 6 nodes. However, N_1 will be removed from G_4 because N_1 is only connected with N_2. Also, those edges which are connected with N_2 will also disappear. Generally, for $i \in \{1, 2, \cdots, n\}$, once we delete N_i from graph G, those edges connected with N_i will be removed from G. Of course, if N_i is deleted, those isolated nodes will be also removed.

Algorithm 2. Adjust D_n and calculate scores for remaining nodes when deleting a node.

1: Calculate the set I_i which contains the orders of those nodes that are only connected with N_i.
2: Judge whether $(n - |I_i|)$ equals 1 and calculate the following $D_{n-|I_i|-1}$.
3: Directly adjust the tth column elements of D_n for $t \in \{1, 2, \cdots, n\} \cap T_i \backslash I_i$.
4: Directly reserve the tth column elements of D_n for $t \in \{1, 2, \cdots, n\} \backslash T_i$.
5: Calculate the expected $D_{n-|I_i|-1}$ by deleting the tth rows as well as the tth column elements of D_n.
6: Calculate scores for the remaining $n - |I_i| - 1$ nodes according to Equation (8) using $D_{n-|I_i|-1}$.

As shown in Algorithm 2, we describe the steps that analyze how matrix D_n will change and calculate scores for the remaining nodes after deleting N_i.

In this Section, we demonstrate how to manage our hash-based graph and rank when faced with addition and deletion operations by giving fast calculation method of the iteration matrix. Incidentally, the operation of modifying a node is actually such a process that we first delete (Algorithm 2) a node and then add (Algorithm 1) a node. We will not elaborate this process due to limited space.

6   Conclusion

This paper builds a hash-based graph using restricted Hamming distance and proposes an undirected hash-based rank algorithm to calculate importance score for each node. By analyzing the iterative matrix, we give a full mathematical proof to verify that our algorithm will converge. Moreover, we illustrate the rationality of our algorithm. At last, we demonstrate how to manage our hash-based graph and rank by performing the minimal change strategy after adding and deleting a node, which can dynamically and fast compute the score and rank for each node in the updated graph.

Acknowledgements. This work is supported by the Innovation Group Project of the National Natural Science Foundation of China No. 61821003 and the National Key Research and Development Program of China under grant No. 2016YFB0800402 and the National Natural Science Foundation of China No. 61672254. Thanks for Jay Chou, a celebrated Chinese singer whose songs have been accompanying the author.

References

1. Baeza, P.B.: Querying graph databases. In: Proceedings of the 32nd ACM SIGMOD-SIGACT-SIGART Symposium on Principles of Database Systems, PODS 2013, 22–27 June 2013, New York, NY, USA, pp. 175–188 (2013). https://doi.org/10.1145/2463664.2465216
2. Cai, H., Huang, Z., Srivastava, D., Zhang, Q.: Indexing evolving events from tweet streams. In: ICDE, pp. 1538–1539 (2016)
3. Cuzzocrea, A., Jiang, F., Leung, C.K.: Frequent subgraph mining from streams of linked graph structured data. In: Proceedings of the Workshops of the EDBT/ICDT 2015 Joint Conference (EDBT/ICDT), 27 March 2015, Brussels, Belgium, pp. 237–244 (2015). http://ceur-ws.org/Vol-1330/paper-37.pdf
4. Gao, S., Cheng, X., Wang, H., Chia, L.: Concept model-based unsupervised web image re-ranking. In: ICIP, pp. 793–796 (2009)
5. Ge, S.S., Zhang, Z., He, H.: Weighted graph model based sentence clustering and ranking for document summarization. In: ICIS, pp. 90–95 (2011)
6. Hua, Y., Jiang, H., Feng, D.: FAST: near real-time searchable data analytics for the cloud. In: International Conference for High Performance Computing, Networking, Storage and Analysis, SC 2014, 16–21 November 2014, New Orleans, LA, USA, pp. 754–765 (2014). https://doi.org/10.1109/SC.2014.67
7. Lei, Y., Li, W., Lu, Z., Zhao, M.: Alternating pointwise-pairwise learning for personalized item ranking. In: Proceedings of the 2017 ACM on Conference on Information and Knowledge Management, pp. 2155–2158. ACM (2017)
8. Liu, Y., et al.: Deep self-taught hashing for image retrieval. IEEE Trans. Cybern. **49**(6), 2229–2241 (2019)
9. Michaelis, S., Piatkowski, N., Stolpe, M. (eds.): Solving Large Scale Learning Tasks, Challenges and Algorithms - Essays Dedicated to Katharina Morik on the Occasion of Her 60th Birthday. LNCS (LNAI), vol. 9580. Springer, Cham (2016). https://doi.org/10.1007/978-3-319-41706-6
10. Mihalcea, R.: Graph-based ranking algorithms for sentence extraction, applied to text summarization. Unt Sch. Works **170–173**, 20 (2004)
11. Page, L., Brin, S., Motwani, R., Winograd, T.: The pagerank citation ranking: Bringing order to the web. Technical report, Stanford InfoLab (1999)
12. Richter, F., Romberg, S., Hörster, E., Lienhart, R.: Multimodal ranking for image search on community databases. In: MIR, pp. 63–72 (2010)
13. Wang, Y., Zhu, L., Qian, X., Han, J.: Joint hypergraph learning for tag-based image retrieval. IEEE Trans. Image Process. **PP**(99), 1 (2018)
14. Yang, J., Jie, L., Hui, S., Kai, W., Rosin, P.L., Yang, M.H.: Dynamic match Kernel with deep convolutional features for image retrieval. IEEE Trans. Image Process. **27**(11), 5288–5302 (2018)
15. Zhou, K., Liu, Y., Song, J., Yan, L., Zou, F., Shen, F.: Deep self-taught hashing for image retrieval. In: Proceedings of the 23rd Annual ACM Conference on Multimedia Conference, MM 2015, 26–30 October 2015, Brisbane, Australia, pp. 1215–1218 (2015). https://doi.org/10.1145/2733373.2806320

Information Extraction and Retrieval

Two-Encoder Pointer-Generator Network for Summarizing Segments of Long Articles

Junhao Li and Mizuho Iwaihara[✉]

Graduate School of Information, Production and Systems, Waseda University,
Kitakyushu, Japan
lijunhao@toki.waseda.jp, iwaiahra@waseda.jp

Abstract. Usually long documents contain many sections and segments. In Wikipedia, one article can usually be divided into sections and one section can be divided into segments. But although one article is already divided into smaller segments, one segment can still be too long to read. So, we consider that segments should have a short summary for readers to grasp a quick view of the segment. This paper discusses applying neural summarization models including Seq2Seq model and pointer generator network model to segment summarization. These models for summarization can take target segments as the only input to the model. However, in our case, it is very likely that the remaining segments in the same article contain descriptions related to the target segment. Therefore, we propose several ways to extract an additional sequence from the whole article and then combine with the target segment, to be supplied as the input for summarization. We compare the results against the original models without additional sequences. Furthermore, we propose a new model that uses two encoders to process the target segment and additional sequence separately. Our results show our two-encoder model outperforms the original models in terms of ROGUE and METEOR scores.

Keywords: Text summarization · Deep learning · Seq2Seq ·
Pointer generator network · Multi-encoder

1 Introduction

Wikipedia, started from 2001, has been recognized as the world's largest encyclopedia. Wikipedia is collaboratively edited and maintained by a great number of collaborators, which assures reliability regardless of its openness [5].

With the number of Wikipedia articles growing quickly, the information that Wikipedia contains is invaluable. But generally, usage of Wikipedia is limited to mainly for human readers. The most important work that computer systems can support in using Wikipedia resource is searching and showing interesting articles to readers [13]. With the continuously growing number of articles in Wikipedia, search queries have to be carefully constructed. In addition, segments in one article sometimes become longer, which is causing difficulties for human readers to grasp a quick view of one segment. Summarizing the whole article may provide one solution. But the summary of one whole article should contain information significant throughout the whole article, where information regarding the target segment could be dropped out. Thus we think it is necessary to focus on segment summarization.

© Springer Nature Switzerland AG 2019
J. Shao et al. (Eds.): APWeb-WAIM 2019, LNCS 11641, pp. 299–313, 2019.
https://doi.org/10.1007/978-3-030-26072-9_23

There are mainly two approaches to text summarization: Extractive and abstractive.

Extractive methods involve selection of related phrases and sentences from the source document to generate a new summary. Techniques involve ranking by the relevance of sentences or phrases in order to choose only those most relevant to the meaning of the source. One widely used algorithm is TextRank [17] that uses a graph-based ranking algorithm to extract important sentences from a single document.

Abstractive methods attempt to develop an understanding of the main concepts in a document and then express those concepts in clear natural language. They use linguistic methods to examine and interpret the text and then to find new concepts and expressions to best describe it by generating a new shorter text that conveys the most important information from the original text document [19]. Recently, deep learning methods have shown promising results for text summarization. One of the significant models is based on a sequence-to-sequence attention model [9].

In this paper, we focus on summarization methods that involve deep learning models where both approaches are utilized. We need to extract additional information from the whole article, to provide background information for summarizing the target segment. However, the whole article is sometimes too long containing noisy information. To overcome this problem, we consider several extractive methods for shortening the article into an additional sequence for the model. We propose several ways to combine the additional sequence with the target segment, as the input for neural summarization. Furthermore, we propose a new model that uses two encoders to process the target segment and additional sequence separately. Our evaluation results show that additional sequences can indeed improve summarization quality in terms of ROGUE and METEOR scores, and our two-encoder model outperforms the original models.

The rest of this paper is organized as follows. Section 2 covers related work. Section 3 describes summarization models we use. Section 4 discusses three approaches for generating additional sequences. Section 5 is the results of our experiments and Sect. 6 discusses the results. Section 7 concludes this paper.

2 Related Work

Long short-term memory (LSTM) was first proposed by Hochreiter and Schmidhuber [18] and improved in 2002 by Felix Gers' team [6]. LSTM networks are well-suited for classifying, processing and making predictions based on time series data. LSTMs were developed to deal with the exploding and vanishing gradient problems that are encountered when training traditional RNNs. Later Graves et al. [7] proposed a neural network model using bidirectional LSTM (Bi-LSTM) for phoneme classification, showing great improvement over the original LSTM.

Neural network-based sequence-to-sequence (Seq2Seq) learning has achieved remarkable success in various NLP tasks, including but not limited to machine translation and text summarization. The Seq2Seq model is first proposed by Ilya et al. [2], which is an encoder-decoder model such that the encoder converts the input sequence into one context vector that contains all information in the input sequence. Then the decoder generates the output sequence based on the context vector.

Bahdanau et al. [4] were the first to introduce attention mechanism to the Seq2Seq model to release the burden of compressing of entire source into a fixed-length vector as context. Instead, they proposed to use a dynamically changing context vector h_t^* in decoding process. e_i^t is the attention energy of hidden states of encoder at timestep i in decoder step t. a^t is the attention weight distribution. v^T, W_h, W_s and b_{atten} are learnable parameters.

$$e_i^t = v^T \tanh(W_h h_i + W_s s_t + b_{attn}) \tag{1}$$

$$a^t = \text{softmax}(e^t) \tag{2}$$

$$h_t^* = \sum_i a_i^t h_i \tag{3}$$

Rush, Chopra, et al. [1] were the first to apply a Seq2Seq and attention model on text summarization, achieving state-of-the-art performance on DUC-2004, Gigaword, and two sentence-level summarization datasets.

Based on Rush [1], Luong et al. [14] proposed a multi-task Seq2Seq model to solve different NLP problems using the same model at the same time. Their model settings include one encoder to many decoders, many encoders to one decoder, and many encoders to many decoders. Their results show that it is a promising model for solving multi-task problems. Several results that use multiple encoders are even better than only one encoder.

Vinyals et al. [15] proposed a Seq2Seq model called pointer network which produces an output sequence consisting of elements from the input sequence applying soft attention distribution of Bahdanau et al. [4]. The pointer network has been utilized to create hybrid approaches for NMT, language modeling and summarization.

See et al. [2] proposed a pointer-generator network model that combines a pointer network with the original Seq2Seq + attention model. Their model allows both copying words through pointing and generating words from a fixed vocabulary. In their model, they use the Bahdanau attention mechanism [4] in Eqs. (1), (2) and (3) to calculate attention distribution a^t and context vector h_t^*. Their generation probability $p_{gen} \in [0, 1]$ for timestep t is calculated from the context vector, decoder state s_t and the decoder input x_t.

$$p_{gen} = \sigma(w_{h^*}^T h_t^* + w_s^T s_t + w_x^T x_t + b_{ptr}), \tag{4}$$

where $w_{h^*}^T$, w_s^T, w_x^T and b_{ptr} are learnable parameters and σ is the sigmoid function. Then the p_{gen} is used as a soft switch to choose between generating a word from the vocabulary by sampling from p_{vocab} or copying a word from the input sequence by sampling from the attention distribution a^t.

$$P_{vocab} = \text{softmax}(V'(V[s_t, h_t^*] + b) + b') \tag{5}$$

$$P(w) = p_{gen}P_{vocab}(w) + (1 - p_{gen})\sum_{i:w_i=w} a_i^t \tag{6}$$

Note that if w is an out-of-vocabulary (OOV) word, then $p_{vocab}(w)$ is zero. Similarly, if w does not appear in the source document, then $\sum_{i:w_i=w} a_i^t$ is zero. Their model has the ability to produce OOV words which is one of the primary advantages compared to the original Seq2Seq model.

3 Neural Summarization Models

In this section, we describe the baseline sequence-to-sequence model, pointer-generator network, and our proposing two-encoder pointer-generator network models.

3.1 Sequence-to-Sequence Attentional Model

We utilize the sequence-to-sequence attentional model whose input is only target segment text as our baseline model. This Seq2Seq model is a direct application of See et al. [2]. Figure 1 shows the structure of this model. The words w of the target segment are fed into the encoder (one single-layer Bi-LSTM) one by one in encoding timestep i, producing a sequence of encoder hidden states h. Then the decoder (one single-layer LSTM) receives the word embedding of the word from the previous step (when training, the previous word is from the reference segment summary) on each step t and has decoder state s_t.

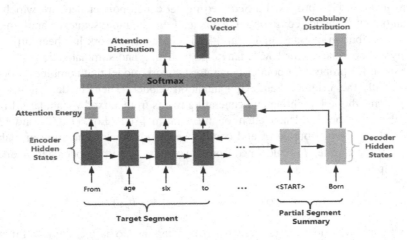

Fig. 1. Sequence-to-sequence attentional model

Then the attention distribution a^T is calculated by Bahdanau attention [4] in (1) and (2). The attention distribution can be considered as a probability distribution over the source words, which helps the decoder decide where to focus to produce the next word.

Next the attention distribution is used to produce a weighted sum of the encoder hidden states, denoted as context vector h_t^* in (3). The context vector can be viewed as a fixed-length representation of what has been read from the target segment for this step.

Then the context vector is concatenated with the decoder state s_t and fed through two linear layers to produce the vocabulary distribution p_{vocab} in (5). p_{vocab} is a probability distribution over all words in the vocabulary. In this model the final distribution $P(w)$ to predict words w is:

$$P(w) = P_{vocab}(\mathrm{w}) \tag{7}$$

The loss for training of decode step t is the negative log likelihood of the target word w_t^*:

$$loss_t = -\log P(w_t^*) \tag{8}$$

The overall *loss* for the whole sequence is:

$$loss = \frac{1}{T}\sum_{t=0}^{T} loss_t \tag{9}$$

The execution of this model has two input settings. One is the baseline that uses only the target segment, and the other uses concatenation of the target segment and compressed whole article as input.

3.2 Pointer-Generator Network

We utilize the pointer-generator network with a coverage mechanism proposed by See et al. [2], which allows both copying words by pointing from the source text and generating words from a fixed vocabulary. Figure 2 shows the structure of this model.

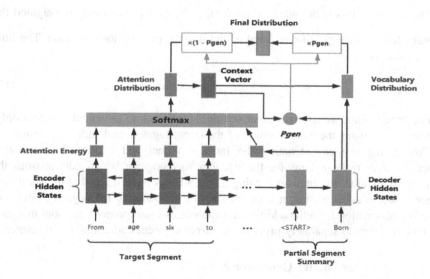

Fig. 2. Pointer-generator network

For the purpose of solving the common repetition problem for sequence-to-sequence models, they proposed coverage mechanism. They define the sum of attention distributions over all previous decoder timesteps as coverage vector c^t:

$$c^t = \sum_{t'=0}^{t-1} a^{t'} \tag{10}$$

The coverage vector c^t can be considered as a distribution over the source document words that represents the degree of coverage that those words have received from the attention mechanism so far. Note that c^0 is a zero vector because on the first decoding timestep, none has been covered from the source document. Then the coverage vector is used as extra input to the attention mechanism. Changing the formula to calculate the attention energy from (1) to:

$$e_i^t = v^T tanh\left(W_h h_i + W_s s_t + W_c c_i^t + b_{attn}\right) \tag{11}$$

Here, W_c is a learnable parameter vector of same length as v. Adding the coverage vector to the attention mechanism lets the attention mechanism's current decision consider its previous decisions summarized in the coverage vector c^t.

Then the attention distribution a^t and context vector h_t^* are calculated as in (2) and (3). The generation probability $p_{gen} \in [0, 1]$ for timestep t is calculated as in (4). Then the final probability distribution is given in (6).

They also defined a coverage loss to penalize repeatedly attending to the same locations:

$$covloss_t = \sum_i \min\left(a_i^t, c_i^t\right) \tag{12}$$

The coverage loss is bounded as: $covloss_t \leq \sum_i a_i^t = 1$. Then they reweighted the coverage loss by hyperparameter λ and added it to the primary loos function. The final loss function is:

$$loss_t = - \log P\left(w_t^*\right) + \lambda\, covloss_t \tag{13}$$

For experiments, we use two input settings. One is using the target segment only, and the other is using the concatenation of the target segment and additional sequence.

Considering just concatenating the target segment and additional information sequence might make it hard for the decoder to recognize which part is from the segment. In addition, the encoder is using a Bi-LSTM, which means that the additional sequence would affect the target segment's result. Considering the negative effects of the noise information from the additional sequence, we also propose a model that uses different encoders to separately process the target segment and additional sequence.

3.3   Two-Encoder Pointer-Generator Network

We extend the pointer-generator network with two encoders, to process the target segment and the additional sequence separately, so that interaction of the two inputs occurs only at the attention layer. Figure 3 shows the structure of our model.

There are two hidden state sequences produced by two different encoders. The target segment encoder produces hidden states h_i^1, and the additional sequence encoder produces hidden states h_i^2. Having two hidden state sequences makes it not suitable for the original Bahdanau attention mechanism. So we modify the attention to fit for our purpose. We calculate the attention energy for two hidden state sequences separately as:

$$e1_i^t = v1^T \tanh\left(W1_h h_i^1 + W1_s s_t + W1_c c_i^t + b1_{attn}\right) \qquad (14)$$

$$e2_i^t = v2^T \tanh\left(W2_h h_i^1 + W2_s s_t + W2_c c_i^t + b2_{attn}\right) \qquad (15)$$

where $e1_i^t$ is the target segment's attention energy and $e2_i^t$ is the additional sequence's attention energy. $v1^T$, $W1_h$, $W1_s$, $b1_{attn}$, $v2^T$, $W2_h$, $W2_s$ and $b2_{attn}$ are learnable parameters.

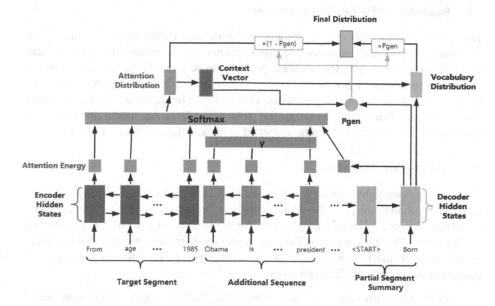

Fig. 3. Two-encoder pointer-generator network

Then we concatenate the two attention energy sequences into one by a learnable parameter γ (1.0 as initial value) to calculate the whole attention weight distribution a^t:

$$a^t = \text{softmax}([e1^t, \gamma e2^t]) \qquad (16)$$

Then we calculate the context vector using attention weight distributions from (3) by the following:

$$h_t^* = \sum\nolimits_{i=0}^{k} a_i^t h_i^1 + \sum\nolimits_{i=k+1}^{m+k} a_i^t h_i^2, \qquad (17)$$

where k is the length of the hidden state sequence of the target segment encoder, and m is the length of the hidden state sequence of the additional sequence encoder.

Then the remaining step is similar to the previous pointer-generator network model. We use the same calculation to obtain the final distribution, and the loss is still using (12).

4 Techniques for Additional Sequences

For generating a summary of a target segment in a long article, the crucial point is how to incorporate contents outside of the target segment. Such long articles would make it hard to train RNN-based models. In addition, we cannot ignore that there would also be a large volume of noise information outside of the target segment. So in this section, we discuss three single-document extractive methods to extract significant sentences from the whole article, to generate additional input sequences.

4.1 Basic Method – Leading Sentences

The basic method simply chooses leading two sentences in each segment of the article, and concatenates these sentences into one sequence as the additional sequence. Here, leading two sentences are experimentally performing best among leading 1–3 sentences.

The experiments in See et al. [2] shows that only using leading three sentences performs better than using longer inputs. This is based on observations that important information is often written in the leading sentences. Thus, we utilize this method as our baseline method for extracting important sentences from the target article.

4.2 TextRank + Word2vec

TextRank, proposed by Mihalcea and Tarau [12], is a graph-based ranking algorithm inspired by PageRank [16]. TextRank is mainly used for keyphrase extraction and important sentence extraction. In our work, we apply TextRank for sentence extraction.

To apply TextRank and word2vec on sentence extraction, we first build a graph associated with the text, where the graph vertices are representing sentences in the document to be ranked. The main steps are:

1. Split the target document into sentences $= [S_1, S_2, S_3, \ldots, S_m]$, and construct graph $G = (V, E)$. Then split sentences into words and remove stop words to obtain $S_i = [t_{(i,1)}, t_{(i,2)}, \ldots, t_{(i,n)}]$, where $t_{(i,j)} \in S_{1i}$ is a candidate keyword.
2. Using word2vec to convert words in a sentence to vectors and calculate the average vector of the sentence:

$$vec(S_i) = \frac{1}{n} \sum_{j=1}^{n} word2vec(t_{i,j}) \tag{18}$$

3. Calculate the cosine similarity between sentences as the edge weights in graph G:

$$Similarity(S_i, S_j) = cos(vec(S_i), vec(S_j)) \tag{19}$$

4. Calculate the score S_1 of each sentence, by the following iterative method:

$$WS(S_i) = (1 - d) + d \sum_{S_j \in In(S_i)} \frac{w_{ji}}{\sum_{S_k \in Out(S_j)} w_{jk}} WS(S_j) \qquad (20)$$

5. Choose the top-10 highly scored sentences, and sort the sentences by their positions in the document. Concatenate these sentences as our additional sequence.

In Step 5, we select ten sentences, to let each additional sequence have at least 200 words.

4.3 TextTeaser

TextTeaser is a feature-based extractive summarization algorithm, which selects sentences that possess best score among others. Its features include title, sentence length, sentence position and keyword frequency. We apply TextTeaser with its default settings.

Additional sequences by all the three methods are truncated into leading 200 words, which is maximum input length for the encoder to show reasonable performance in our experimental environments.

5 Experiments

5.1 Datasets

We collected 26,318 pages from English version of Wikipedia as our dataset for the experiments. There are 81,124 segments and 50,056 sections altogether in the dataset. We convert all the words to lower case and the numbers to "TAG_OF_NUM".

Since there is no human labeled segment summary in Wikipedia articles, and writing this number of segment summaries is not realistic, we construct reference summaries as follows. First we utilize word2vec to calculate the average vector of the segment title and body sentences of the target segment. Then we use cosine similarity to select the body sentence most similar to the segment title. The concatenation of one body sentence most similar to the segment title and the title itself is used as the reference summary of the target segment. Since the words of the target tile are not necessary occurring in the segment body, the ability of abstractive summarization is necessary to generate the perfect solution.

5.2 Settings

For all experiments, we follow the settings in See et al. [2]. Our models have 256-dimensional hidden states and 128-dimensional word embeddings. For the pointer-generator network models, we use a vocabulary of 50k words for both source and target.

The word embeddings we are using are not pre-trained. They are learned from scratch during training. We train using Adagrad [10] with learning rate 0.15 and an initial accumulator value of 0.1. We use gradient clipping with a maximum gradient norm of 2 without using any form of regularization.

For training and testing, we truncated the target segment and additional sequence into leading 400 words and limit the length of the summary to 50 words. According to See et al. [2], truncating the input sequence could raise the performance of the model.

We trained our models on a single 1080Ti GPU with a batch size of 32. For testing, we utilized the beam search algorithm to produce segment summaries of beam size 8.

To compare the performance between models, we trained our models separately using the additional sequence which is generated by either leading sentences, TextRank or TextTeaser. We also performed experiments with different lengths of target segments to see how the length affects the results.

5.3   Results

Our results are shown in Tables 1 and 2. We evaluate our models with the standard ROUGE metric [3], reporting the F_1 scores for ROUGE-1, ROUGE-2 and ROUGE-L, which respectively measure the word-overlap, bigram-overlap, and longest common sequence between the reference summary and the summary to be evaluated. According to See et al. [2], ROUGE tends to reward safe strategies such as selecting the first-appearing content, or preserving original phrasing, which may lead to extractive systems obtains higher ROUGE scores on average. Therefore, we also evaluate our systems with the METEOR metric, which rewards not only exact word matches, but also matching stems, synonyms and paraphrases.

The column "Additional Sequence Method" in Table 1 shows whether additional sequences are used or not, and which extractive summarization algorithm is used to generate an additional sequence from the rest of the article. For the Seq2Seq attentional model and pointer-generator network, the additional sequence is concatenated after the target segment. For the case of "None," only truncated target segment is the input to the model.

In the results of the no-additional sequence methods, we can see that the pointer-generator network outperforms the traditional Seq2Seq attentional model. The model trains faster and takes less iterations. Thus, in succeeding experiments on additional sequence methods, we only use the pointer-generator network.

We now examine the methods that simply concatenate the additional sequence after the target segment. The results show that our idea of additional sequences from the entire article is improving the performance. Among them, sequences that are extracted by TextRank show the best results. Therefore, for our model that processes two sequences separately, we only use TextRank for extraction of additional sequences.

The last row of Table 1 is the result of our proposing model, two-encoder pointer-generator network. As we can see from the scores, our model out-performed all the other models including the original pointer-generator network with additional sequences by TextRank. Thus, our model that uses two encoders to process the target segment and the additional sequence separately appears to be effective.

Table 1. Results on segment summarization

Models	Additional sequence method	ROUGE			METEOR	
		1	2	L	Exact	+stem/syn/para
TextRank	None	44.99	33.20	38.84	26.72	27.46
Seq2Seq attentional model	None	36.70	19.33	29.90	17.08	17.65
Pointer-generator network	None	42.17	28.01	35.82	23.26	23.91
	Basic	42.70	29.02	36.00	25.20	25.73
	TextRank	43.96	30.23	37.31	26.33	26.86
	TextTeaser	43.13	29.43	36.38	25.46	25.98
Two-encoder pointer-generator network	TextRank	**44.69**	**31.04**	**38.17**	**27.17**	**27.69**

We also evaluate TextRank as an extractive method to extract summary from the target segment. The first row shows that TextRank is slightly better than our model by the ROGUE scores. We consider a possible reason such that the major part of our golden standard in the dataset includes one sentence from the target segment. However, in terms of the METEOR scores, which considers flexible matches by stems, synonyms and paraphrases, our proposing method outperforms TextRank.

Additionally, we performed experiments on how the length of a target segment affects the results. We divided the testing dataset into three parts, by the length of the target segments: Less than 200 words, between 200 and 400 words and over 400 words. Then we apply our trained two-encoder pointer-generator network model from the last experiment, to produce the results. Table 2 shows the results.

The results show that, compared to the overall average, our model performs better when the target segment is less than 200 words and slightly better for the case between 200 and 400 words. But it becomes worse when the length exceeds 400 words. It implies that our model is more capable to handle shorter segments, which is one of the existing common problems for RNNs and LSTMs such that longer documents are harder to process. Also, shorter segments mean a smaller number of sentences, giving a higher possibility to locate the reference sentence in the segment.

Table 2. Experiment on length of target segments

Length	ROUGE			METEOR	
	1	2	L	Exact	+stem/syn/para
Overall	44.69	31.04	38.17	27.17	27.69
(0, 200]	53.86	40.65	47.30	33.83	34.26
(200, 400]	45.39	33.19	40.70	28.73	28.93
(400, ∞)	29.86	15.65	23.68	16.53	17.20

Considering that there has no perfect evaluation method for text summarization, we examined several summaries by the four different methods through human eyes. Table 3 shows results of two sample segments in articles "Asa T. Newhall" and "Daniel Johnston." We can see that the Seq2Seq model is clearly showing the worst quality, failing to generate a large number of out-of-vocabulary (OOV) words, which the pointer-generator network is capable of.

Table 3. Summary examples on articles "Asa T. Newhall" (left) and "Daniel Johnston." (right)

Reference summary: TAG_OF_NUM state representative election in TAG_OF_NUM. newhall was elected as a democrat to serve in the massachusetts house of representatives of TAG_OF_NUM.	TAG_OF_NUM. on march TAG_OF_NUM, TAG_OF_NUM, johnston released his first comic book published by boom!.
Seq2Seq Attention model: state representative in the TAG_OF_NUM election [UNK] received TAG_OF_NUM [UNK] TAG_OF_NUM votes coming in second behind republican candidate john [UNK] gonzalez who received TAG_OF_NUM [UNK] TAG_OF_NUM [UNK] and just ahead of fellow democrat michael [UNK] jay who received TAG_OF_NUM [UNK] TAG_OF_NUM [UNK]	and jonathan [UNK] on march TAG_OF_NUM [UNK] TAG_OF_NUM [UNK] johnston released his first comic [UNK] at [UNK] published by [UNK] [UNK] [UNK] comic book [UNK] with the album and an warren's [UNK] has collaborated with skateboarding and clothing company supreme on numerous collections [UNK] of clothing and various [UNK].
Pointer-Generator Network without Additional Sequence: TAG_OF_NUM election in the TAG_OF_NUM election newhall received TAG_OF_NUM , TAG_OF_NUM votes coming in second behind republican candidate john w. blaney who received TAG_OF_NUM , TAG_OF_NUM votes, and just ahead of fellow democrat michael f. phelan who received TAG_OF_NUM , TAG_OF_NUM [UNK]	on march TAG_OF_NUM , TAG_OF_NUM , johnston released his comic book, published by boom! [UNK] the comic book ties-in with the album and an ios app. johnston has collaborated with skateboarding and clothing company supreme on numerous collections
Pointer-Generator Network with TextRank additional Sequence: congress in TAG_OF_NUM newhall was elected as a democrat to serve in the massachusetts house of representatives of TAG_OF_NUM.	and on march TAG_OF_NUM , TAG_OF_NUM , johnston released his first comic book, at sxsw, published by boom! studios.
Two-Encoder Pointer-Generator Network (proposed): TAG_OF_NUM election in TAG_OF_NUM newhall was elected as a democrat to serve in the massachusetts house of representatives of TAG_OF_NUM.	TAG_OF_NUM on march TAG_OF_NUM ,TAG_OF_NUM , johnston released his first comic book, at sxsw, published by boom! studios.
TextRank: newhall was elected as a democrat to serve in the massachusetts house of representatives of TAG_OF_NUM.	on march TAG_OF_NUM, TAG_OF_NUM, johnston released his first comic book published by boom!

• *Green denotes segment title in the reference summary*

Both Seq2Seq attentional model and pointer-generator network are generating long summaries containing excessive words. On the other hand, the methods with additional sequences are generating compact summaries of around 20 words. This can be explained as by incorporating attentions from an additional sequence, words related to the entire article are strongly emphasized, so that such words and their surrounding words are likely to be chosen. On the other hand, without additional sequences, words in the target segment are likely to be given even weights.

6 Discussions

6.1 How Additional Sequences Work?

It is clear from Table 1 that incorporating additional sequences tends to achieve higher scores than without them, no matter by simple concatenation or by using two encoders. We point out two possible explanations for this observation.

Firstly, Wikipedia articles usually have multiple segments, and these segments are not isolated from each other. There exists a main topic of the article, represented by its title, and all the segments should be related to the main topic. Without an additional sequence, it is difficult to find words related to the main topic. In the example of "Asa T. Newhall" in Table 3, the main subject is "newhall" and "election" is a part of the main topic of this article. However, the Seq2Seq Attention model fails to find "newhall," and the methods without additional sequences list words such as "republican" and "gonzalez," which are distant from the main topic.

Secondly, an additional sequence can be regarded as an extractive summary of the entire article. By supplying the target segment and the additional sequence, words in the target segment that are related to the significant words of the whole article can be emphasized.

As an extension, we can consider about generating additional sequences from external related documents. Wikipedia articles need to include references indicating the source of each statement. News articles are often cited as references. Such external reference documents can supplement words that are related to the target segment but not occurring in the segment. By incorporating reference sources that have high similarity with the target segment into the additional sequence, words that are appropriate but not appearing in the article could be introduced, augmenting vocabulary for generating summaries.

6.2 Why Processing Separately Performs Better?

Rows 5 and 7 in Table 1 show that using two encoders to process an additional sequence and target segment separately scores higher than concatenation. We discuss reasoning for this.

As we can see in Table 2, the results are getting worse as the target segment becomes longer. We consider this is due to the common problem of RNNs on long sequences. In our task, concatenating the target segment and additional sequence creates an even longer input sequence, which makes learning by one LSTM-based encoder harder. On the other hand, our two-encoder model uses two LSTMs to process the target segment and additional sequence separately, which can reduce the input length. However, additional sequences have to be generated from long documents such as one whole article. To control the length of additional sequences, extractive summarization such as TextRank that is scalable to long documents is appropriate.

Using several LSTMs to process portions of input text has been discussed. Tan et al. [11] are using a hierarchical encoder-decoder framework for text summarization, including word encoders and sentence encoders. Introducing such an encoder-decoder framework hierarchically organized along the document structure is a possible future direction. However, segment summarization requires uneven weighting between the target segment and the remaining segments.

7 Conclusions

In this paper, we proposed a neural summarization method targeted on generating a summary of one segment within a relatively long document, where contextual information from the whole document has to be reflected to estimate important parts within the target segment. We proposed the two-encoder pointer-generator network, which generates a context vector from two encoders which separately acquire attention energy sequences from the target segment and an additional sequence which is extracted from the whole article.

We performed experiments on three different models on Wikipedia articles, consisting of Seq2Seq + attention, pointer-generator network, and our two-encoder pointer-generator network. We also considered three different ways to extract additional sequences from the whole article. The results show that our approach of processing an additional sequence and the target segment separately by two encoders is effective. There still exist rooms for improvement. One is that our model still suffers from long segments. However, the direction of multiple LSTMs has potential for dealing with longer segments or articles, if an appropriate hierarchical segmentation is employed.

References

1. Rush, A.M., Chopra, S., Weston, J.: A neural attention model for abstractive sentence summarization. In: Empirical Methods in Natural Language Processing (2015)
2. See, A., Liu, P.J., Manning, C.D.: Get to the point: summarization with pointer-generator networks. In: Annual Meeting of the Association for Computational Linguistics (2017)
3. Lin, C.-Y.: Looking for a few good metrics: automatic summarization evaluation-how many samples are enough? In: NACSIS/NII Test Collection for Information Retrieval (NTCIR) Workshop (2004)
4. Bahdanau, D., Cho, K., Bengio, Y.: Neural machine translation by jointly learning to align and translate. Comput. Sci. (2014)
5. Nguyen, D.P.T., Matsuo, Y., Ishizuka, M.: Exploiting syntactic and semantic information for relation extraction from Wikipedia. In: Text-Mining and Link-Analysis (TextLink 2007) (2007)

6. Gers, F.A., Schraudolph, N.N., Schmidhuber, J.: Learning precise timing with LSTM recurrent networks. J. Mach. Learn. Res. **3**(Aug), 115–143 (2002)
7. Graves, A., Schmidhuber, J.: Framewise phoneme classification with bidirectional LSTM networks. In: IEEE International Joint Conference on Neural Networks, vol. 4, pp. 2047–2052 (2005)
8. Hu, M., Sun, A., Lim, E.-P.: Comments-oriented blog summarization by sentence extraction. In: Proceedings of the Sixteenth ACM Conference on Information and Knowledge Management, pp. 901–904. ACM, Lisbon (2007)
9. Sutskever, I., Vinyals, O., Le, Q.V.: Sequence to sequence learning with neural networks. In: NIPS (2014). 2, 3, 7
10. Duchi, J., Hazan, E., Singer, Y.: Adaptive subgradient methods for online learning and stochastic optimization. J. Mach. Learn. Res. **12**, 2121–2159 (2011)
11. Tan, J., Wan, X., Xiao, J.: Proceedings of the 55th Annual Meeting of the Association for Computational Linguistics, pp. 1171–1181 (2017)
12. Mihalcea, R., Tarau, P.: TextRank: bringing order into text. In: Proceedings of the 2004 Conference on Empirical Methods in Natural Language Processing (2004)
13. Volkel, M., Krotzsch, M., Vrandecic, D., Haller, H., Studer, R.: Semantic Wikipedia. In: Proceedings of the WWW 2006, pp. 585–594 (2006)
14. Luong, M.-T., Le, Q.V., Sutskever, I., Vinyals, O., Kaiser, L.: Multi-task sequence to sequence learning. In: ICLR (2016)
15. Vinyals, O., Fortunato, M., Jaitly, N.: Pointer networks. In: Neural Information Processing Systems (2015)
16. Page, L., et al.: The PageRank citation ranking: bringing order to the web. Stanford Info Lab (1999)
17. Mihalcea, R., Tarau, P.: TextRank: bringing order into texts (2004)
18. Hochreiter, S., Schmidhuber, J.: Long short-term memory. Neural Comput. **9**(8), 1735–1780 (1997)
19. Gupta, V., Lehal, G.S.: A survey of text summarization extractive techniques. J. Emerg. Technol. Web Intell. **2**(3), 258–268 (2010)

Enhancing Joint Entity and Relation Extraction with Language Modeling and Hierarchical Attention

Renjun Chi, Bin Wu, Linmei Hu[✉], and Yunlei Zhang

Beijing Key Laboratory of Intelligent Telecommunications Software
and Multimedia, Beijing University of Posts and Telecommunications,
Beijing, China
chirenjun@gmail.com,
{wubin, hulinmei, yunlei0518}@bupt.edu.cn

Abstract. Both entity recognition and relation extraction can benefit from being performed jointly, allowing them to enhance each other. However, existing methods suffer from the sparsity of relevant labels and strongly rely on external natural language processing tools, leading to error propagation. To tackle these problems, we propose an end-to-end joint framework for entity recognition and relation extraction with an auxiliary training objective on language modeling, i.e., learning to predict surrounding words for each word in sentences. Furthermore, we incorporate hierarchical multi-head attention mechanisms into the joint extraction model to capture vital semantic information from the available texts. Experiments show that the proposed approach consistently achieves significant improvements on joint extraction task of entities and relations as compared with strong baselines.

Keywords: Entity recognition · Relation extraction · Joint model · Language modeling objective · Hierarchical attention

1 Introduction

The goal of the entity recognition and relation extraction is to detect pairs of entities and identify the semantic relationships between them. As shown in Fig. 1, the entities "Thomas" and "Chisholm" are detected from the sentence "Composer Thomas is a native of Chisholm, Minn.", and their relation is labeled as "Live_in". The extracted result from this sentence is represented by a triplet $\{Thomas_{e1}, Live\_in_r, Chisholm_{e2}\}$. It is critical for information extraction task to understand massive text corpora, and has been widely used in knowledge completion, auxiliary question answering system and so on. Therefore, it has attracted considerable attention from researchers over the past decades.

Traditional methods deal with this task in a pipeline manner, extracting entities first and then identifying their relationships. This separate framework ignores the correlation between entities and relationships, and between relationships. For example, the "Live_in" relationship often corresponds to two types of entities, Person and Location, and vice versa. In addition, the results of entity recognition generated by pipeline ways

J. Shao et al. (Eds.): APWeb-WAIM 2019, LNCS 11641, pp. 314–328, 2019.
https://doi.org/10.1007/978-3-030-26072-9_24

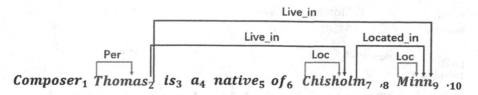

Fig. 1. A standard example sentence for the task. "Live_in" is a relation in the predefined relation set (Green indicates entity arcs and blue indicates relation arcs). (Color figure online)

may deteriorate the performance of relational classification and lead to error propagation. Recent studies [14, 18, 24, 26] focus on joint extraction methods to capture the inherent linguistic dependencies between relations and entity arguments to resolve error propagation. However, these methods still suffer from the following drawbacks:

First, most of the existing joint methods are feature-based structured systems. They require complex feature engineering and heavily rely on additional NLP tools, such as POS-tagging and dependency parsing. Therefore, the performances of these joint models largely depend on the quality of the features obtained from the NLP tools. However, these features are not always accurate for various languages and contexts.

Second, relevant labels in the dataset are sparse. For instance, in the dataset of CoNLL 2003 NER, only 17% of tokens represent entities. Sequence labeling models are able to learn this bias in the distribution of labels without much incentive to learn more general language features from the available texts.

At last but not least, one limitation of most existing joint models is that only a single relation can be assigned to a token. However, most of the sentences in natural language contain more than one triplet. As shown in the example in Fig. 1, there are three triplets in the sentence, i.e. $\{Thomas_{e1}, Live\_in_r, Chisholm_{e2}\}$, $\{Thomas_{e1}, Live\_in_r, Minn_{e2}\}$, $\{Chisholm_{e1}, Located\_in_r, Minn_{e2}\}$.

In this paper, we propose a novel hierarchical attention-based end-to-end bidirectional long short-term memory network integrated with language modeling objective to address the shortcomings described above. Figure 2 shows the overall architecture of our model. We employ Bidirectional Long Short-term Memory (Bi-LSTM), which is powerful to model long range dependencies, to encode the input sentences globally. Hierarchical attention is introduced to capture important semantic information in sentences. We model the relation extraction task as a multi-label head selection problem, allowing multiple relations for each entity in one input sentence. More importantly, language model with auxiliary training objective learns to predict the surrounding words of each word in the data set, stimulating the framework to learn more abundant of semantic synthesis features without additional training data. Consequently, our main contributions in this paper can be summarized as follows.

- We propose an end-to-end joint model, which performs the two tasks of entity recognition and relation extraction simultaneously, and that can handle multiple relations together.

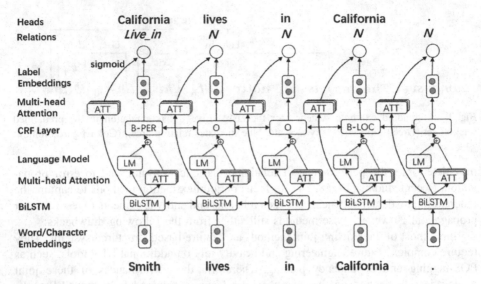

Fig. 2. Our model for joint entity and relation extraction with Language Model (LM) comprises (i) two embedding layers, (ii) a Bi-LSTM layer, (iii) a LM layer, (iv) a CRF layer, (v) a relation extraction layer and (vi) two multi-head attention layers.

- We introduce hierarchical attention mechanism which can not only allow the model to learn relevant information in different representation subspaces, but also capture the most important semantic information in sentences.
- We integrate language model with auxiliary training objective to learn more abundant language features in sentences, making full use of the available data. This strategy obtains remarkable results.

2   Related Work

Entity recognition and relation extraction are two fundamental tasks in information extraction, and have appealed many research interests. Early entity and relation extraction methods are mainly in pipeline ways, regarding this task as two separate tasks, named entity recognition (NER) [1, 5, 6, 25] and relation extraction (RE) [8, 10–12], which ignores inherent linguistic dependencies between entities and relations and results in error propagation.

Feature-based joint models have been proposed to simultaneously solve the entity recognition and relation extraction tasks [13, 18, 24, 26]. Different from the pipeline method, the joint learning framework uses a single model to perform both NER and RE tasks. It can effectively integrate information about entities and relationships. It has been proved that joint learning methods can achieve better performance in this task. However, these feature-based methods rely on the availability of NLP tools (e.g., POS taggers) or manually designed features, leading to additional complexity and poor performance in different application and language settings where the NLP tools are not reliable.

With the revival of neural networks in recent years, several neural network architectures have been successfully applied to overcome this feature engineering problem. Specifically, [19] solves a simpler entity classification problem (EC, assuming that the entity boundary is given) rather than NER. They replicate the context around the entity and provide the entity pairs to the relational extraction layer. [28] applies a bidirectional tree-structured RNN to capture grammatical information. [29] investigates RNNs with attention without taking into account the correlations of relation labels. [16] propose a neural method comprised of a sequence-based LSTM for entity identification and a separate tree-based dependency LSTM layer for relation classification. [2] convert the joint task to a sequence labeling problem, achieving joint decoding for entities and relations in one task. These methods do not infer other potential entities and relations within the same sentence. Our model enables simultaneous extraction of multiple relations from the same input and takes better advantage of the available training data.

3   Methodology

In this section, we describe in detail our proposed end-to-end joint entity and relation extraction model with language modeling objective and hierarchical attention. As shown in Fig. 2, our model consists of three main parts: (i) the sentence encoder part consisting of a character layer, a word embedding layer and a Bi-LSTM layer, (ii) hierarchical multi-head attention layers, (iii) the CRF and softmax decoder part, and a head selection layer, (iv) the auxiliary training objective on language modeling.

3.1   Sentence Encoder

As shown in Fig. 3, the module first converts the words of the input sentence into low-dimensional vectors. Next, Bi-LSTM extracts the feature vectors of the input sentence according to the low-dimensional vectors.

Vector Representation. The original input of our model is a series of tokens (i.e., sentence), $w = (w_1, w_2, \ldots, w_T)$. When using neural networks, we usually convert words into low-dimensional vectors, as in other literatures [10, 20]. By looking up the corresponding representations by pre-trained word embedding, each input word is converted into a vector. In addition, we introduce character-level components to construct alternative representations for each word, with individual characters of the word mapped to character embedding and passed through the bidirectional LSTM. Finally, word and character embedding are concatenated to form the final vector representation, which is subsequently fed into the Bi-LSTM layer to extract sequential information.

Word Embedding. Word embedding is the distributed representation of words. We first transform each word into a vector to provide lexical semantic features. Specifically, each word in the word sequence is mapped into a dense, low-dimensional and real-

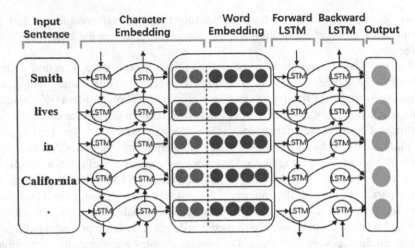

Fig. 3. Sentence encoder via Bi-LSTM

valued vector, through a lookup operation. We make use of pre-trained word embeddings, each dimension of which expresses the potential characteristic of words. It has been verified that this kind of distribution of word representations is able to capture grammatical, semantic information and similarity significantly [22, 23]. The word embedding is encoded by column vectors in the embedding matrix $W \in \mathbb{R}^{d^w \times |V|}$, where d^w is the dimension of word embedding, and $|V|$ is the size of the vocabulary. Each column $W_i \in \mathbb{R}^{d^w}$ corresponds to the embedding of the i-th word in the vocabulary.

Character Embedding. We adopt character-level embedding to implicitly capture morphological features (e.g. prefixes and suffixes). Individual characters of each word are mapped to character embedding and passed through the Bi-LSTM to obtain the character-based vector representation of the word.

By using the dynamic weighting mechanism to combine the resulting vector representation with conventional word embedding, we can convert a sentence into a matrix $E = [e_1, e_2, \ldots, e_T]$, where $e_i \in \mathbb{R}^{d^w + d^c}$ and T is the length of sentence. The matrix E is subsequently fed into the Bi-LSTM network. This mechanism adaptively controls the balance between character level and word level features. More importantly, it allows the model to learn character-based patterns and process previously invisible words, while still making full use of word embedding.

Bi-LSTM Network. In recent years, end-to-end models based on neural networks have been widely used in natural language processing tasks. In this paper, we study an end-to-end model to extract entity pairs and their relations, as shown in Fig. 2. Bi-LSTM layer is used to encode the input statements.

It has been shown that the LSTM coding layer is effective in capturing the semantic information of each word and is able to learn long-term dependencies. But LSTM can only predict the output of the next moment based on the timing information of the previous moment. However, in our problem, the output of the current moment is not only related to the previous state, but also related to the future state. Therefore, we adopt Bi-LSTM to encode the input. After the embedding layer, there are two parallel LSTM layers: forward LSTM layer and backward LSTM layer. The two LSTM components move in the opposite direction of the sentence to construct context-related representations of each word. Each LSTM takes the hidden state from the previous step as input, as well as the vector representation from the current step, and outputs a new hidden state. Hidden representations from both directions are joined to obtain the context-specific representation of each word conditioned on the entire sentence in both directions:

$$\overrightarrow{h_i} = LSTM\left(w_i, \overrightarrow{h_{i-1}}\right) \tag{1}$$

$$\overleftarrow{h_i} = LSTM\left(w_i, \overleftarrow{h_{i+1}}\right) \tag{2}$$

$$h_i = \left[\overrightarrow{h_i}; \overleftarrow{h_i}\right] \tag{3}$$

Next, the joined representation passes through a feedforward layer, mapping the components into a joint space and allowing the model to learn features based on both context directions:

$$d_i = tanh(W_d h_i) \tag{4}$$

where W_d is a weight matrix, and tanh is used as the non-linear activation function.

3.2   Hierarchical Attention

We apply the idea of hierarchical attention to the joint extraction task, as illustrated in Fig. 2. The first-level attention enables the model to capture the possible impact of particular words on the target entities, extracting more abstract higher-level features from the previous layer's output matrix. The second one can focus on words that have a decisive impact on automatically identifying the parts of input sentences associated with relationship classification.

Recently, multi-head attention [27] has become a hotspot of research in neural networks, and has achieved good results in various tasks. Multi-head attention uses separately normalized parameters to apply self-attention multiple times on the same inputs, the advantage of which allows the model to learn relevant information in different representation subspaces. The self-attention updates the result of previous layer d_i by performing a weighted sum over all tokens, weighted by their importance for modeling token i. Using separate affine transformations with ReLU activation, each input is projected to key k, value v, and query q. The attention weights a_{ijh} of head h between tokens i and j are calculated using scaled dot product attention:

$$a_{ijh} = \sigma\left(\frac{q_{ih}^T k_{jh}}{\sqrt{d_k}}\right) \tag{5}$$

$$o_{ih} = \sum_j v_{jh} \odot a_{ijh} \tag{6}$$

with $\odot$ denoting element-wise multiplication and σ indicating a softmax along the j-th dimension.

The outputs of the individual attention heads are connected into o_i. Both the two layers of attention use residual connections between the output of the multi-head attention and its input. Layer normalization, denoted by $LN(\cdot)$, is then applied to the output.

$$o_i = [o_{i1}; \ldots; o_{ih}]$$

$$f_i = LN(d_i + o_i)$$

3.3 Decoder Layer

For the NER tasks, we adopt the BIO (Each NER entity has sub-tags for Beginning, Inside, Outside) encoding scheme. In Fig. 1, the B-PER tag is assigned to the beginning token of a "person" (PER) entity. For the prediction of entity labels, we use: (i) the softmax method for the EC task (assuming given entity boundaries) or (ii) the CRF method for the NER task where we identify the type and boundary of each entity.

Softmax Decoder. For softmax output architecture, we greedily detect the entity type of the token (i.e., independent prediction). Under the condition of vector f_i, the model directly predicts the normalized distribution of all possible labels for each word:

$$P(y_i|f_i) = softmax(W_a f_i)$$
$$= \frac{e^{W_{o,k} f_i}}{\sum_{\tilde{k} \in K} e^{W_{o,\tilde{k}} f_i}} \tag{7}$$

where K is the set of all possible labels, and $W_{o,k}$ is the k-th row of output weight matrix W_o. The model is optimized by minimizing categorical cross-entropy, which is equivalent to minimizing the negative log-probability of the correct labels:

$$\mathcal{L}_{EC} = -\sum_{i=1}^{T} log(P(y_i|f_i)) \tag{8}$$

Although this architecture returns predictions based on all words in the input, the tags are still independently predicted. In spite of the independent distribution of tag types in the EC tasks is reasonable, this is not the case when there is a strong correlation between adjacent tags.

CRF Decoder. For the NER task using BIO scheme, there are strong dependencies between subsequent tags (for example, B-PER and I-LOC tags cannot be sequential). Explicitly modeling these connections may be beneficial. Driven by this intuition, we use linear-chain CRF [15] to perform NER tasks. In the CRF setup, we adopt Viterbi algorithm which allows the network to find the best path through all possible tag sequences [3, 9, 33]. During the training period, we minimize the cross-entropy loss $\mathcal{L}_{NER}$:

$$\mathcal{L}_{NER} = -s(y) + log \sum_{\tilde{y} \in \tilde{Y}} e^{s(\tilde{y})} \tag{9}$$

where s(y) is the score for a given sequence y and Y is the set of all possible label sequences. The entity tags are then fed into the relationship extraction layer as label embeddings (see Fig. 2), assuming that knowledge of entity types is useful for predicting the relationships between the entities involved.

Head Selection. We model the relation extraction task as a multi-label head selection problem [7, 9]. In our model, each word w_i can be involved in multiple relations with other words. For instance, in the example illustrated in Fig. 1, "Chisholm" could be involved not only in relation "Live_in" with the token "Thomas", but also in relation "Located_in" with the token "Minn". The goal of this task is to predict the vector of the head $\tilde{y}_t$ and the vector of the corresponding relation $\tilde{r}_t$ for each word w_i. Through a single layer neural network, we compute the score $s(w_j, w_i, r_k)$ of word w_j to be the head of w_i given a relation label r_k. The corresponding probability is defined as:

$$P(w_j, r_k | w_i; \theta) = \sigma(s(w_j, w_i, r_k)) \tag{10}$$

where $\sigma(.)$ is the sigmoid function. During training, we minimize the cross-entropy loss $\mathcal{L}_{rel}$ as:

$$\mathcal{L}_{rel} = \sum_{i=0}^{n} \sum_{j=0}^{m} -logP(y_{i,j}, r_{i,j} | w_i; \theta) \tag{11}$$

where m is the number of associated heads (and thus relations) per word w_i. During decoding, the most probable heads and relations are selected using threshold-based prediction. The ultimate goal of joint tasks is calculated as:

$$\mathcal{L}_{JOINT}(w; \theta) = \mathcal{L}_{EC} + \mathcal{L}_{rel} \tag{12}$$

or

$$\mathcal{L}_{JOINT}(w; \theta) = \mathcal{L}_{NER} + \mathcal{L}_{rel} \tag{13}$$

3.4   Language Modeling Objective

Although each token in the input has a desired tag, the entity labels are sparse. For example, in the CoNLL 2003 NER dataset, there are 8 possible tags. But 83% of the tokens has the tag O, indicating that no named entity was detected. The model illustrated above is able to learn the unbalanced distribution of labels without obtaining much additional information from the majority labels.

Therefore, we recommend the specific part of the structure be optimized into a language model, as shown in Fig. 4. For each token, the network is optimized to predict the previous word, the current tag and the next word in the sequence. This additional training objective makes better use of each available word and encourages the model to learn more general language features for accurate composition. This goal is also common for any sequence labeling task and data set, because it does not require additional annotated training data.

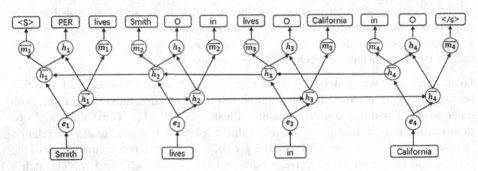

Fig. 4. Structure of language model with an additional language modeling objective, performing NER on the sentence "Smith lives in California". Arrows above variables indicate the directionality of the component (forward or backward).

In order to implement the above objective, we predict the next word in the sequence only based on the hidden representation $\overrightarrow{h_t}$ from the forward-moving LSTM. Similarly, the previous word in the sequence is predicted based on $\overleftarrow{h_t}$ from the backward-moving LSTM. This architecture avoids the problem of using the correct answers as input for language modeling components, while the complete framework is still optimized to predict tags based on the entire sentence.

Firstly, the hidden representation of forward and backward LSTM is mapped to a new space by using a non-linear layer:

$$\overrightarrow{m_t} = tanh\left(\overrightarrow{W}_m \overrightarrow{h_t}\right) \tag{14}$$

$$\overleftarrow{m_t} = tanh\left(\overleftarrow{W}_m \overleftarrow{h_t}\right) \tag{15}$$

where $\overrightarrow{W}_m$ and $\overleftarrow{W}_m$ are weight matrices. These representations are then passed through the softmax layer to predict the previous and the last words:

$$P\left(w_{t+1}|\overrightarrow{m_t}\right) = softmax\left(\overrightarrow{W}_q \overrightarrow{m_t}\right) \tag{16}$$

$$P\left(w_{t-1}|\overleftarrow{m_t}\right) = softmax\left(\overleftarrow{W}_q \overleftarrow{m_t}\right) \tag{17}$$

Then, by calculating the negative log-likelihood of the next word in the sequence, the objective function of the two components is constructed as the goal of conventional language modeling.

$$\overrightarrow{\mathcal{L}_{LM}} = -\sum_{t=1}^{T-1} log\big(P\big(w_{t+1}|\overrightarrow{m_t}\big)\big) \tag{18}$$

$$\overleftarrow{\mathcal{L}_{LM}} = -\sum_{t=2}^{T} log\big(P\big(w_{t-1}|\overleftarrow{m_t}\big)\big) \tag{19}$$

Finally, these additional targets are combined with the training target $\mathcal{L}_{JOINT}$ from Eqs. (12) or (13) to generate a new cost function $\tilde{\mathcal{L}}$:

$$\tilde{\mathcal{L}} = \mathcal{L}_{JOINT} + \gamma\big(\overrightarrow{\mathcal{L}_{LM}} + \overleftarrow{\mathcal{L}_{LM}}\big) \tag{20}$$

where γ is a parameter that is used to control the importance of language modeling objectives compared with sequential tagging objectives.

4 Experiments

The experiments are intended to provide evidence that incorporating language modeling with hierarchical attention-based Bi-LSTM network can not only address the problem of error propagations and sparse relevant labels, but also make full use of semantic information in sentences. For this reason, we first introduce the data set and evaluation indicators used in the experiment. Next, we demonstrate the parameter settings in our experiments. Then we compare the performance of our model with several traditional methods. Finally, we evaluate the effects of hierarchical attention and language model objective.

4.1 Implementation Details

Dataset. The experiment was conducted on the CoNLL04 data set [4], which has been widely used by many researchers. For the CoNLL04 EC task (assuming boundaries are given), we use the same splits as in [18, 19]. We also perform 10-fold cross-validation to evaluate our model in the same data set, similar to [26]. In order to obtain comparable results that are not affected by input embedding, we use the pre-trained word embedding of previous works [20, 22, 23].

Evaluation Metrics. We use two types of assessments, as is shown in Table 1, (i) Strict (S): if both the entity boundaries and entity type are correct, we score the entity as correct; (ii) Relaxed (R): a multi-token entity is considered correct if at least one correct type is assigned to the tokens comprising the entity, assuming that the boundaries are known, to compare to previous works. In all cases, a relationship is considered correct when both the relationship type and the argument entities are correct.

Table 1. Comparison of our method with the state-of-the-art in terms of F1 score.

Methods	Features	Eval.	Entity	Relation	Overall
Gupta et al. [18]	√	R	92.40	69.90	81.15
Gupta et al. [18]	×	R	88.80	58.30	73.60
Adel and Schütze [19]	×	R	82.10	62.50	72.30
Bekoulis et al. (2018)	×	R	93.04	67.99	80.51
Our model EC	×	R	**94.30**	**73.15**	**83.72**
Miwa and Sasaki [26]	√	S	80.70	61.00	70.85
Bekoulis et al. (2018)	×	S	83.61	61.95	72.78
Our model	×	S	**86.51**	**62.32**	**74.71**

In our experiment, we used standard Precision (Prec), Recall (Rec) and F1 scores to evaluate the results. We create a validation set by randomly sampling 10% data from test set and use the remaining data as evaluation based on suggestion [13]. We run 10 times for each experiment and then report the average results.

Parameter Setting. For most hyperparameters, we follow the settings of [9], in order to facilitate direct comparison with previous work. For word-level and character-level components, the dimensions of these two embeddings are 50 and 25 respectively. The size of LSTM hidden layer is set to 64 in each direction. To reduce the complexity of computation in these experiments, the language modeling objective predicted only the 7,500 most frequent words, with an extra token covering all the other words.

We used early stop in all experiments. Training was stopped when performance on the development set had not improved for 30 epochs. We use the Adam optimizer with learning rate 0.001, and we fixed the hyperparameter (i.e. dropout values, best epoch, learning rate) on the validation set. Performance on the development set was also used to select the best model, which was then evaluated on the test set.

During the development process, we found that the application of dropout to embedding can improve the performance of almost all data sets. Therefore, during the training period, elemental dropout is applied to each input embedding with a probability of 0.5. Since next word prediction is not a major part of the evaluation, we set the value of γ, which controls the importance of the language modeling objective, to 0.1 for all experiments in the entire training process.

Baselines. We compare our approach with several classical methods of joint extraction. To avoid any outlier results due to randomness in model initialization, each configuration is trained using 10 different random seeds, and the average results are presented herein. We trained, developed and tested each data set using previously established splits.

- Miwa and Sasaki [26] propose a history-based structured learning approach.
- Gupta et al. [18] proposed a simple approach of piggybacking candidate entities to model the label dependencies from relations to entities.
- Adel and Schütze [19] introduced globally normalized convolutional neural networks for joint entity classification and relation extraction.
- Bekoulis et al. [3] used Adversarial training (AT) to improve the robustness of neural network methods.

The proposed models are: (i) Our model, (ii) Our model EC (predicts only entity classes). The $\sqrt{}$ and $\times$ symbols indicate whether the models rely on external NLP tools. We include different evaluation types (S and R).

4.2   Experimental Results

Table 1 shows our experimental results. The name of the model is displayed in the first column, the second column indicates whether the models rely on external NLP tools, while the type of evaluation is listed in the third column. The last three columns show the F1 results of the two subtasks and their average performance. Bold values indicate the best results among models that use only automatically extracted features.

For the model with EC task, we use the relaxed evaluation similar to [18, 19] on the EC task. Our model is superior to the latest model that does not depend on the features extracted manually (increased by 3.21%). Moreover, compared to the model of Gupta et al. [18] that relies on complex features, our model performs within a margin of 2.57% in terms of overall F1 score.

For the model with NER tasks, our model performs within a margin of 1.93% in terms of overall F1 score, compared to the model of Bekoulis et al. (2018). What's more, the F1 score of our model increased by about 3.86% compared to the one with manual-extracted features [26], indicating that the end-to-end model with automatic-extracted features is more effective than models with hand-made features. The method based on neural networks can fit the data well and thus can learn the common features of training set well.

4.3   Effect of Hierarchical Attention and Language Model

This subsection analyzes how hierarchical attention and the language modeling objective affect our model. To demonstrate the effectiveness of these two techniques, we respectively report the results of four different systems: (1) baseline (EC): we first adopt the end-to-end model alone as a baseline, (2) baseline + LM: the baseline combined with language modeling objective, (3) baseline (EC) + ATT: the baseline regularized using hierarchical multi-head attention, (4) baseline + ATT + LM: applying hierarchical attention mechanism and the language modeling objective in Sect. 3. The results of comparison experiments are presented in Table 2.

Table 2. Ablation tests on the joint task of NER (or EC) and RE.

Settings	Entity			Relation			Overall F1
	Prec.	Rec.	F1	Prec.	Rec.	F1	
Baseline EC	93.54	91.69	92.61	72.88	67.94	68.77	80.6
Baseline EC+ATT	**95.43**	**93.23**	**94.30**	**76.37**	**71.76**	**73.15**	**83.72**
Baseline	84.32	85.86	85.08	59.89	60.91	60.39	72.73
Baseline+LM	84.48	**86.28**	85.37	59.95	**62.79**	61.34	73.35
Baseline+ATT	85.21	85.40	85.30	63.34	61.19	62.24	73.77
Baseline+LM+ATT	**87.12**	85.91	**86.51**	**63.53**	62.32	**62.91**	**74.71**

In Table 2, we not only report the results of F1, but their precision and recall scores. When hierarchical attention is introduced, the results are better than the baseline model. We can conclude that hierarchical attention can effectively capture semantic information. Furthermore, we notice that after integrating language modeling, both the baseline model and the baseline model with hierarchical attention achieve higher performance. This demonstrates that exploring language modeling objective to obtain more semantic features can significantly increase the precision and recall for the joint extraction task.

The bottom line of Table 2 shows the performance of integrating both attention mechanism and language modeling to the baseline model. The architecture outperforms the baseline in all benchmarks, improving accuracy and recall, and increasing the absolute improvement of the CoNLL04 test set by 1.98% over the baseline model.

5 Conclusions

In this paper, we propose an end-to-end neural model, which combines hierarchical attention mechanism and language modeling to extract entities and their relations simultaneously. Hierarchical attention is utilized to capture pivotal words in sentences. More importantly, auxiliary training objective on language modeling is served to learn to predict the surrounding words of each word in the data set, providing additional semantic features for tasks in an unsupervised way. The experimental results show the effectiveness of the proposed joint method for entity recognition and relation extraction. In the future work, we aim to explore the effectiveness of entity pre-training for the entity recognition module. Although our model can make full use of the correlation between entities and relations, the association between two corresponding entities still requires refinement in next works.

Acknowledgment. This work is supported by the National Key Research and Development Program of China (2018YFC0831500), National Natural Science Foundation of China (No. 61772082), National Natural Science Foundation of China (No. 61806020), the Fundamental Research Funds for the Central Universities and Big Data Research Foundation of PICC.

References

1. Nadeau, D., Sekine, S.: A survey of named entity recognition and classification. Lingvisticae Investigationes **30**(1), 3–26 (2007)
2. Zheng, S., Wang, F., Bao, H., Hao, Y., Zhou, P., Xu, B.: Joint extraction of entities and relations based on a novel tagging scheme. In: Proceedings of the 55th Annual Meeting of the Association for Computational Linguistics, pp. 1227–1236 (2017)
3. Bekoulis, G., Deleu, J., Demeester, T., Develder, C.: Adversarial training for multi-context joint entity and relation extraction. In: Empirical Methods in Natural Language Processing, pp. 2830–2836 (2018)
4. Roth, D., Yih, W.: A linear programming formulation for global inference in natural language tasks. In: Proceedings of the Eighth Conference on Computational Natural Language Learning (CoNLL 2004) at HLT-NAACL 2004, pp. 1–8. Association for Computational Linguistics, Boston, USA (2004)

5. Chiu, J.P., Nichols, E.: Named entity recognition with bidirectional LSTM-CNNs. Trans. Assoc. Comput. Linguist. **4**(1), 357–370 (2016)
6. Liu, L., et al.: Empower sequence labeling with task-aware neural language model. In: Thirty-Second AAAI Conference on Artificial Intelligence, pp. 5253–5260 (2018)
7. Zhang, X., Cheng, J., Lapata, M.: Dependency parsing as head selection. In: Proceedings of the 15th Conference of the European Chapter of the Association for Computational Linguistics, Valencia, Spain, pp. 665–676 (2017)
8. Verga, P., Strubell, E., McCallum, A.: Simultaneously self-attending to all mentions for full-abstract biological relation extraction. In: 16th Annual Conference of the North American Chapter of the Association for Computational Linguistics: Human Language Technologies, pp. 872–884 (2018)
9. Bekoulis, G., Deleu, J., Demeester, T., Develder, C.: Joint entity recognition and relation extraction as a multi-head selection problem. Expert Syst. Appl. **114**, 34–45 (2018)
10. Le, H.-Q., Can, D.-C., Vu, S.T., Dang, T.H., Pilehvar, M.T., Collier, N.: Large-scale exploration of neural relation classification architectures. In: 2018 Conference on Empirical Methods in Natural Language Processing, pp. 2266–2277 (2018)
11. Feng, J., Huang, M., Zhao, L., Yang, Y., Zhu, X.: Reinforcement learning for relation classification from noisy data. In: Thirty-Second AAAI Conference on Artificial Intelligence, pp. 5779–5786 (2018)
12. Gao, T., Han, X., Liu, Z., Sun, M.: Hybrid attention-based prototypical networks for noisy few-shot relation classification. In: Thirty-Third AAAI Conference on Artificial Intelligence (2019)
13. Ren, X., et al.: Cotype: joint extraction of typed entities and relations with knowledge bases. In: Proceedings of the 26th International Conference on World Wide Web, pp. 1015–1024. International World Wide Web Conferences Steering Committee (2017)
14. Wang, S., Zhang, Y., Che, W., Liu, T.: Joint extraction of entities and relations based on a novel graph scheme. In: 27th International Joint Conference on Artificial Intelligence, pp. 4461–4467 (2018)
15. Lafferty, J., McCallum, A., Pereira, F.C.: Conditional random fields: probabilistic models for segmenting and labeling sequence data. In: Proceedings of the 18th International Conference on Machine Learning, pp. 282–289 (2001)
16. Miwa, M., Bansal, M.: End-to-end relation extraction using LSTMs on sequences and tree structures. In: Proceedings of the 54th Annual Meeting of the Association for Computational Linguistics, pp. 1105–1116 (2016)
17. Katiyar, A., Cardie, C.: Investigating LSTMs for joint extraction of opinion entities and relations. In: Proceedings of the 54th Annual Meeting of the Association for Computational Linguistics, pp. 919–929 (2016)
18. Gupta, P., Schütze, H., Andrassy, B.: Table filling multi-task recurrent neural network for joint entity and relation extraction. In: Proceedings of COLING 2016, the 26th International Conference on Computational Linguistics: Technical Papers, pp. 2537–2547 (2016)
19. Adel, H., Schütze, H.: Global normalization of convolutional neural networks for joint entity and relation classification. In: Proceedings of the 2017 Conference on Empirical Methods in Natural Language Processing, pp. 1723–1729. Association for Computational Linguistics, Copenhagen, Denmark (2017)
20. Xie, R., Liu, Z., Jia, J., Luan, H., Sun, M.: Representation learning of knowledge graphs with entity descriptions. In: AAAI, pp. 2659–2665 (2016)
21. Vaswani, A., Bisk, Y., Sagae, K., Musa, R.: Supertagging with LSTMs. In: Proceedings of the 2016 Conference of the North American Chapter of the Association for Computational Linguistics: Human Language Technologies, pp. 232–237 (2016)

22. Mikolov, T., Chen, K., Corrado, G., Dean, J.: Efficient estimation of word representations in vector space. arXiv preprint arXiv:1301.3781 (2013)
23. Pennington, J., Socher, R., Manning, C.: Glove: global vectors for word representation. In: Proceedings of the 2014 Conference on Empirical Methods in Natural Language Processing (EMNLP), pp. 1532–1543 (2014)
24. Li, Q., Ji, H.: Incremental joint extraction of entity mentions and relations. In: Proceedings of the 52nd Annual Meeting of the Association for Computational Linguistics, pp. 402–412 (2014)
25. Lample, G., Ballesteros, M., Subramanian, S., Kawakami, K., Dyer, C.: Neural architectures for named entity recognition. In: Proceedings of the 2016 Conference of the North American Chapter of the Association for Computational Linguistics: Human Language Technologies, pp. 260–270 (2016)
26. Miwa, M., Sasaki, Y.: Modeling joint entity and relation extraction with table representation. In: Proceedings of the 2014 Conference on Empirical Methods in Natural Language Processing (EMNLP), pp. 1858–1869 (2014)
27. Vaswani, A., et al.: Attention is all you need. In: Advances in Neural Information Processing Systems, pp. 5998–6008 (2017)
28. Li, F., Zhang, M., Fu, G., Ji, D.: A neural joint model for entity and relation extraction from biomedical text. BMC Bioinform. 18(1), 198 (2017)
29. Katiyar, A., Cardie, C.: Going out on a limb: joint extraction of entity mentions and relations without dependency trees. In: Proceedings of the 55th Annual Meeting of the Association for Computational Linguistics, Vancouver, Canada, pp. 917–928 (2017)
30. Bekoulis, G., Deleu, J., Demeester, T., Develder, C.: An attentive neural architecture for joint segmentation and parsing and its application to real estate ads. Expert Syst. Appl. 102, 100–112 (2018)

An Unsupervised Learning Approach for NER Based on Online Encyclopedia

Maolong Li[1], Qiang Yang[4], Fuzhen He[1], Zhixu Li[1,2]($\boxtimes$), Pengpeng Zhao[1], Lei Zhao[1], and Zhigang Chen[3]

[1] Institute of Artificial Intelligence, School of Computer Science and Technology, Soochow University, Suzhou, China
mlli17@stu.suda.edu.cn, {zhixuli,ppzhao,zhaol}@suda.edu.cn
[2] IFLYTEK Research, Suzhou, China
[3] State Key Laboratory of Cognitive Intelligence, iFLYTEK, Hefei, China
zgchen@iflytek.com
[4] King Abdullah University of Science and Technology, Jeddah, Saudi Arabia
qiangyanghm@hotmail.com

Abstract. Named Entity Recognition (NER) is a core task of NLP. State-of-art supervised NER models rely heavily on a large amount of high-quality annotated data, which is quite expensive to obtain. Various existing ways have been proposed to reduce the heavy reliance on large training data, but only with limited effect. In this paper, we propose a novel way to make full use of the weakly-annotated texts in encyclopedia pages for exactly unsupervised NER learning, which is expected to provide an opportunity to train the NER model with no manually-labeled data at all. Briefly, we roughly divide the sentences of encyclopedia pages into two parts simply according to the density of inner url links contained in each sentence. While a relatively small number of sentences with dense links are used directly for training the NER model initially, the left sentences with sparse links are then smartly selected for gradually promoting the model in several self-training iterations. Given the limited number of sentences with dense links for training, a data augmentation method is proposed, which could generate a lot more training data with the help of the structured data of encyclopedia to greatly augment the training effect. Besides, in the iterative self-training step, we propose to utilize a graph model to help estimate the labeled quality of these sentences with sparse links, among which those with the highest labeled quality would be put into our training set for updating the model in the next iteration. Our empirical study shows that the NER model trained with our unsupervised learning approach could perform even better than several state-of-art models fully trained on newswires data.

Keywords: Named entity recognition · Data augmentation · Enhanced self-training

© Springer Nature Switzerland AG 2019
J. Shao et al. (Eds.): APWeb-WAIM 2019, LNCS 11641, pp. 329–344, 2019.
https://doi.org/10.1007/978-3-030-26072-9_25

1   Introduction

Named Entity Recognition (NER) aims at recognizing and classifying phrases referring to a set of named entity types [5], such as *PERSON, ORGANIZATION*, and *LOCATION* in text. A formal definition of NER is given in Definition 1 below.

Definition 1. *Let $T = \{t_1, t_2, ..., t_e\}$ denote a set of named entity types. Given a sequence of tokens $S = <w_1, w_2, ..., w_N>$, the task of NER is to output a list of tuples $<I_s, I_e, t>$, where $I_s, I_e \in [1, N]$ are the start and the end indexes of a named entity mention in S, and $t \in T$ is the corresponding entity type of the mention.*

Given an example text "United Nations official Ekeus heads for Baghdad", after NER we could have: *"[ORG United Nations] official [PER Ekeus] heads for [LOC Baghdad]."*, where three named entities: *Ekeus* is a person, *United Nations* is an organization and *Baghdad* is a location.

As a core task of Natural Language Processing (NLP), NER is crucial to various applications including question answering, co-reference resolution, and entity linking, etc. Correspondingly, plenty of efforts have been made in the past decades to developing different types of NER systems. For instance, many NER systems are based on feature-engineering and machine learning, such as Hidden Markov Models (HMM) [26], Support Vector Machines (SVM) [11], Conditional Random Fields (CRF) [8], with many handcrafted features (capitalization, numerals and alphabets, containing underscore or not, trigger words and affixes, etc.). However, these features will be useless when adopted to totally different languages like Chinese.

Later work replace these manually constructed features by combining a single convolution neural network with word embeddings [3]. After that, more and more deep neural network (DNN) NER systems are proposed [25]. However, it is well known that DNN models require a large scale annotated corpus for training. The existing open labeled data for NER model training are mostly from the newswire data [10,22], which have few mistakes on grammar or morphology. As a result, the models trained on these gold-standard data usually have a bad performance on noisy web texts. Besides, the NER model trained on such a general corpus could not work well on specific domains, because the contexts of entity names in specific domains are quite different from those in newswires. To get a domain-specific NER model, extra domain-specific labeled corpus for training are required, which, however, are also expensive to achieve.

To deal with the challenges above, many recent work tend to use online encyclopedias for NER model training. As the largest open-world knowledge base, Wikipedia contains a large quantity of weakly-annotated texts with inner links of entities, as well as a well-structured knowledge base of various domains. Someone uses tags and alias in the Wikipedia to build a gazetteer type feature which they use in addition to standard features for CRF model [15], while the others generate NER training data by transforming the inner links to Named Entity (NE) annotations through mapping the Wikipedia pages to entity types [14,16].

However, all the existing work only consider to use the weakly-annotated texts with limited number of inner links of entities in online encyclopedias for training, but ignore other possible ways to fully use the data and knowledge contained in online encyclopedias.

In this paper, we propose a novel way to make full use of the weakly-annotated texts in online encyclopedia pages for unsupervised NER learning, which could greatly reduce the amount of manual annotation without hurting the precision of the NER model. The terms or names in encyclopedia pages are linked to the corresponding articles only when they are **first** mentioned in an encyclopedia article, which makes the first paragraph of the encyclopedia pages contain a lot more links than the other paragraphs. Here we roughly divide the sentences of encyclopedia pages into two parts simply according to the density of inner url links contained in each sentence. While a relatively small number of sentences with dense links could be used directly for training the NER model in the first step, the left sentences with sparse links are then smartly selected for gradually promoting the model in several self-training iterations. Given the limited number of sentences with dense links, a data augmentation method is proposed, which could generate a lot more training data with the help of the structured knowledge of online encyclopedias to greatly augment the training effect. For those sentences with sparse links, we propose an iterative self-training method to make full use of them, where a graph model is utilized to help estimate the labeled quality of these sentences, among which those with the highest labeled quality would be put into our training set for updating the model in the next iteration.

The contributions of this paper are listed as follows:

1. We propose a novel unsupervised learning approach for NER based on online encyclopedias, which provides an opportunity to train the NER model without using any manually-labeled data.
2. We propose a so-called data augmentation method to greatly augment the training effect with the help of structured knowledge of online encyclopedias.
3. We also propose an iterative self-training method to make full use of these sparse links, where a graph model is utilized to help select sentences for updating the model.

Our empirical study shows that the NER model trained with our unsupervised learning approach could perform even better on Wikipedia texts than several state-of-art models fully trained on newswires data.

The remainder of this paper is organized as follows: Sect. 2 reviews the related work. Section 3 provides a framework of our methods. Section 4 introduces our data augmentation method for NER training data and Sect. 5 explains how to enhanced the self-training by a graph-based model. After reporting our experical study in Sect. 6, we conclude our paper in Sect. 7.

2   Related Work

In the following of this section, we first introduce several state-of-art supervised NER models based on DNN and some semi-supervised NER models using encyclopedia data. After that, we also present some existing work using self-training models. Finally, we cover some related work on data augmentation technics.

Supervised DNN-Based NER Model. Collobert et al. firstly propose to use a single convolution neural network architecture to output a host of language processing predictions including NER, which is an instance of multi-task learning with weight-sharing [2]. The feature vectors of the architecture are constructed from orthographic features (e.g.,capitalization of the first character), dictionaries and lexicons. Later work [3] replaces manually constructed feature vectors with word embeddings (a distributed representations of words in n-dimensional space).

After that, the Bidirectional Long Short-Term Memory (BI-LSTM) or BI-LSTM with Conditional Random Field (BI-LSTM-CRF), now widely used, is applied to NLP benchmark sequence tagging data sets, along with different word representations [6,9]. Although the DNN models have a great performance, they need plenty of training data annotated by experts, which is expensive and time-consuming to achieve.

Semi-supervised NER Using Encyclopedia Data. Online encyclopedias, like Wikipedia, have been widely used to generate weakly labeled NER training data. The main idea is to transform the hyperlinks into NE tags by categorizing the corresponding Wikipedia pages into entity types. Some methods categorize the pages based on manually constructed rules that utilize the category information of Wikipedia [16]. Such rule-based entity type mapping methods have high precision, but low coverage. To achieve a better performance, Nothman et al. use a classifier trained by the extra manually labeled Wikipedia pages with entity types [14].

Self-training Model. Self-training is a semi-supervised learning technique where there is only a small amount of labeled data but a large number of unlabeled data. It is a widely used approach for various NLP tasks, such as NER [7,13], part-of-speech (POS) [20] and parsing [12]. The basic idea of self-training is to select a number of instances in the testing set according to some selection strategy, to augment the original training set. The most popular strategy is to select those instances whose predictions given by the baseline models own high confidence values. Kozareva et al. use two classifiers for labeling the entity names and then one external classifier for voting the labeled results [7]. This method only selects the most confidentially predicted instances and the external classifier for voting may not be reliable. A recall-oriented perceptron is used in [13] for NER, along with self-training on Wikipedia, but they select Wikipedia articles at random and still need annotated sentences.

Data Augmentation. Data augmentation refers to a kind of methods for constructing iterative optimization or sampling algorithms via the introduction of unobserved data or latent variables [19]. So far, data augmentation has been proved effective in most image related tasks by flipping, rotating, scaling or cropping the images [23]. When it comes to NLP tasks, data augmentation methods need to be designed specifically [21,24]. Xu et al. propose a method by changing the direction of relations for relation classification task [24]. They split a relation into two sub-paths: *subject-predicate* and *object-predicate*, and then change the order of the two sub-paths to obtain a new data sample with the inverse relationship. Wang et al. propose a novel approach for automatic categorization of annoying behaviors [21]. They replace the words in tweet with its k-nearest-neighbor (knn) words, found by the cosine similarity of word embeddings. When it comes to Chinese word embeddings, the entity mention often meets Out-of-Vocabulary (OOV) problem.

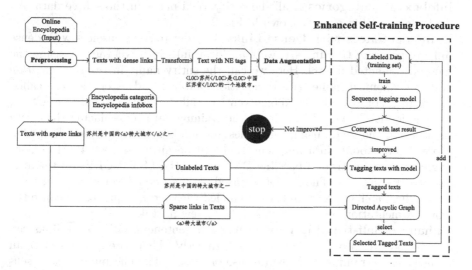

Fig. 1. The framework of the approach

3 Framework

We propose an unsupervised learning approach for NER based on online encyclopedias. Figure 1 depicts an overview of our unsupervised learning approach. Given a number of encyclopedia texts for a specific purpose, we roughly divide all the sentences into two parts: those with dense links, and those with sparse links. For sentences with dense links, we transform their links into NE tags, and then do data augmentation with the help of the structured knowledge contained in online encyclopedia. After that, we could have a set of labeled data for training a baseline DNN model (BI-LSTM-CRF). Next, we use this DNN model to perform NER tagging on a set of sentences with sparse links. We then utilize a graph model to help estimate the labeled quality of these sentences with sparse links, among which those with the highest labeled quality would be put into our training set for updating the NER model in the next iteration. We repeat this self-training step until the performance of the retrained model could not be improved on the test set. Some details about the framework are given below.

- **Online Encyclopedia Preprocessing.** We select the descriptions (paragraphs before *Contents*) of the encyclopedia articles in one domain as source data (e.g. *Geography and places* in Wikipedia). We classify these encyclopedia texts into sentences with dense links and those with sparse links according to a score equation $S_q(n_l, n) = \frac{n_l}{n}$, where n_l and n are the number of linked words and all words (except stop words) in the sentence respectively. Here we take *topn* sentences ordered by their S_q scores as sentences with dense links, which are then taken as the initial training set D_i, while the left sentences with sparse links as the unlabeled data set D_u. Besides, we extract the **infobox** and **categories** of all the entity Wiki pages in the source data. An example of a Wiki page is shown in Fig. 2.

- **Using Sentences with Dense Links.** In order to use sentences with dense links to generate training set, we transform links into NE tags by using the method presented in [16]. Here we call the entity name in texts as *mention* and the canonical name with description, infobox and categories are called *entity.* (e.g: For an entity Wiki page (entity) called *People's Republic of China*, its mention is *China*). We augment the training data by replacing the entity mentions in the texts, using categories and infobox derived from Wikipedia. More details about data augmentation will be given in Sect. 4. After data augmentation, we train a baseline DNN model (BI-LSTM-CRF) on the augmented training set. The model is based on character level instead of word level, where a character sequence is labeled with a tag sequence, because the word segmentation on web texts still has a few mistakes.

- **Enhanced Self-training Procedure.** The sentences with sparse links can not be added into training data set directly. Therefore, we develop an enhanced self-training to make full use of these links. The main idea of self-training is to augment the initial training set by labeling the unlabeled data with the baseline model, and then use some strategies to select part of these auto-labeled instances to add into the training set. The two steps are conducted for several iterations until the performance of the retrained model could not be further improved on the test set. The performance of the self-training algorithms strongly depends on how to select automatically labeled data in each iteration of the iterative procedure. Therefore, We explore extra metric as an auxiliary measurement to help select auto-labeled instances. This method, called Enhanced Self-training, will be expressed thoroughly in Sect. 5.

4   Data Augmentation

The most effective way to improve the performance of a deep learning model is to add more data to the training set. But in most circumstances, it is difficult

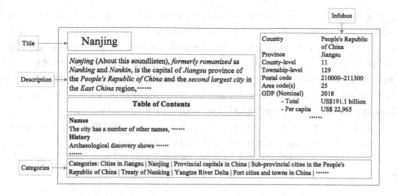

Fig. 2. An example Wiki page of Nanjing, where the italic parts have inner-links.

or expensive to get plenty of extra annotated data. As an alternative, we could also develop some methods to enhance the data we have already had, which are named as data augmentation [21,23,24]. The existing data augmentation method used in [21] is not suitable for our model, since our model is based on character level and it is meaningless to replace characters in the Chinese sentence.

We propose a novel data augmentation approach, which could generate a set of sentences from one sentence, by properly replacing each entity mention in the sentence with some other entity mentions of the same entity category. For example, we may replace the *Gusu District* in the sentence "The population of Suzhou Gusu District is 950,000." with *Wuzhong District*, as the example sentence S_1 shown in Fig. 3. An important assumption behind our approach is that: sentences in a right presentation form but with incorrect knowledge (which is inconsistent with the real-world case) could also be a sample for training NER. For example, the mention *Suzhou* in sentence "The population of Suzhou Gusu District is 950,000." is replaced with *Hangzhou*, as the example sentence S_2 shown in Fig. 3.

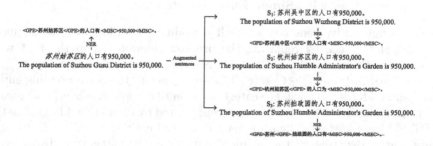

Fig. 3. The NER result of Stanford NER Tagger on augmented multi-entity sentences.

Our approach, however, may also generate bad sentences if we replace an entity mention with another entity which has a different relation with other entities in the sentence, such as the example sentence S_3 shown in Fig. 3. To reduce this side effect, we conduct the data augmentation differently according to the number of the entities in text. For single-entity sentences, we could replace the entity mention simply. But for multi-entity sentences, we find some entity pairs, having the same relation with the pairs in sentences, and replace mentions with mentions of the new entity pairs, because the new entities with same relation often have some common attributes with the replaced entities, which keeps the context of entities in augmented sentences correct. More details are shown as follows.

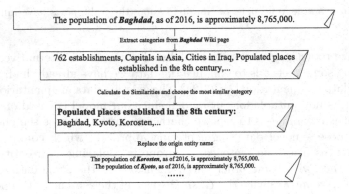

Fig. 4. The example of one **single-entity sentence**: The entity name in the example sentence is **Baghdad** and the categories in the corresponding Wiki page are in the second blocks. Then, we find the most similar category is **Populated places established in the 8th century**. There are 35 entities in this category. Finally, we replace the **Baghdad** with the top-n most similar entity names in the category, like **Kyoto**, **Korosten**.

4.1   Augmenting with Single-Entity Sentences

Given a single-entity sentence s which contains a mention m_o, we augment the original sentence by replacing the original mention m_o in the text with its related entities' mentions. First, we get the most similar category $c_{sim} = \text{argmax}_c\ CosSim(c_i, s)(c_i \in Cate(e_o))$, where e_o means the corresponding entity of m_o, and $Cate(e)$ means all the categories of entity e, and $CosSim(x, y)$ means the cosine similarity of two texts, which represented by their embeddings, like the BERT hidden states [4]. Then, we get the related entities set $E_r = Ent(c_{sim})$, where $Ent(c)$ denoting all the acquired entities in the category c. However, if none of words in the sentence is contained in the text of category, we set the $E_r = \emptyset$. Note that if we augment the sentence with all the entities in E_r, many similar sentences will be added into the training data set. This will have a bad influence on the model since the augmented sentences bring in a few noises. To deal with this situation, we get the final entities E_{rep} by setting a max-replacing number n_{max} through Eq. 1 below:

$$E_{rep} = \begin{cases} E_r & \|E_r\| \leq n_{max} \\ TOP(n_{max}, E_r, e_o) & \|E_r\| > n_{max} \end{cases} \tag{1}$$

where $TOP(n, E, e)$ gets the n entities most similar to e from an entities set E. The similarity of two entities is calculated by the cosine similarity using the first paragraphs of their Wikipedia articles. After that, we replace the original mention with the mentions of entities in E_{rep}. The single-entity sentence example is illustrated in Fig. 4.

Unfortunately, it still brings noises into the training set when we replace the entity mention with the name of a entity, which is not at the same level with the original entity, as the sentence S_3 shown in Fig. 3. To reduce these mistakes, we should also take the attributes of the entity into consideration, which is in our future work plan.

4.2 Augmenting with Multi-entity Sentences

We call a triple denoted by $(subject, relation, object)$ as a Relation Triple, where $subject$ and $object$ are entities, $relation$ denotes the relation between the two entities. Given a multi-entity sentence, we first find all the relation triples T_o in the sentence. For a triple $t_i = (s_i, r_i, o_i) \in T_o$, we get the corresponding replacing relation triple set in two conditions: (1) We get the relation triple set T_1 by only replacing the s or o in one triple. (2) We get the relation triple set T_2 by replacing both s and o of one triple. The two relation triple sets are expressed by Eq. 2, where E_{s_i} is the related entities of s_i, so is E_{o_i}. Then, the replacing relation triple set of t_i is $T_{new} = T_1 \cup T_2$.

$$\begin{aligned} T_1 &= \{t|t = (s,r,o), r = r_i, (s \in E_{s_i}, o = o_i) \vee (s = s_i, o \in E_{o_i})\} \\ T_2 &= \{t|t = (s,r,o), r = r_i, s \in E_{s_i} \wedge o \in E_{o_i}\} \end{aligned} \tag{2}$$

After that, we replace the original mentions of s_i, o_i with mentions of s, o in new relation triples T_{new}. The multi-entity sentence example is illustrated in Fig. 5.

5   Enhanced Self-training Approach

In addition of the sentences with dense links, we also need to find a method to make full use of the sentences with sparse links. Self-training is a useful approach to semi-supervised learning when we only have a small amount of labeled data but a large number of unlabeled data.

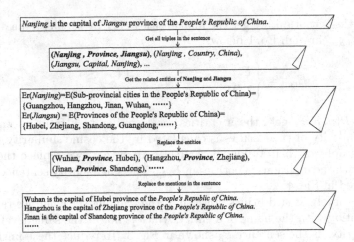

Fig. 5. The example of one **multi-entity sentence**: The triples in the sentence is in the second block. Then, we give the example of getting the related entities of the subject and object in the triple **(Nanjing, Province, Jiangsu)**. After that, we get the new relation triples illustrated in the forth block. Finally, we replace the original mentions in the sentence with the mentions of entities in new triples.

5.1   Self-training

The basic self-training algorithm is described in Algorithm 1. The initial training data set D_{init} is generated from Wikipedia pages by transforming the dense links into NE tags, expressed in Sect. 3. Our baseline model (BI-LSTM-CRF) is trained on the D_{init}. Then, we use the baseline model to predict a batch size n_b of samples selected from unlabeled data set D_u, which consists of the sentences with sparse links. The F_b is the F-measure of baseline model on the test set.

Algorithm 1. An Self-training Algorithm

Input   : D_{init}, D_u, M, T, where D_{init} denote the initial training set, D_u
          denote the unlabeled data set, M denote the baseline model, T
          denote the max iteration step.

Output: D_l, a set of labeled data.

$D_l \leftarrow D_i, f_{last} \leftarrow F_b, f_{new} \leftarrow -1$;
while $T > 0$ and $f_{last} > f_{new}$ **do**
   $\quad D_l \leftarrow D_l \cup Select(M, D_u, n_b)$;
   $\quad M \leftarrow Train(D_l)$;
   $\quad f_{new} \leftarrow Test(M)$;
   $\quad T \leftarrow T - 1$;
end
return D_l;

The *Select* function is the core of the algorithm, which uses some strategies to select high quality labeled data from D_u. However, traditional self-training solutions usually only depend on heuristics such as model confidence for instance selection. Such a selection strategy may not provide a reliable selection [1]. To handle this problem, we develop an enhanced self-training approach.

5.2 Enhanced Self-training

We propose a selection approach by combining a graph model with the confidence-based selection function to help evaluate the quality of the auto-labeled instances. When there are few or even no mentions linked in the sentences of unlabeled data, it is difficult to transform the inner links into NE annotations. For example, the sentence "Beijing is an important world capital and **global power city**." has linked words: *global power city*, except *Beijing*. Nevertheless, we can still use the sparse links to help estimate the labeled quality of these sentences due to our model based on character level, where it only tags the characters referred to entity names. Specifically, we consider that the labeled sentences are in high quality when the most probable word segmentation contains the entity names. Therefore, we use a directed acyclic graph (DAG) to conduct the word segmentation, where we can ensure some certain words are segmented. We calculate a score $S_p(y, E, W) = Conf(y) + \frac{|E \cap W|}{|E|}$ of the labeled instances, where $Conf(y)$ denotes the confidence of the result y predicted by model, which are the output of the model, and E denotes entity mentions set in the sentence tagged by model, and W denotes the words of the sentences segmented by DAG. In each self-training iteration, we select labeled sentences with *topn* highest scores to add into the training set.

We build a DAG for all possible word segmentations (see an example in Fig. 6) where the node denotes one word and the directed path from *start* to *end* denotes a possible word segmentation of the sentence. In addition, linked words in the sentences with sparse links are definitely contained in the DAG. Each node has a probability calculated by the word frequency. Then, we use dynamic programming to find the most probable word segmentation by calculating the Π of all nodes in the path [17]. Besides, if the linked words in selected sentences are mentions and not tagged by the BI-LSTM-CRF model, we will transform the links into NE annotations and then add them into the training set.

6 Experiments

We conduct a series of experiments to evaluate our proposed approaches for NER training on real-world data collections.

6.1 Datasets and Metrics

Datasets. Since our NER model is unsupervised, all the training data are derived from Wikipedia. We randomly select 8K sentences from the Wiki pages in

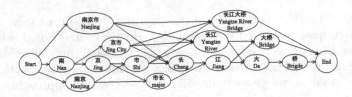

Fig. 6. The example of the DAG for sentence: "Yangtze River Bridge of Nanjing". After using dynamic programming on the DAG, we get a word segmentation: "Nanjing", "Yangtze River Bridge".

Geography and extract 1K sentences as initial labeled data set named GEO, 7K sentences as unlabeled data set. We manually label 500 sentences in Wikipedia for testing the model. Besides, we evaluate the data augmentation method on the MSRA NER data set, which has 46.4K sentences for training, 4.4k sentences for test [10].

Metrics. Standard precision (P), recall (R) and F1-score (F1) are used as evaluation metrics. We use the conll eval script in CoNLL 2003 [18]. The hyperparameters of our model are set as follows: $drop = 0.5$, $learningrate = 0.005$, $optimizer = adam$ and $batchsize = 32$.

6.2 Approaches for Comparison

- **BI-LSTM-CRF.** It is the baseline method with sequence tagging model (BI-LSTM-CRF) [6], which is trained on the initial training set transformed by sentences with dense links. We conduct the Chinese NER task on character level without word segmentation.
- **BI-LSTM-CRF+DA.** This method is similar to the first one but it uses the data augmentation (DA for short) on the initial training set and it is trained on the augmented data set.
- **BI-LSTM-CRF+EST.** In this method, we use the enhanced self-training (EST for short) approach on the basis of the baseline model to train the model but without data augmentation.
- **BI-LSTM-CRF+EST+DA.** On the basis of the baseline, we add enhanced self-training method and data augmentation to train our model.

6.3 Experimental Results

Data Augmentation. We conduct the experiments on the MSRA NER data set to evaluate the effect of DA. We select different numbers of sentences from the dataset in turn. We set $n_{max} = 10$ for DA experiments, and the change in size of training data is shown in Table 1. We can see that DA effectively augments the training set from the sentence level, character level and entity level.

As illustrated in Fig. 7, we can see the results of improvements with different size of training data. The improvement increases gradually in the beginning and then progressively decreases, since the diversity of data changes from less to more

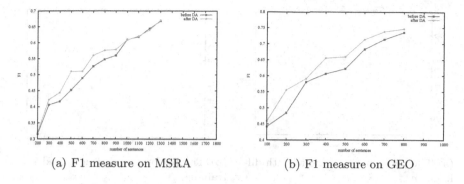

(a) F1 measure on MSRA   (b) F1 measure on GEO

Fig. 7. The improvement of performance of baseline model after DA

Table 1. The changing of number of sentence, character and entity after DA

Type		Number				
Sentence	Before DA	0.1k	0.2k	0.3k	0.4k	0.5k
	After DA	1.4k	2.8k	4.3k	5.4k	7.2k
Character	Before DA	4.3k	8.5k	13k	17k	21k
	After DA	79k	162k	253k	314k	422k
Entity	Before DA	0.3k	0.6k	1k	1.2k	1.7k
	After DA	8.2k	17k	25k	30k	43k

and becomes stable when applying DA for the small data set. This means that DA has limited impact on the performance of model when the number of diverse sentences reaches a certain size. Also, we find that MSRA represents the similar results. Apart form that, the DA on Wikipedia data has a better improvement than on MSRA because there are errors in process of linking the mentions to entities on MSRA.

We also conduct experiments on using different values of max-replacing number n_{max}. As shown in Fig. 8(a), the performance shows a declining trend in the beginning since the augmented sentences are inadequate and also brings in noises. After falling to the lowest point, as the number of sentences increases, DA begins to have a positive effect. However, if the n_{max} is set too large, the DA will have little improvement on the performance, even hurt it, because of excess similar augmented sentences along with some noises in the training data. Also, the model will take a longer training time on the immoderately augmented training data.

Enhanced Self-training. We conduct two self-training approaches to train the baseline model without DA, shown in Fig. 8(b). The two approaches have approximate performance at the beginning because these selected labeled sentences have high confidence, which means the model has already learned the features of them. After two iterations, the traditional ST model begins to select the left sentences with low confidence, leading to a bad performance. Different

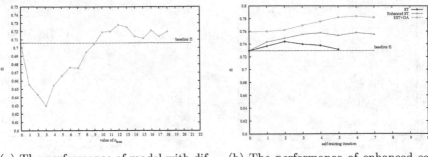

(a) The performance of model with different n_{max}

(b) The performance of enhanced self-training

Fig. 8. The performances of the model combined with different methods

from the traditional ST model, our enhanced ST model takes the sparse links into consideration to select and revise the left sentences. Therefore, it still has a positive influence in two more iterations, up to the best on the $4 - th$ iteration. **Combination of EST and DA.** Figure 8(b) shows that the performance of combining enhanced self-training and data augmentation, is better than the performance of using each of the two. The baseline model has a 3% F1 improvement after using DA and EST. Besides, unlike traditional self-training, the performance of our model will not decrease after several self-training iterations.

7   Conclusions

In this paper, we propose a method to make full use of the weakly-annotated texts in encyclopedia pages for exactly unsupervised NER learning. We propose a so-called data augmentation approach to greatly augment the training effect with the help of structure knowledge of online encyclopedias containing sentences with dense links. In addition, we put forward an iterative self-training method to make fully use of sentences with sparse links by combining the NER model where a graph model is utilized to help select sentences for updating the model. Our empirical study shows that the NER model trained with our unsupervised learning approach could perform even better than several state-of-art models.

In the future, we will improve the data augmentation method by considering the attributes of the entities. Besides, we will find an efficient crowdsourcing approach to reduce the annotation cost as much as possible.

Acknowledgments. This research is partially supported by National Natural Science Foundation of China (Grant No. 61632016, 61572336, 61572335, 61772356), and the Natural Science Research Project of Jiangsu Higher Education Institution (No. 17KJA520003, 18KJA520010).

References

1. Chen, M., Weinberger, K.Q., Blitzer, J.: Co-training for domain adaptation. In: Advances in Neural Information processing systems, pp. 2456–2464 (2011)
2. Collobert, R., Weston, J.: A unified architecture for natural language processing: deep neural networks with multitask learning. In: Proceedings of the 25th International Conference on Machine Learning, pp. 160–167. ACM (2008)
3. Collobert, R., Weston, J., Bottou, L., Karlen, M., Kavukcuoglu, K., Kuksa, P.: Natural language processing (almost) from scratch. J. Mach. Learn. Res. **12**(Aug), 2493–2537 (2011)
4. Devlin, J., Chang, M.W., Lee, K., Toutanova, K.: Bert: pre-training of deep bidirectional transformers for language understanding. arXiv preprint arXiv:1810.04805 (2018)
5. Grishman, R., Sundheim, B.: Message understanding conference-6: a brief history. In: COLING 1996 Volume 1: The 16th International Conference on Computational Linguistics, vol. 1 (1996)
6. Huang, Z., Xu, W., Yu, K.: Bidirectional LSTM-CRF models for sequence tagging. arXiv preprint arXiv:1508.01991 (2015)
7. Kozareva, Z., Bonev, B., Montoyo, A.: Self-training and co-training applied to Spanish named entity recognition. In: Gelbukh, A., de Albornoz, Á., Terashima-Marín, H. (eds.) MICAI 2005. LNCS (LNAI), vol. 3789, pp. 770–779. Springer, Heidelberg (2005). https://doi.org/10.1007/11579427_78
8. Lafferty, J., McCallum, A., Pereira, F.C.: Conditional random fields: probabilistic models for segmenting and labeling sequence data (2001)
9. Lample, G., Ballesteros, M., Subramanian, S., Kawakami, K., Dyer, C.: Neural architectures for named entity recognition. arXiv preprint arXiv:1603.01360 (2016)
10. Levow, G.A.: The third international Chinese language processing bakeoff: word segmentation and named entity recognition. In: Proceedings of the Fifth SIGHAN Workshop on Chinese Language Processing, pp. 108–117 (2006)
11. Li, Y., Bontcheva, K., Cunningham, H.: SVM based learning system for information extraction. In: Winkler, J., Niranjan, M., Lawrence, N. (eds.) DSMML 2004. LNCS (LNAI), vol. 3635, pp. 319–339. Springer, Heidelberg (2005). https://doi.org/10.1007/11559887_19
12. McClosky, D., Charniak, E., Johnson, M.: Effective self-training for parsing. In: Proceedings of the Main Conference on Human Language Technology Conference of the North American Chapter of the Association of Computational Linguistics, pp. 152–159. Association for Computational Linguistics (2006)
13. Mohit, B., Schneider, N., Bhowmick, R., Oflazer, K., Smith, N.A.: Recall-oriented learning for named entity recognition in Wikipedia. CMU-LTI-11-012, Language Technologies Institute School of Computer Science Carnegie Mellon University 5000 Forbes Ave., Pittsburgh, p. 15213 (2011)
14. Nothman, J., Ringland, N., Radford, W., Murphy, T., Curran, J.R.: Learning multilingual named entity recognition from Wikipedia. Artif. Intell. **194**, 151–175 (2013)
15. Radford, W., Carreras, X., Henderson, J.: Named entity recognition with document-specific KB tag gazetteers. In: Proceedings of the 2015 Conference on Empirical Methods in Natural Language Processing, pp. 512–517 (2015)
16. Richman, A.E., Schone, P.: Mining wiki resources for multilingual named entity recognition. In: Proceedings of ACL 2008: HLT, pp. 1–9 (2008)
17. Sun, J.: 'jieba' Chinese word segmentation tool (2012)

18. Tjong Kim Sang, E.F., De Meulder, F.: Introduction to the CoNLL-2003 shared task: language-independent named entity recognition. In: Proceedings of the Seventh Conference on Natural language Learning at HLT-NAACL 2003-Volume 4, pp. 142–147. Association for Computational Linguistics (2003)
19. Van Dyk, D.A., Meng, X.L.: The art of data augmentation. J. Comput. Graph. Stat. **10**(1), 1–50 (2001)
20. Wang, W., Huang, Z., Harper, M.: Semi-supervised learning for part-of-speech tagging of Mandarin transcribed speech. In: IEEE International Conference on Acoustics, Speech and Signal Processing, ICASSP 2007, vol. 4, pp. IV–137. IEEE (2007)
21. Wang, W.Y., Yang, D.: That's so annoying!!!: A lexical and frame-semantic embedding based data augmentation approach to automatic categorization of annoying behaviors using# petpeeve tweets. In: Proceedings of the 2015 Conference on Empirical Methods in Natural Language Processing, pp. 2557–2563 (2015)
22. Weischedel, R., et al.: OntoNotes release 4.0. LDC2011T03. Linguistic Data Consortium, Philadelphia (2011)
23. Wong, S.C., Gatt, A., Stamatescu, V., McDonnell, M.D.: Understanding data augmentation for classification: when to warp? arXiv preprint arXiv:1609.08764 (2016)
24. Xu, Y., et al.: Improved relation classification by deep recurrent neural networks with data augmentation. arXiv preprint arXiv:1601.03651 (2016)
25. Yadav, V., Bethard, S.: A survey on recent advances in named entity recognition from deep learning models. In: Proceedings of the 27th International Conference on Computational Linguistics, pp. 2145–2158 (2018)
26. Zhou, G., Su, J.: Named entity recognition using an HMM-based chunk tagger. In: Proceedings of the 40th Annual Meeting on Association for Computational Linguistics, pp. 473–480. Association for Computational Linguistics (2002)

Pseudo Topic Analysis for Boosting Pseudo Relevance Feedback

Rong Yan[1,2](✉) and Guanglai Gao[1,2]

[1] College of Computer Science, Inner Mongolia University,
Hohhot, People's Republic of China
{csyanr,csggl}@imu.edu.cn
[2] Inner Mongolia Key Laboratory of Mongolian, Information Processing Technology,
Hohhot, People's Republic of China

Abstract. Traditional Pseudo Relevance Feedback (PRF) approaches fail to mode real-world intricate user activities. They naively assume that the first-pass top-ranked search results, i.e. the pseudo relevant set, have potentially relevant aspects for the user query. It is make the major challenge in PRF lies in how to get the reliability relevant feedback contents for the user real information need. Actually, there are two problems should not be ignored: (1) the assumed relevant documents are intertwined with the relevant and the non-relevant content, which influence the reliability of the expansion resource and can not concentrate in the real relevant portion; (2) even if the assumed relevant documents are real relevant to the user query, but they are always semantic redundance with various forms because the peculiarity of natural language expression. Furthermore, it will aggravate the 'query drift' problem. To alleviate these problems, in this paper, we propose a novel PRF approach by diversifying feedback source, which main aim is to converge the relatively single semantic as well as diversity relevant information from the pseudo relevant set. The key idea behind our PRF approach is to construct an abstract pseudo content obtained from topical networks modeling over the set of top-ranked documents to represent the feedback documents, so as to cover as diverse aspects of the feedback set as possible in a small semantic granularity. Experimental results conducted in real datasets indicate that the proposed strategies show great promise for searching more reliable feedback source by helping to achieve query and search result diversity without giving up precision.

Keywords: Pseudo Relevance Feedback · Topical network ·
Diversity · Latent Dirichlet Allocation · Community detection

1 Introduction

Information Retrieval (IR) research traditionally assumes that a given query can be associated with a single underlying user intent or information need. However, human behavior patterns is usually in a concise way to express his (her)

© Springer Nature Switzerland AG 2019
J. Shao et al. (Eds.): APWeb-WAIM 2019, LNCS 11641, pp. 345–361, 2019.
https://doi.org/10.1007/978-3-030-26072-9_26

intention, which will make the query disambiguation even worse, especially the short query. As web technology enhances the ability to publish and spread information rapidly, a large of information are available for retrieval, which make the enabling effective information search is an intractable problem. Automatic Query Expansion (AQE) technique settles this problem well by augmenting new features with a similar meaning [6]. As for a classical AQE technique, Pseudo Relevance Feedback (PRF) assumes that the first-pass top-ranked documents, being called the pseudo relevant set, are relevant to the user query, and then uses them to remedy the under-represented information need expression. Numerous previous studies show the effectiveness of PRF to improve the recall for IR [6]. However, this feedback scheme, which keeps considering whether the feedback document is good enough or being real relevant to the user query, has to face the inherent constrained problem of PRF [26].

1.1 Motivation

In general, our work is based on two observations. First, the traditional PRF algorithms consider mere the document itself as the feedback unit, and ignore the connotation of the words behind the document. Recently, to get high-quality of the feedback source from text content analysis point of view is prevalently applied in PRF researches. As for a classic text mining technique, topic modeling has attracted significant attention to PRF researches in the last decade. Generally, the work on topic analysis in PRF are revolved around two sides. One is to evaluate the reliability of the feedback terms in particular topics at the collection level [10,18,24,27]. Another is to evaluate the reliability of the feedback topics at the document content level [16,26]. The aim of topic modeling is to find groups of related terms by considering similarities between how those terms co-occur across a collection. Each topic acts as a latent concept to represent different semantic aspects of the document, which makes the document more comprehensively, as well as the terms in the topics to be a reality. Nonetheless, the intrinsic relationships between the topics are ignored to some extent, except the topic structures (e.g., hierarchical structures or correlated topics) and a partly superficial high clustering of terms being mined to explain the semantic relations between each term pair at the collection level. Consequently, it fails to obtain the better semantic similarity between the documents.

Thus, uncovering the internal coupling structure, which is hidden between each topic pair and within each topic, is the vital reason in our work. E.g.: considering two documents d_1 and d_2 in Fig. 1, we assume that they have been modeled with different topic number: Fig. 1(a) with 3 topics and Fig. 1(b) with 4 topics (the order of each term t denotes its rank in the specific topic T). Each topic is described by a fixed number of terms. As shown in Fig. 1, we can see that each topic has different semantic contribution for the document and each topic are aggregated by some terms with different ranks. Both the size of each term contribution for the semantic representation in the topic and the size of each topic contribution for the semantic representation in the document are enslaved to the topic number. The description of topic T_3 with term set $\{t_{31}, t_{32}, t_{33}\}$ in

Fig. 1(a) but with term set $\{t_{21}, t_{32}, t_{31}\}$ in Fig. 1(b). Also, overlapping terms are appeared in different topics but with different role or semantic contribution, e.g. term t_{13} appears simultaneously in topic T_1 and T_2 in Fig. 1(a).

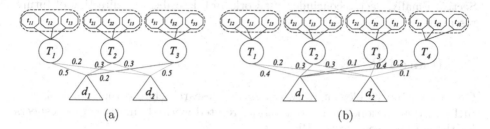

(a)   (b)

Fig. 1. Illustration of our motivation

In Fig. 1(a) and (b), although both documents have been modeled by different topic number, each document almost has the same description of topical terms with different ranks: $d_1 = \{t_{11}, t_{12}, t_{13}, t_{21}, t_{22}\}$, $d_2 = \{t_{21}, t_{22}, t_{13}, t_{31}, t_{32}, t_{33}\}$. If we select these topical terms, then it will get a relatively stable semantic representation for the document. Selecting representative topical terms for the document is the first motivation of this paper, rather than sensitively constraining each document to the topic number.

Second, incomplete specification of the user's information need is the vital issue for IR. To avoid pushing redundant contents, as well as providing diversification but relevant to the user, will benefit for the user retrieval satisfaction, thus Search Result Diversification (SRD) [1,17] technique is coming out as an alternative solution. Numerous studies [2,20,23] have focused on promoting the diversity by using diversify the query through PRF. The main strategy is through contextualizing user's information need to select less redundancy but large coverage for the user query so as to improve the performance of SRD. Recently, Yan et al. [26] attempted to diversify the pseudo relevant set by using text content analysis to enhance the performance of PRF. It is a new effort for PRF using SRD, and we believe that SRD will provide a new perspective to insight and mine the essence of the PRF. In this work, we will follow this thread to research PRF, and this is the second motivation of this paper.

In this paper, we shift the emphasis from the term in the text to the topical term in the topic, and endow the selected representative topical terms to represent the initial document content. The main contributions of this paper comprise the following.

1. Offering an intuitive explanation for the internal structure of the community structure to topic description for supplementing and enriching the contribution of each term for the topic peculiarity, in order to obtain the most representative topical terms to explain the semantic of the document.
2. Ranking and trimming the candidate terms as the selected representative terms, in order to create a succinct feature to form a new document feature.

The remainder of this paper is organized as follows. Section 2 reviews related work about this research. Section 3 demonstrates the details of our proposed pseudo topic content analysis framework. Section 4 presents how to integrate topical information into PRF, and we present the experiment and evaluation in Sect. 5. Finally, Sect. 6 concludes our work and describes the direction of future research.

2   Related Work

Our work is most relevant to two bodies of research efforts: topic analysis and SRD. In this section, we describe a brief related work about these two respects for the specific scenario of PRF.

2.1   Topic Analysis

To distinguish the quality and reliability of the feedback source is the vital issue for PRF. Topic analysis in PRF is an alternative approach for finding more reliable relevant information from the particular topics, and can be roughly classified into the local mode and the globe mode.

In the local mode, the topics are merely extracted from the pseudo relevant set, which main goal is to evaluate the reliability of the feedback terms from the relevant topics [10, 18, 24, 27]. In the globe mode, the topics are extracted from the whole collection, which main goal is to evaluate the reliability of the feedback topics from the whole collection [16, 26]. However, to identify more reliability relevant topic is superior to identify more reliability relevant terms, because it is hard to get an optimal topic number for the collection, much less a specific task. There are some other work, like Chen et al. [7] attempted to tailor different topic set to different document to give a more completed topic representation for the document in order to alleviate the coarse topic representation. Li et al. [13] organized both the local topics and the global topics to represent the local and the global semantics of the collection so that it could assign the more related topics for the document to tackle the 'forced topic' problem.

However, previous work that integrating topic model in IR focused mainly on finding the most relevant topics, and rarely concerning the inherent inner-relationship within the topics. In this work, we assume that if a term considered as a selected representative term, then the topic would be an aggregation of this kind of term. That is to say, the importance of this kind of term will be changed with the topic number, but in any case they must together collaborate with each other to explain the semantic of the topic, and it is. Meanwhile, the content of the document is stable no matter how the topic number changes. That is the crux of our work and starting point. We emphasis on the unadulterated representation of the document content by mining the most representative topical terms.

2.2   SRD

It is quite clear that promoting relevance alone is insufficient for IR when the user query is ambiguous or the retrieval results are excessive redundant. The relevance being overemphasized in standard PRF approaches are seldom taken into account the user satisfaction, the 'relevance' does not mean the true sense after all. SRD is an effective technique to address the user's real complex information need expression, which aim is to enhance satisfactory results by providing diversified results as early in ranking as possible from a given set of retrieval documents.

In recent years, several approaches have been introduced to tackle the SRD problem by promoting documents with high novelty and coverage of multiple query aspects. In Santos et al. [17] resent work, a comprehensive survey about SRD is described in detail. Nevertheless, the diversified object in SRD is still an individual document being acts as the source for diversity. Note that although the semantic content expression of the document is always the only one core topic, but the different parts of the document are not identical, which may be express the subtopic of the main topic for the document. In particular, in [2, 23], the PRF technique is used to alleviate the redundancy of the expansion terms and find diversified query for SRD. Recently, some work [26] used SRD in PRF research, which aim was to obtain a new diversified pseudo relevant set as the expansion object. However, the content description of the document in [26] was inappropriate by lacking explicitly consideration about the topical information within the topic, which resulted in the incorrect similarity between the documents, and eventually affected the quality of the feedback resource. In this paper, another important work is to calculate the similarity between the documents from the topical information point of view, and it will describe in next section in detail.

3   Pseudo Topic Content Analysis Framework

In this paper, how to calculate the similarity between the documents is the core work. Thus, devising an appropriate document content representation becomes the basis for solving the problem. Different from the general topic content analysis, we enrich the text content analysis through the pseudo topic content analysis from two levels, including topical community inner structure level and complex relationships within the topic level. In this section, we elaborate on our approach to analysis the pseudo topic content for the document. The overall framework is shown in Fig. 2, with details presented as follows.

3.1   Generate Topical Networks

We first give some notations. As for a text collection D of N documents with W unique words. We denote D with K topics as $T = \{T_1, T_2, \cdots, T_K\}$. Then, there are K topical networks, being denote as $G = \{G_1, G_2, \cdots, G_K\}$. Each topical

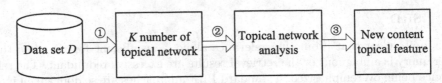

Fig. 2. Pseudo topic content analysis framework

network is denoted as $G_i = (V_i, E_i)$, V_i denotes the nodes set with $n_i = |V_i|$ number nodes as $V_i = \{v_{i,1}, v_{i,2}, ..., v_{i,n}\}$. In each edges set E_i, $e_{i,j}$ denotes the link relation of each node pair (v_i, v_j) in V_i with $m_i = |E_i|$ number edges. C denotes the community set, $M = |C|$ denotes the community number, and $|C_i|$ denotes the node number of C_i community. w_i denotes the weight of node v_i. The word distribution set θ for topic set Z, where θ_z is the word distribution for topic z. The topic distribution set ϕ_c $(c \in C)$ for community set C.

Figure 3 shows the topology of the topical network that we intend to generate. For generating meaningful con-relationships within the topic and between the topics, there are two levels that we plan to create, including the word level and the topic level. We discuss the details about what and how the topical network construction in Sect. 3.3.

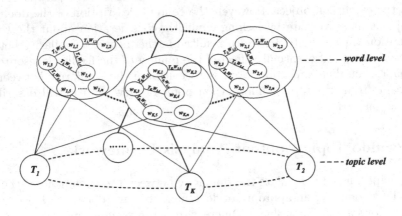

Fig. 3. The topology of the topical network

3.2 Topical Network Analysis

In our framework, one of the hardest problems is how to obtain the features by using analysis of the topical network to characterize the document content. Note that although 'the feature' with richness in the semantic, the importance of each term in the topic and the importance of each topic in the document are quite different. Actually, they are interweaving each other.

Wei et al. [25] proposed a topic-based summarization approach to model the documents into a set of topics and each topic was characterized by a set of keywords by using word-topic and document-topic two distributions. Similar to their work, Yan et al. [26] ranked the topics and the terms in the topic by using the importance of the topic and the term respectively, to abstractly represent the document content. Furthermore, Smith et al. [21] proposed a relationship-enriched topic model to visualize the relationships between the terms in the topics. However, they only characterized the correlation between each term pair within each topic on the document level but not for a specific topic, without considering the correlation between the topics. In a word, all these work rarely concern about the internal structure within the topic, but it is the vital for the topic-based representation.

The key insight of our work is to consider the coherent internal relationships which is concealed within the topic and between the topics at the community level. A community is a group of subjects with similar behavior and internal agreement [11]. Recently, community detection algorithms are utilized to extract topic features. In recent work of [8,9,14], discovering the internal structure of network community have been studied in detail in order to obtain deep knowledge of network properties. Two types of internal communities have been proposed. One is the leader community, where a few nodes regarded as the leaders of the community that have highly associated with nearly all or a substantial number of other nodes. Another is the self-organizing community, where all the nodes are associated with similar degrees that have nearly equal status. Inspired by these, the discovered community structure can be used for generating more meaningful con-relationships within the topic and between the topics. Actually, the same scene is occurred in topical network community. In the description of each topic, the semantic expression is implemented through a small number of terms that have large proportion, to accomplish coordination and semantic correlation with the most other terms that act as supplementary to the description, but with weaker correlation between them. That is to say, a few terms play an important role in keeping the topical network communities dense and more clustered, furthermore acting a core link to the other community members. But few researchers focus on the internal structure within the topic community and the correlation between different topic communities in topic analysis. In this paper, we will focus on this scene and offer an intuitive explanation for the internal structure of the community to topic description for supplementing and enriching the contribution of each term for the topic peculiarity. To the best of our knowledge, our paper presents the very first work for analyzing the internal coupling structure in topic modeling from much more fine property of complex networks point of view. We believe that the coherence analysis of the topics and communities can benefit each other, and bridge the gap between the standard topic statistical analysis and text content analysis.

3.3 Generate New Content Topical Feature

In essence, it is the term, which to keep the topic dense and be aggregated together, being to explain the semantic of the document in topic modeling. But the truth is that there is an inherent diversity among individual topics, the same term may possess different explanation or meaning in different topics, which always gives different interpretation to represent the semantic meaning for the topic. Thus, the following work is how to evaluate the reliability of each term and how to obtain the effective terms that are used to explain the semantic of the document. Generally speaking, the features that characterized the semantic of the document are bringed up at two levels, including the word level and the topic level. We will take a graph analytical approach by mining the topical terms to construct a pseudo content space for the document. Then, we will describe the details from the following two stages point of view. In this work, we apply the popular topic modeling method Latent Dirichlet Allocation (LDA) to model the real dataset.

First Stage Ranking. The first stage ranking is to identify the terms and the co-occurrences between the terms at the word level. Generally, the weight of each edge is used to reveal its peculiarity that has actually practical significance in different network, and this prior information will benefit to detect communities. Therefore, we should not only consider the topology of the network, but also the practical significance of the connection between each edge pair. In order to get the more accurately measurement of the correlation intension of each word pair in the topic, we firstly give the definition of the weight of each edge pair that showed by the following equations:

$$w(v_i, v_j) = \frac{w^{'}(v_i, v_j)}{w^{'}_{avg}} \tag{1}$$

$$w^{'}_{avg} = \frac{\sum_{v_i v_j \in E_i(G_i)} w^{'}(v_i, v_j)}{|E_i(G_i)|} \tag{2}$$

$$w^{'}(v_i, v_j) = \mu_1 sim\_con(v_i, v_j) + \mu_2 sim\_word(v_i, v_j) \tag{3}$$

$$sim\_con(v_i, v_j) = \frac{|N(v_i, v_j)|^2}{min\{|N(v_i)|, |N(v_j)|\}^2} \tag{4}$$

where $N(v_i, v_j)$ denotes the set of the public neighbor node with node v_i and v_j, and $N(v_i)$ denotes the set of neighbor node with node v_i. sim_con [15] denotes the topology similarity weight between each node pair. sim_word denotes the actual weight value between each node pair, and we use word2vec[1] to accomplish it. The parameter μ_1 and μ_2 is linear combination for $w^{'}(v_i, v_j)$, and $\mu_1 + \mu_2 = 1$.

As for the weight of each word node $v \in G_i$, we will use Ref. [26] to calculate, as it shows in Eq. (5).

$$w(v) = \phi_{i,m} \cdot log \frac{\phi_{i,m}}{K\sqrt{\prod_{j=1}^{K} \phi_{j,m}}} \tag{5}$$

---

[1] http://code.google.com/p/word2vec.

Second Stage Ranking. The second stage is to extract latent topic content at the topic level. Just like the weight of the word node in each topical network, the weight of the topic node in the whole topical network is restricted through each topic content coverage, topic variance and the topology of the whole topical network. Then, the weight of each topic node is calculated by the following equations:

$$w(T_i) = \lambda_1 T\_cov_i \cdot T\_var_i + (1 - \lambda_1) sim\_contopic(v_{Ti}, v_{Tj}) \qquad (6)$$

$$T\_cov_i = \frac{\sum_{j=1}^{N} N_j \cdot \Theta_{j,i}}{\sum_{j=1}^{N} N_j} \qquad (7)$$

$$T\_var_i = \sqrt{\frac{\sum_{j=1}^{N} N_i \cdot (\Theta_{j,i} - T\_cov_i)^2}{\sum_{j=1}^{N} N_i}} \qquad (8)$$

where T_cov_i and T_var_i denote each topic content coverage and topic variance, respectively. $sim\_contopic(v_{Ti}, v_{Tj})$ is the topology similarity weight between each topic pair [15].

In this work, we apply BGLL [4] algorithm to detect the communities, in order to get the more detailed correlation between each term pair which has the same semantic behavior. BGLL is a fast modularity optimization method for yielding a smaller weighted network by repeatedly optimizing the modularity. We choose EQ (Extended Modularity) metric to measure the quality of community detection because the built topical network communities without community labeling. The EQ [19] value of each topical community is computed by Eq. (9):

$$EQ = \frac{1}{2} \sum_{i=1}^{M} \left(\sum_{v \in C_i} \sum_{w(\neq v) \in C_i} \frac{1}{O_v O_w} \left[A_{vw} - \frac{k_v k_w}{2m} \right] \right) \qquad (9)$$

where O_v is the number of communities that the node v belongs to the final detected communities. A is the adjacency matrix of the original network, and m is the total number of edges in the original network.

In Fu et al. [9] work, they used the ratio between the degree variance of the given community and the expected degree variance of the corresponding random model to quantify a community. We can use it as a yardstick to extract the topical terms as mentioned in Sect. 3.3 in the two stage ranking procedures. However, the node degree is just only one of the characteristics of the node attribute description in the actual network. Furthermore, it also includes the actual meaning of nodes in a specific network, e.g. the strength of the node. In our constructed network, the strength of the node refers to the intensity of the description of the topic semantic. Thus, we do some adjustment for the ration in [9], as shown in the Eq. (10):

$$\rho = \frac{VAR_{real}}{VAR_{rand}} \qquad (10)$$

where VAR_{real} and VAR_{rand} denote the the degree and the strength variance of the topical communities and random communities, respectively. When $\rho > 1$

means that the nodes in the given community have stronger role for keeping the topical community more clustered, indicating more latent semantic representation for the document.

After these two rankings, the document will be represented by the topical term nodes set and with complex association relationships among them to embody the semantic content itself. There, we will use the characteristic of the terms itself and the relationships between each term pair to estimate the content similarity between each document pair, and it will explain in Sect. 4 in detail.

4   Integrating Topical Information into PRF

In the work of [26], it is empirically to set the parameter to determine the quantity of the topical terms to accomplish the document brief description. It would undermine the result of the document understanding, which lost the relationships that hidden behind between the topical terms and could lead to an unreliable similarity between each document pair. As an alternative strategy, in order to obtain the ability of the word node in the document much objectively, we evaluate the term node from the whole topical networks point of view. In this section, we will describe how to get the pseudo relevant set in detail. Algorithm 1 summarizes the strategy. We choose the classical metric called marginal relevance (MMR) [5] algorithm to realize the re-ranking of the first-pass retrieval results. After the re-ranking, the top-ranked documents will be considered not only relevant to the user query, but also have the smallest semantic redundancy each other. The MMR algorithm focuses on identifying diverse search results based on document content, and how to calculate the similarity calculation between each document pair is the core for MMR. As mentioned in Sect. 3.3, each document is depicted in a new document content space. We will combine three metrics, including the node weighted degree WD, the weighted aggregation coefficient to WC and Betweenness WB, to evaluate expression ability of the term node in topical network. The term node integrated characteristic value $Z(v)$ is defined as Eq. (11):

$$Z(v) = \lambda_1 * WD(v) + \lambda_2 * WC(v) + \lambda_3 * WB(v) \qquad (11)$$

where, the weighted degree $WD(v)$ of node v is defined as the number of $v's$ neighbors. W_{vu} is the weight between node v and u. It reflects the joint strength of v to the other adjacent nodes directly. The weighted aggregation coefficient $WC(v)$ of node v is the connectivity density and strength in the local network. The node Betweenness $WB(v)$ of the node v is the ratio of the number of shortest paths that pass through v to all the shortest paths in the network. It reflects the influence of the node on the information flow of the network. λ_1, λ_2 and λ_3 are the adjustable parameter. $\lambda_1 + \lambda_2 + \lambda_3 = 1$ and $0 < \lambda_1, \lambda_2, \lambda_3 < 1$.

So far, we have extract the $z(v)$ value for each node v in the topical network space, i.e., every document has been represented in the pseudo content space to be a pseudo content vector. We denote each document as $d = \{Z_1, Z_2, \cdots, Z_n\}$.

Algorithm 1. Obtaining the pseudo relevant set procedure

Input:
    Original data set D;
    The user query q;
Output:
    The pseudo relevant set P_{pse};
1: Initialization the topic number: $K = 10$;
2: Select a test data set R_{test}, calculate the Perplexity 3 value for different topic number, step sets 10;
3: Regard the lowest perplexity corresponding topic number as the optimal topic number K;
4: **for** each topical network $i \in K$, G_i **do**
5:    $i=1$;
6:    **for** each word node $v \in G_i$ **do**
7:      Compute the weight of (v, v') $(v' \in G_i, v \neq v')$ using Eq. (1);
8:      Compute $w(v)$ using Eq. (5);
9:    **end for**
10:   Compute the weight of each topical network node using Eq. (6);
11:   $i = i + 1$;
12: **end for**
13: Evaluate each network G_i and each node v in G_i using Eq. (9) and Eq. (10) for re-ranking;
14: Let q to retrieval D, the first passed retrieval results denote R;
15: Re-ranking R using the MMR algorithm;
16: **return** R;
17: Select the top-ranked documents as the pseudo relevant set P_{pse};

Thus, we apply Cosine similarity to determine the content similarity of each document pair, as shown in Eq. (12).

$$Sim(d_i, d_j) = \frac{\sum_{k=1}^{n} Z_k(d_i) \times Z_k(d_j)}{\sqrt{\sum_{k=1}^{n} Z_k^2(d_i)} \times \sqrt{\sum_{k=1}^{n} Z_k^2(d_j)}} \qquad (12)$$

5 Experiments and Evaluation

5.1 Experimental Settings and Datasets

In addition, we do use offline LDA so that we can easily extend our work on large collections and avoid the challenge of choosing the topic number.

- **XINHUA:** it contains four years (2002–2005) XINHUA (*Simplified Chinese*) news including several different topics with 308,845 documents in NTCIR8[2]. We use the *TEXT* field for topic modeling. Because XINHUA is in Chinese, we first do some preprocessing, including automatic word segmentation and stop-word removal.

---

[2] http://research.nii.ac.jp/ntcir/index-en.html.

- **OHSUMED:** it contains five years (1987–1991) relatively short abstracts of references from medical journals in the MEDLINE[3] database with 348,566 documents. We use the *abstract* field for topic modeling.

Firstly, in the experiment, we use fivefold cross validation to get the proper topic number for different datasets. We use the *Perplexity* [3] metric to get the optimal topic number for different dataset. Generally, the lower *Perplexity* is, the better the generalization of model has. In the experiment, we obtain the optimal topic number $K = 60$ and $K = 70$ for XINHUA and OHSUMED respectively.

We use the classical probabilistic retrieval model the Lemmur Toolkit[4] for specific implementation. The *request* field of the topics 1–100 for XINHUA, and 1–106 for OHSUMED being used as query set respectively.

5.2   Experimental Evaluation

In this work, we focus on alleviating the information overload problem for the user using PRF strategy by using the diversified pseudo relevant set, but not aiming for diversifying the retrieval results. So the standard retrieval metric MAP, nDCG and P@n will be used in our evaluation. Also, we temporarily ignore the impaction of the low precision of the first-pass results.

Pseudo Content Result. Figure 4 shows the *EQ* value in two datasets. From Fig. 4, we can see that the performs well for each topical network overall. But we still notice that there have some lower *EQ* value for a few topical network because the lack of the association of each word pair in the topical network.

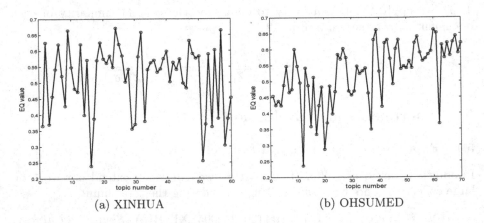

(a) XINHUA   (b) OHSUMED

Fig. 4. EQ value in two datasets

---

[3] http://mlr.cs.umass.edu/ml/machine-learning-databases/ohsumed/.
[4] http://www.lemurproject.org/.

Table 1 shows some examples of the comparison between our topic analysis results and original topic analysis (*top*-20). Figure 5 shows some topical networks descriptions of examples for two datasets.

Table 1. Examples on topic analysis comparisons on different data sets.

XINHUA

Topic	Original topic analysis	Our topic analysis
Topic2	growth\last year\this year\dollar\economic \export\decline\increase\statistic\simultaneous \achieve\consume\occupy\display\quarter\rise \year-on-year\reach\report\reduce	growth\economic\dollar \consume\simultaneous \year-on-year\export\reach \rise\statistic
Topic4	education\college\student\school\children \talents\train\woman\major\study \employment\youth\cultivate\academy \juvenile\job\profession\university\socity	cultivate\children \profession\job\woman \service\employment \juvenile\youth\education \train\talents

OHSUMED

Topic	Original topic analysis	Our topic analysis
Topic4	expression\cells\class\surface\lines\complex \expressed\T-cell\molecules\cells\major\bound \interferon\HLA-DR\murine\molecule\sites \interleukin\distinct\transcripts	sites\complex\distinct \interferon\lines\HLA-DR \molecules\cells
Topic18	hospital\patient\support\study\time\program \costs\nursing\programs\status\admitted \elderly\community\care\patients\survey \recommended\systems\improve\benefits	underwent\duration \radiation\preoperative \tumor\surgery

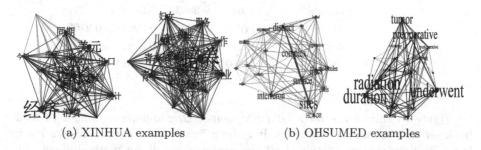

(a) XINHUA examples   (b) OHSUMED examples

Fig. 5. Topic network descriptions of examples for two data sets. (**XINHUA:** *Topic2* (left) and *Topic4* (right); **OHSUMED:** *Topic4* (left) and *Topic18* (right))

It can be concluded from Table 1 and Fig. 5, the results of the topical network analysis are more common sense than the original topic analysis, and can get better representation with fewer terms, i.e., to some extent, the topic semantic representation with the topical network analysis will separate out the terms that have bigger probability.

Retrieval Results and Analysis. Our experiments are conducted under the language modeling framework, and we use a query generative language mode with Dirichlet prior for smoothing (with a hyperparameter of $\mu = 1000$) to produce the first-pass retrieval results. We consider the top-1000 first-pass retrieval results as the diversity target. In the MMR algorithm, we apply Okapi [22] similarity to measure the similarity between the query and document. Because we focus on how the topical information to accomplish the document content understanding, we apply BM25 [12] PRF retrieval model as the baseline. Also, we compare our results with Ref. [26].

Table 2 shows the results of the comparison on two language datasets in detail. We conducted a significance test (t-test) on the improvements by our approach. The result indicates that the improvements are statistically significant (p-value < 0.05). ('*' means significant improvement over other methods). As shown in Table 2, we can see that our proposed method can get better retrieval performance for two language datasets.

Table 2. Performance comparisons on two data sets.

Date set	Method	MAP	nDCG	P@5	P@10
XINHUA	BM25_PRF	0.2830	0.3997	0.5239	0.5056
	LM_PRF	0.2709	0.3920	0.5139	0.4847
	TF-IDF_PRF	0.2815	0.4128	0.5296	0.4817
	Ref. [26]	0.3010	0.4328	0.5452	0.4986
	PRF_Topic	**0.3210***	**0.4418***	**0.5510***	0.5010
OHSUMED	BM25_PRF	0.1731	0.3344	0.3426	0.2772
	LM_PRF	0.1467	0.2893	0.2693	0.2337
	TF-IDF_PRF	0.1702	0.3288	0.3208	0.2703
	Ref. [26]	0.1812	0.3580	0.3487	0.2698
	PRF_Topic	**0.1910***	**0.3728***	**0.3500***	0.2701

Figure 6 exhibits the retrieval performance over different number of feedback set on two language datasets. It is clear from Fig. 6 that the results are in line with previous expectations that the performance of our method is not only insensitive to the number of feedback set, but also the better performance we can obtain with smaller number. Meanwhile, this comparison confirms that the well document understanding can guarantee for getting better feedback resource.

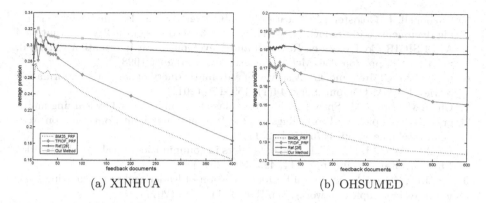

(a) XINHUA   (b) OHSUMED

Fig. 6. Performance over the number of feedback set on two data sets

6   Conclusions and Future Work

We proposed in this paper a new view to PRF, that is able to enrich the topic
and intensively characterize the topic description using community detection
analysis method. In our approach, we attempt to obtain particular terms that
can represent the semantic of the topic well. Besides the challenge of this, we
explore to explain the semantic of the document using a common perspective,
but not confine to find relevant terms in particular topics or to find relevant
topics for the user query, and the whole procedure is realized with off-line mode.

In the future work, we plan to research other similarity equations that can
affect the performance significantly. Also we can apply our proposed method to
document expansion research, the off-line semantic document analysis in document expansion is superior to query expansion after all, and this is just what
our proposed method is good at.

Acknowledgements. This research is jointly supported by the National Natural Science Foundation of China (Grant No. 61866029, 61763034), Natural Science Foundation of Inner Mongolia Autonomous Region (Grant No. 2018MS06025) and Program
of Higher-Level Talents of Inner Mongolia University (Grant No. 21500-5175128).

References

1. Abid, A., et al.: A survey on search results diversification techniques. Neural Comput. Appl. **27**(5), 1207–1229 (2016)
2. Agrawal, R., Gollapudi, S., Halverson, A., Ieong, S.: Diversifying search results. In:
   Proceedings of the 2nd ACM International Conference on Web Search and Data
   Mining, WSDM 2009, Barcelona, Spain , pp. 5–14, February 2009
3. Blei, D.M., Lafferty, J.D.: Correlated topic models. In: Proceedings of the 18th
   International Conference on Neural Information Processing Systems, NIPS 2005,
   pp. 147–154. MIT Press, MA (2005)
4. Blondel, V.D., Guillaume, J.L., Lambiotte, R., Lefebvre, E.: Fast unfolding of
   communities in large networks. J. Stat. Mech: Theory Exp. **P1008**, 155–168 (2008)

5. Carbonell, J., Goldstein, J.: The use of MMR, diversity-based reranking for reordering documents and producing summaries. In: Proceedings of the 21st International ACM SIGIR Conference on Research and Development in Information Retrieval, SIGIR 1998, pp. 335–336. Melbourne, Australia, August 1998
6. Carpineto, C., Romano, G.: A survey of automatic query expansion in information retrieval. ACM Comput. Surv. **44**(1), 159–170 (2012)
7. Chen, M., Jin, X.M., Shen, D.: Short text classification improved by learning multi-granularity topics. In: Proceedings of 22nd International Joint Conference on Artificial Intelligence, pp. 1776–1781 (2011)
8. Fu, J.C., Wu, J.L., Liu, C.J., Xu, J.: Leaders in communities of real-world networks. Phys. A: Stat. Mech. Appl. **444**, 428–441 (2016)
9. Fu, J.C., Zhang, W.X., Wu, J.L.: Identification of leader and self-organizing communities in complex networks. Sci. Rep. **7**(1), 1–10 (2017)
10. Ganguly, D., Jones, J.F.G.: A non-parametric topical relevance model. Inf. Retr. J. 1–31 (2018)
11. Han, X., et al.: Emergence of communities and diversity in social networks. Proc. Nat. Acad. Sci. **114**(11), 2887 (2017)
12. Jones, K.S., Walker, S., Robertson, S.E.: A probabilistic model of information retrieval: development and comparative experiments part 2. Inf. Process. Manag. **36**(6), 809–840 (2000)
13. Li, X.M., Ouyang, J.H., Lu, Y., Zhou, X.T., Tian, T.: Group topic model: organizing topics into groups. Inf. Retr. J. **18**(1), 1–25 (2015)
14. Liu, C.J.: Community ditection and analytical application in complex networks. Ph.D. thesis, Shandong University, Shandong, China (2014)
15. Lv, L.Y., Zhou, T.: Link prediction in complex networks: a survey. Phys. A **390**(6), 1150–1170 (2011)
16. Miao, J., Huang, X., Zhao, J.S.: TopPRF: A probabilistic framework for integrating topic space into pseudo relevance feedback. ACM Trans. Inf. Syst. **34**(4), 1–36 (2016)
17. Santos, R.L.T., Macdonald, C., Ounis, I.: Search result diversification. Found. Trends Inf. Retr. **9**(1), 1–90 (2015)
18. Serizawa, M., Kobayashi, I.: A study on query expansion based on topic distributions of retrieved documents. In: Gelbukh, A. (ed.) CICLing 2013. LNCS, vol. 7817, pp. 369–379. Springer, Heidelberg (2013). https://doi.org/10.1007/978-3-642-37256-8_31
19. Shen, H.W., Cheng, X.Q., Cai, K., Hu, M.B.: Detect overlapping and hierarchical community structure in networks. Phys. A **388**(8), 1706–1712 (2009)
20. Shen, X.H., Zhai, C.X.: Active feedback in ad hoc information retrieval. In: Proceedings of the 28th International ACM SIGIR Conference on Research and Development in Information Retrieval, SIGIR 2005, Salvador, Brazil, pp. 59–66, August 2005
21. Smith, A., Chuang, J., Hu, Y.N., Boyd-Graber, J., Findlater, L.: Concurrent visualization of relationships between words and topics in topic models. In: Proceedings of the Workshop on Interactive Language Learning, Visualization, and Interfaces, ACL 2014, pp. 79–82. ACM Press, New York (2014)
22. Stephen, R.: Okapi at TREC 3. In: Overview of the Third Text Retrieval Conference (TREC 3), pp. 109–125 (1994)
23. Vargas, S., Santos, R.L.T., Macdonald, C., Ounis, I.: Selecting effective expansion terms for diversity. In: Proceedings of the 10th Conference on Open Research Areas in Information Retrieval, OAIR 2013, Lisbon, Portugal, pp. 69–76, May 2013

24. Wang, X.W., Zhang, Q., Wang, X.J., Sun, Y.P.: LDA based pseudo relevance feedback for cross language information retrieval. In: IEEE International Conference on Cloud Computing and Intelligent Systems, CCIS 2012, vol. 3, pp. 1511–1516 (2012)
25. Wei, F.R., et al.: TIARA: a visual exploratory text analytic system. In: Proceedings of the 16th ACM International Conference on Knowledge Discovery and Data Mining, SIGKDD 2010, Washington, DC, USA, pp. 168–168, July 2010
26. Yan, R., Gao, G.L.: Pseudo-based relevance analysis for information retrieval. In: 2017 IEEE 29th International Conference on Tools with Artificial Intelligence, ICTAI 2017, Boston, MA, USA, pp. 1259–1266, November 2017
27. Ye, Z., Huang, J.X., Lin, H.F.: Finding a good query-related topic for boosting pseudo-relevance feedback. J. Assoc. Inf. Sci. Technol. **62**(4), 748–760 (2011)

Knowledge Graph

Leveraging Domain Context for Question Answering over Knowledge Graph

Peihao Tong, Junjie Yao[⊠], Linzi He, and Liang Xu

East China Normal University, Shanghai 200062, China
xherot@gmail.com, junjie.yao@sei.ecnu.edu.cn, helinzihelen@gmail.com,
xulyeng@gmail.com

Abstract. This paper focuses on the problem of question answering over knowledge graph (KG-QA). With the increasing availability of different knowledge graphs in a variety of domains, KG-QA becomes a prevalent information interaction approach. Current KG-QA methods usually resort to semantic parsing, retrieval or neural matching based models. However, current methods generally ignore the rich domain context, i.e., category and surrounding descriptions within the knowledge graphs. Experiments shows that they can not well tackle the complex questions and information needs.

In this work, we propose a new KG-QA approach, leveraging the domain context. The new method designs a neural cross-attention QA framework. We incorporate the new approach with question and answer domain contexts. Specifically, for questions, we enrich them with users' access log, and for the answers, we equip them with meta-paths within the target knowledge graph. Experimental study on real datasets verifies its improvement. The new approach is especially beneficial for domain knowledge graphs.

1 Introduction

Recent years have witnessed an information access paradigm shift, from a proactive search to voice/question oriented automatic answering. A lot of personal assistants, i.e., Siri, Alexa Echo, and Google Home are emerging. In these question answering services, we submit questions and get answers or suggestions. Under the hood, the crucial ingredient is the structured knowledge graph, including a full range of necessary information, constructed from related data sources.

Take the insurance product domain as our motivation scenario, we have constructed a knowledge graph for insurance products (InsKG, abbrv. later in this paper) and set up an online question-answer service on top of it. This insurance product knowledge graph has more than 200k triples, consisting of insurance companies, categories, types, attributes, and terms, etc. With intuitive and comprehensive answers, the mobile question-answer service on top of it has attracted almost 100k input questions from ordinary users since its launch six months earlier.

Question answering over knowledge graph (KG-QA) has attracted many attentions [2, 6, 21, 22]. Common ways of KG-QA include semantic parsing and

© Springer Nature Switzerland AG 2019
J. Shao et al. (Eds.): APWeb-WAIM 2019, LNCS 11641, pp. 365–381, 2019.
https://doi.org/10.1007/978-3-030-26072-9_27

retrieval based. The semantic parsing methods formalize the input question into the logical forms and locate the entities in the target knowledge graph. The retrieval methods conduct IR-based metrics to rank the candidate entities from the knowledge graph. With the recent improvement of deep learning algorithms, some embedding representation and generated models [6] are also introduced to tackle the problem of KG-QA as the matching between the input question and answers.

Though remarkable progress have been achieved, KG-QA is still a challenging problem. The difficulty issues lie at not only the vague question description but also a variety of entity and relationship types within the knowledge graph [19,25].

Table 1. QA cases of current and proposed approaches

Question	Truth	State of the art	Proposed
What types of insurance products does China Life sell?	12, including disease, medicine and accident insurances, etc.	8, including disease, medicine and accident insurances, etc.	Same as ground truth
What categories does CPIC's critical disease insurance belong to?	cancer, disease, investment	cancer, disease	cancer, disease, investment

The returned answers of commonly used and the proposed KG-QA approaches, are listed in Table 1. For factual questions (first question), i.e., related to product attributes, both current and proposed methods can return satisfying answers. However, for some general category related or survey questions (second, third), the current method can not compete with the proposed approach.

By investigating the underlying knowledge graph, we find that these questions usually cover more nodes in the knowledge graph and these corresponding nodes have more connections between them. Due to the complexity of KG-QA and the generality of current methods, they can not uncover or utilize the patterns or more latent structures within the knowledge graph.

In this paper, motivated by these domain characteristics, we design a new neural question-answer matching model to incorporate more domain context features into KG-QA tasks. In the feature extraction layers, we extract the meta-path patterns and push the embedded graph features into the meta-path levels [9]. We also discuss the semantic parsing of input questions with the tree model [20] for better identifying the users' information need. Besides, as an integrated neural generation KG-QA model, we fit the new approach into a cross-attention framework [11], which can model the interactive question-answer matching process.

The contributions of this paper can be listed as follows.

1. We exploit the domain context in KG-QA, improving the QA performance with more KG patterns as the representation features.

2. We utilize the new cross-attention model to match the question and answers in KG-QA task, taking the rich connections between the comprehensive representation of questions and answers.
3. The experimental results demonstrate the effectiveness of our proposed approach, especially in domain knowledge graphs.

2   Related Work

Existing KG-QA models can be categorized into two lines, i.e., common and neural network-based methods. In the general lines, semantic parsing and retrieval based methods are popular ones.

Semantic parsing-based methods compile the input natural language questions into logical forms, with the help of a combinatory grammar [7] or dependency-based compositional semantics [16]. The answers are then returned through SPARQL queries or other well-designed query processing mechanism. These methods are strict and usually take many efforts in the annotation and pre-processing stage [10,14].

The retrieval based methods turn to analyze the dependency of the words in the question and then use the target information extracted from the question to locate candidate answers from the knowledge graph. Final answers are chosen from the candidate set by further quality or relevance evaluation [22].

In contrast, recent deep learning methods usually transform questions and answers into the form of semantic vectors [4,6], and then conduct the matching, ranking operations. Recently, some combined or fused methods are also proposed, and at the same time, some domain features are investigated in the models. [23] developed a semantic parsing framework utilizing convolutional neural networks to process single-relation questions. [21] proposed a model to map the natural language questions into logical forms by leveraging the semantic associations between lexical representations and KG-properties in the latent space.

In [8], the authors discussed three different aspects of the answers. For each aspect of the answer, the proposed model utilized a text-CNN [13] to generate the corresponding question representation. [11] utilized the bidirectional LSTM model to create the representation for the questions which entirely took advantage of the forward and backward information of sentences.

The proposed approach in this paper follows similar fusion paradigm, but we present the question and answer representation separately in a more general way, and also provide more comprehensive question-answer matching model. We further profile the question representation in a tree structure and the knowledge graph representation with meta-path. The new approach presents a more general context feature extraction approach.

3   Domain Context Feature Extraction

3.1   Framework

KG-QA is designed to extract the answer set A from the knowledge graph when the question q is given. It involves input question and the underlying knowledge graph respectively. Input questions are the form of natural language and consisting of a set of words. Knowledge graph is usually regarded as a set of entities and the associated edges, i.e., fact triples, in the form of $(subject, predicate, object)$.

To be specific, we introduce the architecture of the KG-QA framework proposed in this paper. At first, we should identify the topic entity from questions in order to generate candidate answer set in knowledge graph. After candidate answer set is generated, we should rank these candidates to find the candidates which match the given question best. In matching stage, we obtain representations for questions and answers respectively. We adopt the deep learning model tree structure LSTM to understand the question and obtain the representation of questions. For the answers, the knowledge graph embedding method TransE [5] is usually used to obtain the representation for different answer aspects. Finally, the score function is trained to measure the matching score between the representation of questions and candidate answers. The candidate which scores highest will be selected as the final answer. Besides, for the specific-domain knowledge graph, we make some modifications to the general framework. As mentioned in Sect. 1, we carefully design the novel domain context features. Both the answer and question-related context are utilized. In the domain context feature extraction and representation layer, the proposed approach processes input question with two modules, i.e., question representation and answer representation.

- In the extraction of question representation, we enrich the question representation of words by collecting external text resources. After that, we parse the input question with the help of the tree model, which profiles the users' question intents.
- In the stage of answer representation, we adopt the meta-path random walk [9] to capture the context of the targeted entities in the knowledge graph. The extracted paths are used to represent the entities and later answers.

Figure 1 illustrates the framework of our proposed KG-QA approach. We split the framework into offline training and online querying stages. After the domain context feature extraction, in the matching stage, the cross-attention model [11,15] is designed to capture the relations between input question and selected answers. We will provide the matching details in Sect. 4.

3.2 Input Question Representation

Shown in Fig. 1, the question representation has two stages. In offline training, we collect external text resources, to embed and represent the questions. After that, we parse the input question with the help of the tree model, which parses the users' question intents. In the online response, the extracted embedding vectors are used to represent the input questions.

Speaking of the word embedding, we choose word2vec [17, 18] method to train the word vectors, with the help of the specific-domain text corpus. Here we use the users' input question logs. The word embedding vectors for insurance dataset are generated and then used as input to the later question training and online response.

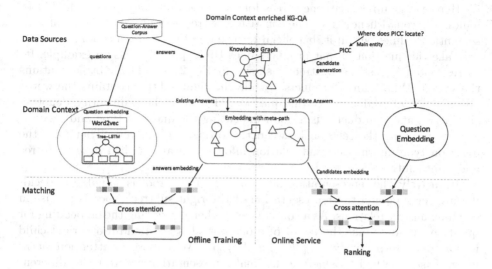

Fig. 1. Domain context for KG-QA tasks

Formally, the question q consists of several words $q = (w_1, w_2, \cdots, w_n)$, where w_i means the ith word in the question. For each word, x_{w_i} denotes the latent representation of the word w_i, i.e., embedded vectors in our case.

We continue to the tree-structured LSTM model, a variant of recursive neural networks [20, 24]. Different from sequential LSTM which only allows strictly sequential information propagation, the tree-LSTM is a network that accepts tree-structured input. Moreover, the architecture of the network will be constructed according to the tree structure of the extracted concept layers from the input question.

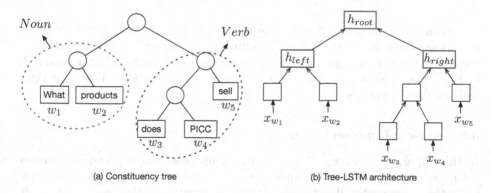

Fig. 2. Constituency tree and the corresponding tree-LSTM

Here we set up constituency trees for each question, and use tree-LSTM to handle the constituency tree. The constituency tree is intuitive to cover different semantic modules within it, beneficial for the question representation.

Take the question "What products does PICC provide?" as an example. Its corresponding constituency tree is shown in Fig. 2. The left subtree contains the words "What" and "products", while the right subtree contains the words "does", "PICC" and "sell". In this question, "PICC" is the main entity, denoting the target, and "products" is a noun indicating the question type and "sell" is a verb, meaning the relation between the target and answers. Therefore, the question type information and relation information are implied in left subtree and right subtree respectively.

Intuitively, the representation of the left subtree's root is valuable in deciding answer type and the representation of the right subtree's root is precise in deciding answer relations. As a consequence, when generating the embedding for questions, we only take the vector of root, root's left child and root's right child into account, respectively h_{root}, h_{left} and h_{right}. We propose an attention mechanism based method to generate the final representation according to different answer aspects which will be explained in Sect. 4 in detail.

After the constituency tree is generated, the tree-LSTM is adopted to handle the questions. The following equations show the underlying tree-LSTM. The leaf units of the tree-LSTM receive the word representation as input, and the internal units take their left and right children's output as the input.

$$i_j = \sigma(W^{(i)}x_j + U_l^{(i)}h_{jl} + U_r^{(i)}h_{jr} + b^{(i)}) \tag{1}$$

$$f_{jk} = \sigma(W^{(f)}x_j + U_{kl}^{(f)}h_{jl} + U_{kr}^{(f)}h_{jr} + b^{(f)}) \tag{2}$$

$$o_j = \sigma(W^{(o)}x_j + U_l^{(o)}h_{jl} + U_r^{(o)}h_{jr} + b^{(o)}) \tag{3}$$

$$u_j = tanh(W^{(u)}x_j + U_l^{(u)}h_{jl} + U_r^{(u)}h_{jr} + b^{(u)}) \tag{4}$$

$$c_j = i_j \odot u_j + f_{jl} \odot c_{jl} + f_{jr} \odot c_{jr} \tag{5}$$

$$h_j = o_j \odot tanh(c_j) \tag{6}$$

When the node is a leaf node, the input h is set to zero. On the contrary, when the node is an internal node, the input x is set to zero. The Eqs. 1, 2 and 3 respectively define the input gate, forget gate and output gate in the tree-LSTM. In Eq. 2, the k denotes a binary index variable which indicates the left or right child of the current node.

3.3   Answer Representation

In this paper, we categorize four different kinds of answer aspects, i.e., answer entity a_e, answer relation a_r, answer type a_t and answer context a_c. Specifically, the answer entity denotes the entity identification in the knowledge graph. Answer relation denotes the relation path from the main entity to answer entity. For each answer in the training or online response stages, we generate latent representation e_e, e_r, e_t and e_c for each aspect. If the relation path only includes one relation, we can directly use the relation representation as e_r. If there are more than one relations in the path, the relation path representation should be calculated in $e_r = \frac{1}{n} \sum_i^n e_{r_i}$.

In the specific-domain knowledge graphs, the relation types are usually limited, which means that it is suitable to capture and summarize the paths. Here we propose an approach which utilizes the meta-path to model the path context in the knowledge graph.

Take the insurance domain as the motivating example, we define a meta-path scheme $p_i = (Company \rightarrow Product \rightarrow Type \rightarrow Product \rightarrow Company)$.

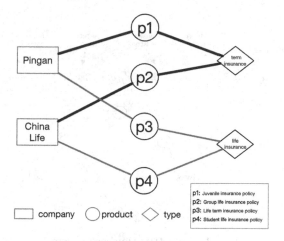

Fig. 3. Meta-path examples (Color figure online)

Under this scheme, we sample several sequences of nodes in the knowledge graph. As the example shown in Fig. 3, the path $(Pingan \rightarrow p_1 \rightarrow term\ insurance \rightarrow p_2 \rightarrow ChinaLife)$ marked with the red solid line is one of the chosen sequence. The meta-path usually implies rich information. The marked path means that the companies $Pingan$ and $ChinaLife$ provide products of the same type. From this path, we find that the companies $Pingan$ and $ChinaLife$ are similar and they would be embedded close in the latent representation space.

We choose several meta-path schemes to form a scheme set as P, based on prior knowledge or expert advice. The meta-path schemes in knowledge graph can be denoted in the form of $(e_1 \rightarrow e_2 \rightarrow \cdots \rightarrow e_n)$. After the selection of meta-path schemes, we utilize them to sample sequences in the knowledge graph. Then these sequences are fed into the Skip-gram model as the input [18], a natural language processing model and are widely applied in embedding field.

4   Question-Answer Matching

We proceed to discuss the question and answer matching process. We first illustrate the cross-attention model used in QA match and then present its training process.

4.1   Cross-Attention

The overview of cross-attention mechanism [12] used in this work is shown in Fig. 4.

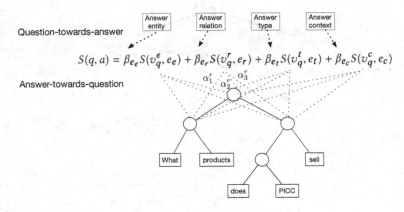

Fig. 4. Cross-attention mechanism

It consists of two attentions, i.e., answer-towards-question and question-towards-answer attention.

1. Given one answer's type, we should re-read the given question to target the corresponding part related to the answer type. As one answer usually has several aspects, we can re-read the question several scans, i.e., answer-towards-question attention.
2. On the other hand, when we concatenate different question parts together, we could re-read the answer to identify more important parts. It is the effect of question-towards-answer attention.

It is vivid that, the Cross-Attention model can be interpreted as a re-reading mechanism to mutually align the question and answers. The following details will reveal the closeness matching process, with the help of cross-attention model.

Answer-Towards-Question Attention. Here we discuss how to combine the h_{root}, h_{left} and h_{right}. The weight of attention is measured by the relevance between each tree node on question representation and answer aspect representation. Final representation should be dynamic according to the different aspect of the answer.

$$\alpha_{ij} = \frac{\exp(w_{ij})}{\sum_{k=1}^{n} \exp(w_{ik})} \tag{7}$$

$$w_{ij} = tanh(W^T[h_j; e_i] + b) \tag{8}$$

Here α_{ij} denotes the attention weight between an answer aspect a_i and tree node representation h_j where $h_j \in \{h_{left}, h_{root}, h_{right}\}$. For each answer aspect, there is a corresponding question representation.

When a question is transformed into a constituency tree, the left subtree's root corresponds to the noun phrase, i.e., question type, while the right subtree's root corresponds to the verb phrase, as a kind of question focus. It means that we may pay more attention to the h_{left} in the choice of the final answer type. We would pay more attention to h_{right} when choosing the answer relation. For answer type, the corresponding question representation can be defined as follow.

$$v_q^t = \alpha_1^{type} h_{left} + \alpha_2^{type} h_{root} + \alpha_3^{type} h_{right} \tag{9}$$

v_q^r, v_q^e and v_q^c are formalized in the same way. Given a candidate answer, we derive the scores of different aspects, which are $S(v_q^e, e_e)$, $S(v_q^r, e_r)$, $S(v_q^t, e_t)$ and $S(v_q^c, e_c)$ respectively, where the scoring function $S(\cdot)$ is the inner product between question representation and aspect representation.

Question-Towards-Answer Attention. Questions should also take varying attention to different answer aspects. As we know, different questions should concentrate on different answer aspects, which leads to question-towards-answer attention. We generate question-towards-answer attention distribution β and β_{e_e}, β_{e_r}, β_{e_t}, β_{e_c} respectively, denoting the attention weight for each answer aspect. This weight distribution reflects the importance of different answers aspects w.r.t. the given question.

$$\beta_{e_i} = \frac{\exp(w_{e_i})}{\sum_{e_k \in e_e, e_r, e_t, e_c} \exp(w_{e_k})} \tag{10}$$

$$w_{e_i} = tanh(W^T[\bar{q}; e_i] + b) \tag{11}$$

$$\bar{q} = \frac{1}{3}(h_{left} + h_{root} + h_{right}) \tag{12}$$

For example, in some QA cases, the answer type may play a key role in determining the final answer. So the corresponding weight β_{e_t} will be larger than other weights. The final score of each answer is summed up by scores from different answer aspects in the question-towards-answer attention. Candidate with the highest score is selected.

$$S(q, a) = \beta_{e_e} S(v_q^e, e_e) + \beta_{e_r} S(v_q^r, e_r) + \beta_{e_t} S(v_q^t, e_t) + \beta_{e_c} S(v_q^c, e_c) \tag{13}$$

4.2 Attention Model Training and Testing

In the training dataset setup, we generate the negative training set for each question. For each QA pair, we construct a candidate answer set for the question. Then we expand the candidate set with incorrect answers. For each question, we generate m negative samples from the entities connected to the primary entity within $k - hops$. For example, if the number of main entity's neighbors within $1 - hop$ is more than m, we will randomly choose m negative samples from the $1 - hop$ neighbors. If the number of $1 - hop$ neighbors is less than m, we will randomly select the negative samples within $2 - hop$. We will expand the hop boundaries until there are enough negative samples.

In the model training procedure, we employ the pairwise training strategy. After generating the training set, the pairwise training loss is defined in the following.

$$\mathcal{L}_{q,a,a'} = \sum_{a \in P_q} \sum_{a' \in N_q} [\gamma + S(q, a') - S(q, a)]_+ \tag{14}$$

Here a denotes the correct answer, a' denotes the wrong answer and γ is the margin which is always a small positive real number within 1.

The $[]_+$ here denotes the function $max(0, z)$. The basic idea of this training strategy is that the score of a question paired with a correct answer is higher than any wrong answer by at least γ. As a result, once the score of correct answers $S(q, a)$ is no higher than the score $S(q, a') + \gamma$, the loss will be counted. The minimized loss objective function is defined as follows.

$$\min \sum_q \frac{1}{|P_q|} \mathcal{L}_{q,a,a'} \tag{15}$$

In the optimization process, we adopt the stochastic gradient descent (SGD) to minimize the learning process with mini-batches utilized. During the testing stage, for each QA pair, we have the question q and the candidate answer set C_q. For every candidate answer $\hat{a} \in C_q$, we calculate the score $S(q, \hat{a})$. Finally, the scores are ranked and the candidate answer with the highest score is selected as the final answer. However, it is worth noting that some questions have more than one correct answers. The strategy mentioned above is improper because only one answer will be returned. We define the margin m and the answers whose gap to the best score is less than m will be put into the final answer set $\hat{A}_q$ as follows:

$$\hat{A}_q = \{\hat{a} | S_{max} - S(q, \hat{a}) < m\} \tag{16}$$

5 Experiments

Here we report the empirical studies of different KG-QA methods on the insKG dataset and the commonly used WebQuestions [1] dataset. The InsKG has limited scheme patterns, i.e., domain context. In contrast, the WebQuestions dataset has a more varying scheme and covers more domains. Therefore, in the WebQuestions experiment, it is noting that our proposed method is without question and answer representation modifications.

5.1 Experimental Setup

Training/Testing. For WebQuestions dataset, there are 5.8k q-a pairs, containing 3,778 training pairs training and 2,032 testing pairs. The questions are collected from Google Suggest API, and the answers are annotated by Amazon Mechanical Turk. The dataset is based on the open-domain knowledge graph Freebase [2] and all the answers can be found in Freebase. For topic entity extraction, with the help of Freebase API [22], we uses the top-1 recommendation as the topic entity.

For InsKG dataset, we randomly select 2k question-answer pairs from the users' question access logs in InsKG. In order to evaluate the performance more comprehensively, we take different training ratio (20%, 40%, 60% and 80%) to split training and testing pairs. As for validation pairs, N-fold cross-validation is used in the training/validation split. The N in our experiment is set to 10.

Parameter Settings. In the InsKG experiment, the word embedding dimension is set to 128 and the hidden size of the LSTM cell is set to 128. The margin here for pairwise training is set to 1. Negative example number is set to 200. For the meta-path, with the help of heuristic algorithm, the method generates 18 meta-path schemes and we keep 12 meta-path schemes according to prior knowledge.

In WebQuestions experiment, the vector dimension of words is set to 256 and thus the tree structure LSTM cell's hidden size is also set to 256. The margin γ for pairwise training is set to 0.8. For every positive answer, we sample 1000 negative examples. For knowledge graph embedding, we set the embedding dimension to 256.

Baselines

- SimpleEmbedding [6]. Bag-of-words model is used to generate question and answer vectors.
- SubgraphEmbedding [3]. It takes the candidate answer's neighbor nodes within 2-hops for embedding.
- Bi-LSTM [11]. It designs a bi-LSTM model to profile the words' forward and backward dependencies.
- TreeLSTM with A-Q. The tree LSTM model is utilized but only with answer-towards-question attention, which is a common attention model.

5.2 Quantitative Study of KG-QA

General Dataset, WebQuestions. KG-QA results on the Webquestions dataset are listed in Table 2.

The NN lines have better ones, but can not compete with LSTM variants. The proposed approach has the best performance, but just with a slight increase.

SimpleEmbedding and SubgraphEmbedding merely get the worst results under the bag-of-word assumptions. Bi-LSTM drops the bag-of-words assumption and considers three different aspects of answers and resorts to the sequence

Table 2. QA result on WebQuestions

Methods	Avg F1
SimpleEmbedding	29.7
SubgraphEmbedding	39.2
Bi-LSTM	42.9
TreeLSTM with A-Q	43.6
Proposed approach	**44.1**

modeling. Tree-LSTM model with answer-towards-question attention also has achieved a better performance than the previous work. The proposed model takes both question-towards-answer and answer-towards-question attention into account. The improvements show the advantage of deep learning models.

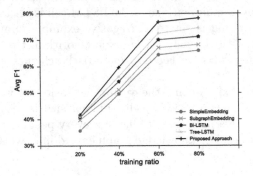

Fig. 5. QA result on InsKG

Specific Domain, InsKG. Results on InsKG are shown in Fig. 5. Similar to the experiments on the general WebQuestions dataset, embedding lines cannot compete with LSTM methods. The proposed TreeLSTM with modifications model has significant improvement. When the training ratio is 60% in Table 3, the improvement is most significant that the proposed method ranks first at 76.8% and followed by common Tree-LSTM at 72.5%. It proves the usefulness of the meta-path in capturing the contextual structure information in the knowledge graph, as well as the value of domain information in modeling vector for answers.

These quantitative QA experiments reveal that the domain context features are valuable for KG-QA, and the proposed domain context KG-QA is better at specific domain knowledge graph usages.

Table 3. QA accuracy on InsKG (training ratio 60%)

Methods	Avg F1
simpleEmbedding	64.4
SubgraphEmbedding	67.1
Bi-LSTM	70.1
Tree-LSTM	72.5
Proposed Approach	**76.8**
Only with W2V	74.6
Only with MP	75.2

5.3   Model Component Analysis

Answer Context Contribution. The KG-QA experiments have revealed the contributions of meta-path features. We continue to analyze the details of different path types. In this experiment on InsKG, we take *"company → product → type → type"*, *"product → type → type → type"* and *"company → product → status"* as the starting meta-path schemes, shown in Fig. 6. Within each meta-path scheme, we compare their corresponding questions' QA performances of Tree-LSTM and Tree-LSTM with meta-path modification methods.

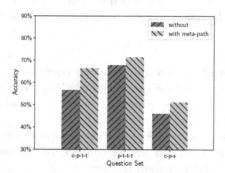

Fig. 6. QA accuracy under different meta-path context

The results show that for some simple/short meta-path scheme, two methods get similar results. In the complex/long meta-path scheme, the method with meta-path feature gains an impressive improvement. The meta-path features are valuable for complex pattern extraction, leveraging the KG-QA performance.

Question Context Contribution. We further analyze the contribution from the question's tree model extraction. Here we classify the questions into fact and relation types based on the question intent. In each category, we compare their QA performance of biLSTM and Tree-LSTM models. The results are shown in Fig. 7.

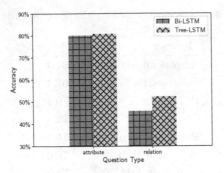

Fig. 7. QA accuracy with/without tree model in questions

We find that common fact/attribute questions have roughly the same performance with/without tree models. It is reasonable that for factual questions, their returned answers usually cover just the entity and directly connected attributes or categories, and the tree model parsing of input question cannot provide additional associations. In contrast, for relationship questions, sometimes their answers cover several entities and the inter-connected edges in the knowledge graph. The tree model is good at profile the question intent/type and question focus/entity topic.

It is admitted that the relationship questions take a small portion of users' input questions. But the tree model's parsing potential contributes to the answering of the complex question.

Cross-Attention Models. To illustrate the contributions of cross-attention mechanism, We continue to take the question "What city does PICC locate?" as the running example. In this question, the answer relation path is "locate", the answer type is denoted as "location.city".

As mentioned above, we consider four answer aspects which are answer entity, answer relation, answer type and answer context. In the heat map, the abscissa of the table is the left subtree's root, root, right subtree's root, and the ordinate is the answer entity, answer relation, answer type and context. The heavier the color, the higher the value of attention weight (Fig. 8).

	Left	Root	Right
Answer entity	0.38	0.32	0.30
Answer relation	0.14	0.23	0.63
Answer type	0.82	0.10	0.08
Answer context	0.28	0.39	0.33

Fig. 8. Cross-attention heat map

As discussed in Sect. 3, we use the constituency tree to parse the input question. The left subtree should play a more important role in predicting the type of the answer while the right subtree is more critical in answer relation prediction. For the answer-towards-question attention, we first focus on the attention distribution for answer type. The results show the left subtree root's weight is more significant, indicating the dominant role that the left subtree plays in the answer type prediction. The right subtree root obtains the most significant value weight in the distribution. For the question-towards-answer attention, the answer type and the answer relation is more important in the final answer selection. It is vivid that the distribution of attention weights fit our expectations and the tree model well uncovers the intents of the input question.

For the question-towards-answer attention, we find the answer type and the answer relation is more important in the final answer decision which is also in line with our intuitive experience.

In conclusion of the experimental study, we find that the proposed approach shows an advantage in both the KG-QA tasks and the question/answer explanation. The introduced meta-path features are beneficial for complex answer processing. The tree structure is good at question understanding. For the KG-QA performance, the new approach is especially useful in domain knowledge graph usages.

6 Conclusion

In this paper, we propose a new neural KG-QA approach by leveraging the domain context, with the help of a cross-attention model. In this model, we parse the question tree and utilize meta-path random walk to enrich representation for answers. We also introduce the cross-attention mechanism, reveal the mutual influences between the questions and answers. Experiments reveal the advantage. The experimental results demonstrate the effectiveness and improvement of the proposed model in specific-domain knowledge graph. Besides the improved KG-QA performance, the introduced domain context is better at capturing the correlation within the knowledge graphs. We are investigating its extensions in the relationship inference and entity analytics in the future work.

Acknowledgement. This work is supported by NSFC 61502169, U1509219 and SHEITC.

References

1. Berant, J., Chou, A., Frostig, R., Liang, P.: Semantic parsing on freebase from question-answer pairs. In: Proceedings of EMNLP, pp. 1533–1544 (2013)
2. Bollacker, K.D., Evans, C., Paritosh, P., Sturge, T., Taylor, J.: Freebase: a collaboratively created graph database for structuring human knowledge. In: Proceedings of SIGMOD, pp. 1247–1250 (2008)

3. Bordes, A., Chopra, S., Weston, J.: Question answering with subgraph embeddings. In: Proceedings of EMNLP, pp. 615–620 (2014)
4. Bordes, A., Usunier, N., Chopra, S., Weston, J.: Large-scale simple question answering with memory networks. arXiv preprint arXiv:1506.02075 (2015)
5. Bordes, A., Usunier, N., García-Durán, A., Weston, J., Yakhnenko, O.: Translating embeddings for modeling multi-relational data. In: Proceedings of NIPS, pp. 2787–2795 (2013)
6. Bordes, A., Weston, J., Usunier, N.: Open question answering with weakly supervised embedding models. In: Calders, T., Esposito, F., Hüllermeier, E., Meo, R. (eds.) ECML PKDD 2014. LNCS (LNAI), vol. 8724, pp. 165–180. Springer, Heidelberg (2014). https://doi.org/10.1007/978-3-662-44848-9_11
7. Cai, Q., Yates, A.: Large-scale semantic parsing via schema matching and lexicon extension. In: Proceedings of ACL, pp. 423–433 (2013)
8. Dong, L., Wei, F., Zhou, M., Xu, K.: Question answering over freebase with multi-column convolutional neural networks. In: Proceedings of ACL, pp. 260–269 (2015)
9. Dong, Y., Chawla, N.V., Swami, A.: Metapath2vec: scalable representation learning for heterogeneous networks. In: Proceedings of KDD, pp. 135–144 (2017)
10. Fader, A., Zettlemoyer, L., Etzioni, O.: Open question answering over curated and extracted knowledge bases. In: Proceedings of KDD, pp. 1156–1165 (2014)
11. Hao, Y., et al.: An end-to-end model for question answering over knowledge base with cross-attention combining global knowledge. In: Proceedings of ACL, pp. 221–231 (2017)
12. Hermann, K.M., et al.: Teaching machines to read and comprehend. In: Proceedings of NIPS, pp. 1693–1701 (2015)
13. Kim, Y.: Convolutional neural networks for sentence classification. arXiv preprint arXiv:1408.5882 (2014)
14. Kwiatkowski, T., Zettlemoyer, L.S., Goldwater, S., Steedman, M.: Inducing probabilistic CCG grammars from logical form with higher-order unification. In: Proceedings of EMNLP, pp. 1223–1233 (2010)
15. Li, H., Min, M.R., Ge, Y., Kadav, A.: A context-aware attention network for interactive question answering. In: Proceedings of KDD, pp. 927–935 (2017)
16. Liang, P., Jordan, M.I., Klein, D.: Learning dependency-based compositional semantics. Comput. Linguist. **39**(2), 389–446 (2013)
17. Mikolov, T., Chen, K., Corrado, G., Dean, J.: Efficient estimation of word representations in vector space. In: ICLR (2013)
18. Mikolov, T., Sutskever, I., Chen, K., Corrado, G.S., Dean, J.: Distributed representations of words and phrases and their compositionality. In: Proceedings of NIPS, pp. 3111–3119 (2013)
19. Saha, A., Pahuja, V., Khapra, M.M., Sankaranarayanan, K., Chandar, S.: Complex sequential question answering: towards learning to converse over linked question answer pairs with a knowledge graph. In: Proceedings of AAAI, pp. 705–713 (2018)
20. Tai, K.S., Socher, R., Manning, C.D.: Improved semantic representations from tree-structured long short-term memory networks. In: Proceedings of ACL, pp. 1556–1566 (2015)
21. Yang, M., Duan, N., Zhou, M., Rim, H.: Joint relational embeddings for knowledge-based question answering. In: Proceedings of EMNLP, pp. 645–650 (2014)
22. Yao, X., Durme, B.V.: Information extraction over structured data: question answering with freebase. In: Proceedings of ACL, pp. 956–966 (2014)
23. Yih, W., He, X., Meek, C.: Semantic parsing for single-relation question answering. In: Proceedings of ACL, pp. 643–648 (2014)

24. Yin, J., Zhao, W.X., Li, X.M.: Type-aware question answering over knowledge base with attention-based tree-structured neural networks. J. Comput. Sci. Technol. **32**(4), 805–813 (2017)
25. Zhang, Y., Dai, H., Kozareva, Z., Smola, A., Song, L.: Variational reasoning for question answering with knowledge graph. In: Proceedings of AAAI, pp. 6069–6076 (2018)

Leveraging Lexical Semantic Information for Learning Concept-Based Multiple Embedding Representations for Knowledge Graph Completion

Yashen Wang[1]([✉]), Yifeng Liu[1], Huanhuan Zhang[1], and Haiyong Xie[1,2]

[1] China Academy of Electronics and Information Technology, Beijing, China
yashen_wang@126.com, yliu@csdslab.net, huanhuanz_bit@139.com,
haiyong.xie@ieee.org
[2] University of Science and Technology of China, Hefei, Anhui, China

Abstract. Knowledge graphs (KGs) are important resources for a variety of natural language processing tasks but suffer from incompleteness. To address this challenge, a number of knowledge graph completion (KGC) methods have been developed using low-dimensional graph embeddings. Most existing methods focus on the structured information of triples in encyclopaedia KG and maximize the likelihood of them. However, they neglect semantic information contained in lexical KG. To overcome this drawback, we propose a novel KGC method (named as TransC), that integrates the structured information in encyclopaedia KG and the entity concepts in lexical KG, which describe the categories of entities. Since all entities appearing in the head (or tail) position with the same relation have some common concepts, we introduce a novel semantic similarity to measure the distinction of entity semantics with the concept information. And then TransC utilizes concept-based semantic similarity of the related entities and relations to capture prior distributions of entities and relations. With the concept-based prior distributions, TransC generates multiple embedding representations of each entity in different contexts and estimates the posterior probability of entity and relation prediction. Experimental results demonstrate the efficiency of the proposed method on two benchmark datasets.

Keywords: Knowledge graph completion · Concept information · Representation learning

1 Introduction

Knowledge Graphs (KGs) are graph-structured knowledge bases, where factual knowledge is represented in the form of relationships between entities. Knowledge Graphs have become a crucial resource for many tasks in machine learning, data mining, and artificial intelligence applications including question answering [34],

© Springer Nature Switzerland AG 2019
J. Shao et al. (Eds.): APWeb-WAIM 2019, LNCS 11641, pp. 382–397, 2019.
https://doi.org/10.1007/978-3-030-26072-9_28

entity linking/disambiguation [7], fact checking [29], and link prediction [44]. In our view, KGs are an example of a heterogeneous information network containing entity-nodes and relationship-edges corresponding to RDF-style triples (h, r, t) where h represents a head entity, and r is a relationship that connects h to a tail entity t.

KGs are widely used for many practical tasks, however, their completeness are not guaranteed. Nonetheless, KGs are far from completion. For instance consider Freebase, a core element in the Google Knowledge Vault project: 71% of the persons described in Freebase have no known place of birth, 75% of them have no known nationality, while the coverage for less frequent predicates can be even lower. Therefore, it is necessary to develop Knowledge Graph Completion (KGC) methods to find missing or errant relationships with the goal of improving the general quality of KGs, which, in turn, can be used to improve or create interesting downstream applications. Motivated by the linear translation phenomenon observed in well trained word embeddings [18], Many Representation Learning (RL) based algorithms [3,15,37,41], have been proposed, aiming at embedding entities and relations into a vector space and predicting the missing element of triples. These models represents the head entity h, the relation r and the tail entity t with vectors $\mathbf{h}$, $\mathbf{r}$ and $\mathbf{t}$ respectively, which were trained so that $\mathbf{h} + \mathbf{r} \approx \mathbf{t}$.

The objects most KGC models handle are encyclopedic KG (e.g., Freebase). Although these models have significantly improved the embedding representations and increased the prediction accuracy, there is still room for improvement by exploiting semantic information in the representation of entities. Generally speaking, semantic information includes concepts, descriptions, lexical categories and other textual information. As discussed in [12,31,35,38], it is essential to utilize lexical KGs to help the machine to understand the world facts and the semantics. That is, the knowledge of the language should be used. Encyclopedic KGs contain facts such as Barack Obama's birthday and birthplace, while lexical KGs could definitely indicate that birthplace and birthday are properties of a person.

Generally, each entity or relation may have different semantics in different triples. For example, in the triple (David_Beckham, place_of_birth, London), David Beckham is a person, while in (David_Beckham, player_of, Manchester_United), David Beckham is a player or athlete. Unfortunately, most recent works represent each entity as a single vector which cannot capture the uncertain semantics of entities. To address the above-mentioned issue, we propose a concept-based multiple embedding model (TransC). TransC fully utilizes the entity *concept* information which represents the domains or categories of entities in lexical KG Probase. Probase is widely used in research about short-text understanding [31,32,39] and text representation [12,36]. Probase uses an automatic and iterative procedure to extract concept knowledge from 1.68 billion Web pages. It contains 2.36 millions of open domain terms. Each term is a concept, an instance, or both. Meanwhile, it provides around 14 millions relationships with two kinds of important knowledge related to concepts: concept-attribute co-occurrence (isAttrbuteOf) and concept-instance co-occurrence (isA). Moreover, Probase provides huge number of high-quality and robust concepts without builds. Therefore, we model each entity as multiple semantic vectors with concept information, and construct multiple concepts of relations from common concepts of related entities.

We utilize the concept-based semantic similarity to incorporate prior probability in the optimization objective. This is because all entities appearing in the head (or tail) with the same relation have some common concepts. Therefore, the prior distribution of the missing element could be derived from the semantic similarity between the missing element and the others. In the "David Beckham" example mentioned above, if the head of (David_Beckham, player_of, Manchester_United) is missing, we can predict the head is an entity with "player" or "athlete" since we know the relation is "player_of" and the tail is "Manchester_United", a football club.

In summary, the contributions of this work are: (i) proposing a novel knowledge base completion model that combines structured information in encyclopedic KG and concept information in lexical KG. To the best of our knowledge, this is the first study aiming at combing the encyclopedic KG and the lexical KG for knowledge graph completion (KGC) task. (ii) showing the effectiveness of our model by outperforming baselines on two benchmark datasets for knowledge base completion task.

2   Related Work

Many knowledge graphs have recently arisen, pushed by the W3C recommendation to use the resource description framework (RDF) for data representation. Examples of such knowledge graphs include DBPedia [1], Freebase [2] and the Google Knowledge Vault [8]. Motivating applications of knowledge graph completion include question answering [5] and more generally probabilistic querying of knowledge bases [11,22]. First approaches to relational learning relied upon probabilistic graphical models [10], such as bayesian networks [28] and markov logic networks [25,26]. Then, asymmetry of relations was quickly seen as a problem and asymmetric extensions of tensors were studied, mostly by either considering independent embeddings [9] or considering relations as matrices instead of vectors in the RESCAL model [23]. Pairwise interaction models were also considered to improve prediction performances. For example, the Universal Schema approach [27] factorizes a 2D unfolding of the tensor (a matrix of entity pairs vs. relations).

Nowadays, a variety of low-dimensional representation-based methods have been developed to work on the KGC task. These methods usually learn continuous, low-dimensional vector representations (i.e., embeddings) for entities and relationships by minimizing a margin-based pairwise ranking loss [14]. The most widely used embedding model in this category is TransE [3], which views relationships as translations from a head entity to a tail entity on the same low-dimensional plane.

Based on the initial idea of treating two entities as a translation of one another (via their relationship) in the same embedding plane, several models have been introduced to improve the initial TransE model. The newest contributions in this line of work focus primarily on the changes in how the embedding planes are computed and/or how the embeddings are combined. For example, the entity translations in TransH [37] are computed on a hyperplane that is perpendicular to the relationship embedding. In TransR [15] the entities and relationships are embedded on separate planes and then the entity-vectors are translated to the relationships plane. Structured Embedding (SE) [4] creates two translation matrices for each relationship and applies them to head and tail entities separately. Knowledge Vault [8] and HolE [21], on the other hand, focus on learning a new combination operator instead of simply adding two entity embeddings element-wise. Take HolE as example, the circular correlation is used for combining entity embeddings, measuring the covariance between embeddings at different dimension shifts.

Semantic information, such as types, descriptions, lexical categories and other textual information, is an important supplement to structured information in KGs. DKRL [42] represents entity descriptions as vectors for tuning the entity and relation vectors. SSP [40] modifies TransH by using the topic distribution of entity descriptions to construct semantic hyperplanes. Entity descriptions are also used to derive a better initialization for training models [16]. With type information, type-constraint model [13] selects negative samples according to entity and relation types. In a similar way, TransT [17] leveraged the type information for the representation of entity. However, TransT have to construct or extend entity concepts from other semantic resources (e.g., WordNet), if there is no explicit concept information in a KG. TKRL [43] encodes type information into multiple representations in KGs with the help of hierarchical structures. It is a variant of TransR with semantic information and it is the first model introducing concept information.

3 Methodology

A typical knowledge graph (KG) is usually a multiple relational directed graph, recorded as a set of relational triples (h, r, t), which indicate relation r between two entities h and t. We model each entity as multiple semantic vectors with concept information to represent entities more accurately. Different from using semantic-based linear transformations to separate the mixed representation, the proposed TransC models the multiple semantics separately and utilizes the semantic similarity to distinguish entity semantics. Moreover, we measure the semantic similarity of entities and relations based on entity concepts and relation concepts.

3.1 Semantic Similarity Based on Concept

As discussed in [17], all of the entities located in the head position (or tail position) with the same relation may have some common entity types or concepts, as

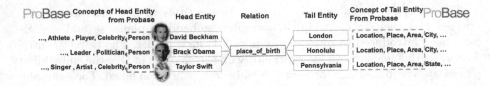

Fig. 1. example showing that the entities in the head or tail of a relation have some common concepts from the lexical KG Probase

shown in Fig. 1. In this example, all the head entities have "Person" concept and all the tail entities have concepts of "Location", "Place" and "Area". Therefore, we could see that, "Person" is the head concept of relation "place_of_birth", and "Location", "Place" and "Area" are the tail concepts of this relation. Based on aforementioned correlation, this paper introduces a concept-based semantic similarity, which utilizes *entity concepts* to construct *relation concepts*. Apparently, each relation (such as "place_of_birth" in Fig. 1) relates two components, and thus each relation r has two concept sets: (i) head concept set C_r^{head}, consisting of concepts of the entities occurring in the head position; and (ii) tail concept set C_r^{tail}, consisting of concepts of the entities occurring in the tail position. From the lexical KG Probase, we could distill entities appearing in the head position of relation r to form the head entity set, denoted as E_r^{head}. Similarly, the tail entity set, denoted as E_r^{tail}, could be constructed in the same way. Moreover, given entity e, we denote its concept set as C_e, consisting the corresponding concepts deriving from Probase by leveraging single instance conceptualization algorithm [12,36,39]. With efforts above, given r relation, the corresponding C_r^{head} and C_r^{tail} could be defined as follows:

$$C_r^{head} = \bigcap_{e \in E_r^{head}} C_e \tag{1}$$

$$C_r^{tail} = \bigcap_{e \in E_r^{tail}} C_e \tag{2}$$

Therefore, the semantic similarity between the relation and the head entity, to measure the distinction of entity semantics with the concept information, is defined as:

$$\text{sim}(r_{head}, h) = \frac{|C_r^{head} \bigcup C_h|}{|C_r^{head}|} \tag{3}$$

Similarly, the semantic similarity between the relation and the tail entity is:

$$\text{sim}(r_{tail}, t) = \frac{|C_r^{tail} \bigcup C_t|}{|C_r^{tail}|} \tag{4}$$

And the semantic similarity between the head entity h and tail entity t is

$$\text{sim}(h, t) = \frac{|C_h \bigcup C_t|}{|C_h|} \tag{5}$$

3.2 Methodology

In our perspective, the prediction probability is a *conditional probability* because except the missing element, rest of the two elements in a triple are known. E.g., when predicting the tail entity for a triple $(h, r, ?)$, we expect to maximize the probability of t under the condition that the given triple satisfies the principle $\mathbf{h} + \mathbf{r} \approx \mathbf{t}$ and the head entity and relation are h and r. Wherein, $\mathbf{h}$, $\mathbf{r}$, and $\mathbf{t}$ denote the embedding representation of h, r, and t respectively. Intuitively, we could denote this conditional probability as $\mathcal{P}(t|h, r, fact)$, meaning that triple $(h, r, ?)$ is a fact, which means that the triple satisfies $\mathbf{h} + \mathbf{r} \approx \mathbf{t}$ principle. According to Bayes theorem [6], $\mathcal{P}(t|h, r, fact)$ could be reformed as follows:

$$\mathcal{P}(t|h, r, fact) = \frac{\mathcal{P}(fact|h, r, t)\mathcal{P}(t|h, r)}{\mathcal{P}(fact|h, r)} \propto \mathcal{P}(fact|h, r, t)\mathcal{P}(t|h, r) \qquad (6)$$

The above-mentioned Eq. 6 consists of two components: (i) $\mathcal{P}(fact|h, r, t)$ is the likelihood that (h, r, t) is a fact, which is estimated by the multiple embedding representations; (ii) $\mathcal{P}(t|h, r)$ is the prior probability of the tail entity t, estimated by the semantic similarity.

Then, we describe how to estimate the prior probabilities Eq. (6). We assume that, the prior distribution of the missing element could be derived from the semantic similarity between the missing element and the others. For example, when we predict t in the triple (h, r, t), the entities with more common or similar concepts belonging to r and h, have higher probability. Hence, the semantic similarity between t and its context $(?, h, r)$ could be utilized to estimate t's prior probability:

$$\mathcal{P}(t|h, r) \propto \text{sim}(r_{tail}, t)^{\alpha_{tail}} \text{sim}(h, t)^{\alpha_{relation}} \qquad (7)$$

wherein, $\text{sim}(r_{tail}, t)$ is the semantic similarity between the relation r and the tail entity t, and $\text{sim}(h, t)$ is the semantic similarity between the head entity h and tail entity t. Furthermore, $\alpha_{tail} \in \{0, 1\}$ and $\alpha_{relation} \in \{0, 1\}$ are the concept similarity weights, because h and r have different impacts on the prior probability of t.

Similarly, the objective of the head entity prediction is

$$\mathcal{P}(h|r, t, fact) \propto \mathcal{P}(fact|h, r, t)\mathcal{P}(h|r, t) \qquad (8)$$

This paper also estimates prior probability Eq. (8) by the concept-based semantic similarity. Similarly, the prior estimation of head entity h is defined as follows:

$$\mathcal{P}(h|r, t) \propto \text{sim}(r_{head}, h)^{\alpha_{head}} \text{sim}(t, h)^{\alpha_{relation}} \qquad (9)$$

Wherein, $\text{sim}(r_{head}, t)$ is the semantic similarity between the relation r and the head entity h, and $\alpha_{head} \in \{0, 1\}$ is also the concept similarity weight, because t and r have different impacts on the prior probability of h. And the objective of the relation prediction is

$$\mathcal{P}(r|h, t, fact) \propto \mathcal{P}(fact|h, r, t)\mathcal{P}(r|h, t) \qquad (10)$$

By the similar derivation, the prior estimation of relation r, i.e., the prior probability Eq. (10) could be estimated by leveraging the concept-based semantic similarity.

$$\mathcal{P}(r|h,t) \propto \text{sim}(r_{head}, h)^{\alpha_{head}} \text{sim}(r_{tail}, t)^{\alpha_{tail}} \tag{11}$$

3.3 Multiple Semantic Vector Representations

We adopt the similar assumption as discussed in [17] to generate multiple semantic vector representations for each entity, to accurately model the ubiquitous rich semantics, while each relation is still represented as a single vector.

Taking the previous TransE as an example. TransE represents each entity as a single vector, trying to describe (or compact) all semantics of the given entity. There is only one vector representation for an entity in TransE. Thus the vector representation is not accurate for any entity semantics, and discards the rich semantic representations of the given entity. To overcome this drawback, the proposed TransC represents each entity concept as a concept vector, and represents each entity as a set of concept vectors, following the assumption that relations have single semantic and entities have multiple semantics [17]. Hence separate representations of entity semantics describe the relationship among a triple more accurately. The likelihood of the vector representations for the triple $\mathcal{P}(fact, h, r, t)$ (in Eq. 6) could be defined as below:

$$\mathcal{P}(fact, h, r, t) = \sum_{i=1}^{|C_h|} \sum_{j=1}^{|C_t|} w_{h,i} w_{t,j} f_r(h_i, t_j) \tag{12}$$

where $|C_h|$ and $|C_t|$ are the number of concepts of head entity h and tail entity t, by leveraging single instance conceptualization algorithm based on Probase [24, 35, 39]; $\{w_{h,1}, \ldots, w_{h,|C_h|}\}$ and $\{w_{t,1}, \ldots, w_{t,|C_t|}\}$ are the distributions of random variables of h and t; $f_r(h_i, t_j)$ is the likelihood of the component with i-th concept vector $\mathbf{h}_i$ of the head entity h and j-th concept vector $\mathbf{t}_j$ of the tail entity t. The previous models represents the head entity h, the relation r and the tail entity t with vectors $\mathbf{h}$, R and $\mathbf{t}$ respectively, which were trained so that $\mathbf{h} + \mathbf{r} \approx \mathbf{t}$, which have been viewed as a principle. This paper also following this principle. Thus, Motivated by the linear translation phenomenon observed in well trained word embeddings [3,18], this paper defines $f_r(h_i, t_j)$ in the form of the widely-used energy function, as follows:

$$f_r(h_i, t_j) = \|\mathbf{h}_i + \mathbf{r} - \mathbf{t}_j\|_l \tag{13}$$

Wherein, $\mathbf{h}_i$, $\mathbf{r}$ and $\mathbf{t}_j$ are the vectors of h, r and t. $l = 1$ or $l = 2$, which means either the l_1 or the l_2 norm of the vector $\mathbf{h}_i + \mathbf{r} - \mathbf{t}_j$ will be used depending on the performance on the validation set.

We model the generating process of semantic vectors as a Dirichlet process [33] like TransG [41]. In training process, the probability that the head entity (or the tail entity) in each triple generates a new (denoted as the superscript *) concept vector, could be computed as follows:

$$P_{head}^{*}(h,r,t) = (1 - \text{sim}(r_{head}, h))\frac{\beta e^{-\|r_{head}\|_1}}{\beta e^{-\|r_{head}\|_1} + P(fact|h,r,t)} \tag{14}$$

This formula means that, if the current set of concepts could accurately represent the head entity h, the new concept semantics may be generated. Wherein, β is the scaling parameter controlling the generation probability [17]. Similarly, the generation probability of new concept vector of the tail entity t could be defined as follows:

$$P_{tail}^{*}(h,r,t) = (1 - \text{sim}(r_{tail}, t))\frac{\beta e^{-\|r_{tail}\|_1}}{\beta e^{-\|r_{tail}\|_1} + P(fact|h,r,t)} \tag{15}$$

3.4   Optimization with Concept Domain Sampling

Recall that we need to sample a negative triple (h', r, t') to compute hinge loss, given a positive triple $(h, r, t) \in \Delta$. The distribution of negative triple is denoted by Δ'. Previous work [3,15,20] generally constructs a set of corrupted triples by replacing the head entity or tail entity with a random entity uniformly sampled from the KG. However, uniformly sampling corrupted entities may not be optimal. Often, the head and tail entities associated a relation can only belong to a specific concept domain or category. E.g., in Fig. 1, the prime candidate domain for head entities is "Person". When the corrupted entity comes from other concept domains, it is very easy for the model to induce a large energy gap between true triple and corrupted one. As the energy gap exceeds some threshold value, there will be no training signal from this corrupted triple. In comparison, if the corrupted entity comes from the same concept domain, the task becomes harder for the model, leading to more consistent training signal.

Motivated by this observation, we propose to sample corrupted head or tail from entities in the same concept domain with a probability P_r and from the whole entity set with probability $1 - P_r$. In the rest of the paper, we refer to the new proposed sampling method as *concept domain sampling*.

With efforts above, we define the For a triple (h, r, t) in the training set Δ, we sample its negative triple $(h', r', t') \notin \Delta$ by replacing one element with another entity or relation. When predicting different elements of a triple, we replace the corresponding elements to obtain the negative triples, wherein the negative triple set could be denoted as $\Delta'_{(h,r,t)}$. With efforts above, we denote the prediction error as $l(h, r, t, h', r', t')$. Therefore, the optimization function could be viewed as the sum of prediction errors with the above-mentioned concept domain sampling, as follows:

$$\sum_{(h,r,t)\in\Delta}\sum_{(h',r',t')\in\Delta'} \max\left\{0, \gamma + l(h,r,t,h',r',t')\right\} \tag{16}$$

The stochastic gradient descent (SGD) strategy [19] is applied to optimize the optimization function in the proposed algorithm. To optimize the parameters in the formulae Eq. (16), we defined the prediction error as follows:

$$l(h,r,t,h',r',t') = \begin{cases} -\ln \mathcal{P}(h|r,t,fact) + \ln \mathcal{P}(h'|r,t,fact) & h' \neq h \\ -\ln \mathcal{P}(t|h,r,fact) + \ln \mathcal{P}(t'|h,r,fact) & t' \neq t \\ -\ln \mathcal{P}(r|h,t,fact) + \ln \mathcal{P}(r'|h,t,fact) & r' \neq r \end{cases} \quad (17)$$

4 Experiments

We evaluate our proposed TransC on several experiments. Generally, The Knowledge Graph Completion (KGC) task could be divided into two non-mutually exclusive sub-tasks: (i) Entity Prediction task, and (ii) Relationship Prediction task. We evaluate our model on both tasks with benchmark static datasets. Moreover, Triple Classification task is also introduced for our comparative analysis.

4.1 Datasets and Baselines

To evaluate entity prediction, link prediction and triple classification, we conduct experiments on the **WN18** (WordNet) and **FB15k** (Freebase) introduced by [3] and use the same training/validation/test split as in [3]. The information of the two datasets is given in Table 1. Wherein, $\#E$ and $\#R$ denote the number of entities and relation types respectively. $\#Train$, $\#Valid$ and $\#Test$ are the numbers of triple in the training, validation and test sets respectively. Concept information of entities in FB15K and WN18 is generated by instance conceptualization algorithm based on Probase [24, 39].

The baselines include **TransE** [3], **TransH** [37], and **TransR** [15], which didn't utilize semantics information. Moreover, four semantic-based models are also included: (i) **TKRL** [43] and **TransT** [17] utilize entity types; (ii) **DKRL** [42] and **SSP** [40] take advantage of entity descriptions. Two widely-used measures are considered as evaluation metrics in our experiments: (i) Mean Rank, indicating the mean rank of original triples in the corresponding probability ranks; HITS@N, indicating the proportion of original triples whose rank is not larger than N. Lower mean rank or higher Hits@10 mean better performance. What's more, we follow [3] to report the filter results, i.e., removing all other correct candidates h in ranking, which is called the "Filter" setting. In contrast to this stands the "Raw" setting.

4.2 Entity Prediction

The Entity Prediction task takes a partial triple $(h, r, ?)$ as input and produces a ranked list of candidate entities as output. Our Entity Prediction task utilizes FB15K dataset and WN18 dataset as benchmark dataset, and utilizes Mean Rank and HITS@10 as evaluation metric.

Table 1. Statistics of FB15k and WN18 used in experiments.

Dataset	#E	#R	#Train	#Valid	#Test
WN18	40,943	18	141,442	5,000	5,000
FB15k	14,951	1,345	483,142	50,000	59,071

Table 2. Evaluation results of entity prediction on FB15k.

	Mean rank		HITS10 (%)	
	Raw	Filter	Raw	Filter
TransE	238	143	46.4	62.1
TransH	212	87	45.7	64.4
TransR	199	77	47.2	67.2
DKRL	181	91	49.6	67.4
TKRL	202	87	50.3	73.4
SSP	**163**	82	57.2	79.0
TransT	199	46	53.3	85.4
TransC(Concept.)	180	64	57.1	84.2
TransC(Multiple.)	207	55	52.7	85.9
TransC	175	**44**	**58.4**	**86.7**

Following [17], the same protocol used in previous studies is utilized here. For each triple (h, r, t), we replace the tail t (or the head h) with the concept domain sampling strategy discussed in Sect. 3.4. We calculate the probabilities of all replacement triples and rank these probabilities in descending order.

As the datasets are the same, we directly reuse the best results of several baselines from the literature [15,17,37,43]. In both "Raw" setting and "Filter" setting, a higher HITS@10 and a lower Mean Rank mean better performance. The optimal-parameter configurations are described as follows: (i) For dataset WN18, the learning rate is 0.001, the vector dimension is 100, the margin is 3, the scaling parameter β is 0.0001, concept similarity weights $\alpha_{head} = \alpha_{tail} = 1$, and $\alpha_{relation} = 0$; (ii) For dataset FB15K, the learning rate is 0.0005, the vector dimension is 150, the margin is 3, the scaling parameter β is 0.0001, concept similarity weights $\alpha_{head} = \alpha_{tail} = 1$, and $\alpha_{relation} = 0$.

The overall entity prediction results on FB15K and WN18 are reported in Tables 2 and 3, respectively. It is worth mentioning that, both the proposed **TransC** and previous [17] utilize multiple semantic vectors for representing the entity. In Table 2, For **TransC**, "Concept." means concept information is used, and "Multiple." means entities are represented as multiple vectors. At the end of the row, "**TransC**" means the complete model combing "Concept." with "Multiple.". From the result, we observe that: **TransC** outperforms all baselines on FB15k with "Filter" setting. E.g., compared with **TransT** and **SSP**, **TransC**

Table 3. Evaluation results of entity prediction on WN18.

	Mean rank		HITS10 (%)	
	Raw	Filter	Raw	Filter
TransE	263	251	75.4	89.2
TransH	401	338	73.0	82.3
TransR	238	225	79.8	92.0
DKRL	202	198	77.4	92.4
TKRL	180	167	80.6	94.1
SSP	168	156	81.2	93.2
TransT	137	130	92.7	**97.4**
TransC	**136**	**125**	**94.5**	96.9

improves the Mean Rank by 4.35% and 46.34%, and improves the HIT@10 by 1.52% and 9.75%, On WN18, **TransT** achieve the best results on metric HIT@10 for "Filter" setting. Beyond that, **TransC** performs the best. This shows that our method successfully utilizes conceptual information, and that multiple conceptual vectors could capture the different semantics of each entity more accurately than the linear transformation of a single entity vector. While **SPP** achieves the best performance for "Raw" setting on the metric mean rank.

Similar to **TransT**, the proposed **TransC** has the largest difference between the results of "Raw" and "Filter" settings on FB15K. Different from using semantic-based linear transformations to separate the mixed representation, the proposed **TransC** models the multiple semantics separately and utilizes the semantic similarity to distinguish entity semantics. In order to capture entity semantics accurately, we could dynamically generate new semantic vectors for different contexts. This indicates the importance of the prior probability $(\mathcal{P}(t|h,r))$, which significantly improves the entity prediction performance.

Because the prior distribution of the missing element could be derived from the semantic similarity between the missing element and the others. For example, when we predict the head entity h in the triple (h,r,t), the entities with more common or similar concepts belonging to the relation r and the tail entity t, have higher probability. Therefore, **TransC** utilizes these similarities to estimate the prior probability resulting in ranking similar entities higher.

Comparing the two approaches, multiple-vector representation (denoted as **TransC(Multiple.)**) and concept information (denoted as **TransC (Concept.)**), the later one is more instrumental for performance. While, **TransC** is better than both of them, demonstrating the necessity of the combination of multiple-vector representation and concept information.

4.3   Relation Prediction

We evaluated **TransC**'s performance on relationship prediction task using the FB15K dataset, following the experiment settings in [3].

Table 4. Evaluation results of relation prediction on FB15K.

	Mean rank		HITS10 (%)	
	Raw	Filter	Raw	Filter
TransE	2.91	2.53	69.5	90.2
TransH	8.25	7.91	60.3	72.5
TransR	2.49	2.09	70.2	91.6
DKRL	2.41	2.03	69.8	90.8
TKRL	2.47	2.07	68.3	90.6
SSP	1.87	1.47	70.9	90.9
TransT	1.59	**1.19**	72.0	94.1
TransC	**1.37**	1.38	**73.6**	**95.2**

We adopt the same protocol used in entity prediction. For relationship prediction, we replaced the relationship of each test triple with all relationships in the KG, with the concept domain sampling strategy, and rank these replacement relationships in descending order. This section utilizes Mean Rank and HITS@10 as evaluation metric.

The optimal-parameter configurations are described as follows: For dataset FB15K, the learning rate is 0.0005, the vector dimension is 150, the margin is 3, the scaling parameter β is 0.0001, concept similarity weights $\alpha_{head} = \alpha_{tail} = 1$, and $\alpha_{relation} = 0$. We train the model until convergence.

The overall entity prediction results on FB15K are reported in Table 4. The experimental results demonstrate that, the proposed **TransC** significantly outperforms all baselines in the most cases: (i) **TransC** achieves the best results on the "Raw" setting; (ii) compared with **TransT** and **SSP**, **TransC** improves the HIT@10 by 1.52% and 9.75%, on "Filter" setting. Compared with **TransT**, which utilized type information, **TransC** improves HITS@1 by 1.17% and Mean Rank by 4.73% in "Raw" setting. **TransT** performs better in relation prediction task than in entity prediction task. This is because, we use the concept-based semantics of both head entity and tail entity to generate relation's semantic.

4.4 Triple Classification

Generally, the triple classification task could be reviewed as a binary classification task, which discriminate whether the given triple is correct or not. We utilize FB15K as the benchmark dataset, and utilize the binary classification accuracy as the evaluation metric. Moreover, We adopt the same strategy for negative samples generating used in [30].

The optimal-parameter configurations are described as follows: For dataset FB15K, the learning rate is 0.0005, the vector dimension is 100, the margin is 3, the scaling parameter β is 0.0001, concept similarity weights $\alpha_{head} = \alpha_{tail} = 1$, and $\alpha_{relation} = 0$. We train the model until convergence.

Table 5. Evaluation results of triple classification on FB15K.

	Accuracy (%)
TransE	85.7
TransH	87.7
TransR	86.4
DKRL	87.1
TKRL	88.5
SSP	90.1
TransT	91.0
TransC	**93.4**

Evaluation results on FB15K are shown in Table 5. **TransC** outperforms all baselines significantly. Compared with **TransT** and **SSP**, **TransC** improves the accuracy by 2.64% and 3.67%. We argue that, this phenomenon is rooted in the methodology that, the proposed **TransC** represents each entity as a set of concept vectors instead of a single vector, which adapting the rich entity semantics significantly and representing entities more accurately.

5   Conclusions

The paper studies aiming at combing the encyclopedic KG and the lexical KG for knowledge graph completion. The paper constructs multiple concepts of relations from entity concepts and designs the concept-based semantic similarity for multiple embedding representations and prior knowledge discovering. In summary, we leverage lexical knowledge base for knowledge graph completion task and propose TransC, a novel algorithm for KGC, which combines structured information in encyclopaedia KG and concept information in lexical KG (e.g., Probase). Different from using semantic-based linear transformations to separate the mixed representation, such as previous TransE, the proposed TransC models the multiple semantics separately and utilizes the semantic similarity to distinguish entity semantics. Empirically, we show the proposed algorithm could make full use of concept information and capture the rich semantic features of entities, and therefore improves the performance on two benchmark datasets over previous models of the same kind. Reasoning with temporal information in knowledge bases has a long history and has resulted in numerous temporal logics. Investigating the work on incorporating temporal information in knowledge graph completion methods, may become the future direction.

Acknowledgement. This work is funded by China Postdoctoral Science Foundation (No. 2018M641436), the Joint Advanced Research Foundation of China Electronics Technology Group Corporation (CETC) (No. 6141B08010102) and Joint Advanced Research Foundation of China Electronics Technology Group Corporation (CETC) (No. 6141B0801010a).

References

1. Bizer, C., et al.: DBpedia - a crystallization point for the Web of Data. Web Semant. Sci. Serv. Agents World Wide Web **7**(3), 154–165 (2009)
2. Bollacker, K., Evans, C., Paritosh, P., Sturge, T., Taylor, J.: Freebase: a collaboratively created graph database for structuring human knowledge. In: SIGMOD Conference, pp. 1247–1250 (2008)
3. Bordes, A., Usunier, N., Garcia-Duran, A., Weston, J., Yakhnenko, O.: Translating embeddings for modeling multi-relational data. In: Advances in Neural Information Processing Systems, pp. 2787–2795 (2013)
4. Bordes, A., Weston, J., Collobert, R., Bengio, Y.: Learning structured embeddings of knowledge bases. In: AAAI Conference on Artificial Intelligence, AAAI 2011, San Francisco, California, USA, August 2011 (2011)
5. Bordes, A., Weston, J., Usunier, N.: Open question answering with weakly supervised embedding models. In: Calders, T., Esposito, F., Hüllermeier, E., Meo, R. (eds.) ECML PKDD 2014. LNCS (LNAI), vol. 8724, pp. 165–180. Springer, Heidelberg (2014). https://doi.org/10.1007/978-3-662-44848-9_11
6. Cornfield, J.: Bayes theorem. Rev. Linstitut Int. Stat. **35**(1), 34–49 (1967)
7. Cucerzan, S.: Large-scale named entity disambiguation based on Wikipedia data. In: Proceedings of the 2007 Joint Conference on Empirical Methods in Natural Language Processing and Computational Natural Language Learning, EMNLP-CoNLL 2007, Prague, Czech Republic, 28–30 June 2007, pp. 708–716 (2007)
8. Dong, X., et al.: Knowledge vault: a web-scale approach to probabilistic knowledge fusion. In: ACM SIGKDD International Conference on Knowledge Discovery and Data Mining, pp. 601–610 (2014)
9. Franz, T., Schultz, A., Sizov, S., Staab, S.: TripleRank: ranking semantic web data by tensor decomposition. In: Bernstein, A., et al. (eds.) ISWC 2009. LNCS, vol. 5823, pp. 213–228. Springer, Heidelberg (2009). https://doi.org/10.1007/978-3-642-04930-9_14
10. Getoor, L., Taskar, B.: Introduction to Statistical Relational Learning. MIT Press, Cambridge (2007)
11. Huang, H., Liu, C.: Query evaluation on probabilistic RDF databases. In: Vossen, G., Long, D.D.E., Yu, J.X. (eds.) WISE 2009. LNCS, vol. 5802, pp. 307–320. Springer, Heidelberg (2009). https://doi.org/10.1007/978-3-642-04409-0_32
12. Huang, H., Wang, Y., Feng, C., Liu, Z., Zhou, Q.: Leveraging conceptualization for short-text embedding. IEEE Trans. Knowl. Data Eng. **30**(7), 1282–1295 (2018)
13. Krompaß, D., Baier, S., Tresp, V.: Type-constrained representation learning in knowledge graphs. In: Arenas, M., et al. (eds.) ISWC 2015. LNCS, vol. 9366, pp. 640–655. Springer, Cham (2015). https://doi.org/10.1007/978-3-319-25007-6_37
14. Lin, Y., Liu, Z., Luan, H.B., Sun, M., Rao, S., Liu, S.: Modeling relation paths for representation learning of knowledge bases. In: EMNLP (2015)
15. Lin, Y., Liu, Z., Zhu, X., Zhu, X., Zhu, X.: Learning entity and relation embeddings for knowledge graph completion. In: Twenty-Ninth AAAI Conference on Artificial Intelligence, pp. 2181–2187 (2015)
16. Long, T., Lowe, R., Cheung, J.C.K., Precup, D.: Leveraging lexical resources for learning entity embeddings in multi-relational data. CoRR abs/1605.05416 (2016)
17. Ma, S., Ding, J., Jia, W., Wang, K., Guo, M.: TransT: type-based multiple embedding representations for knowledge graph completion. In: Ceci, M., Hollmén, J., Todorovski, L., Vens, C., Džeroski, S. (eds.) ECML PKDD 2017. LNCS (LNAI), vol. 10534, pp. 717–733. Springer, Cham (2017). https://doi.org/10.1007/978-3-319-71249-9_43

18. Mikolov, T., Sutskever, I., Chen, K., Corrado, G., Dean, J.: Distributed representations of words and phrases and their compositionality. In: Advances in Neural Information Processing Systems, vol. 26, pp. 3111–3119 (2013)

19. Needell, D., Srebro, N., Ward, R.: Stochastic gradient descent, weighted sampling, and the randomized Kaczmarz algorithm. Math. Program. **155**(1–2), 549–573 (2016)

20. Nguyen, D.Q., Sirts, K., Qu, L., Johnson, M.: STransE: a novel embedding model of entities and relationships in knowledge bases. In: HLT-NAACL (2016)

21. Nickel, M., Rosasco, L., Poggio, T.: Holographic embeddings of knowledge graphs. In: Thirtieth AAAI Conference on Artificial Intelligence, pp. 1955–1961 (2016)

22. Krompaß, D., Nickel, M., Tresp, V.: Querying factorized probabilistic triple databases. In: Mika, P., et al. (eds.) ISWC 2014. LNCS, vol. 8797, pp. 114–129. Springer, Cham (2014). https://doi.org/10.1007/978-3-319-11915-1_8

23. Nickel, M., Tresp, V., Kriegel, H.P.: A three-way model for collective learning on multi-relational data. In: International Conference on International Conference on Machine Learning, pp. 809–816 (2011)

24. Park, J.W., Hwang, S.W., Wang, H.: Fine-grained semantic conceptualization of FrameNet. In: AAAI, pp. 2638–2644 (2016)

25. Raedt, L.D., Kersting, K., Natarajan, S., Poole, D.: Statistical relational artificial intelligence: logic, probability, and computation, vol. 10, no. 2, pp. 1–189 (2016)

26. Richardson, M., Domingos, P.: Markov logic networks. Mach. Learn. **62**(1–2), 107–136 (2006)

27. Riedel, S., Yao, L., McCallum, A., Marlin, B.M.: Relation extraction with matrix factorization and universal schemas. In: HLT-NAACL (2013)

28. Schmidt, D.C.: Learning probabilistic relational models (2000)

29. Shi, B., Weninger, T.: Fact checking in heterogeneous information networks. In: International Conference Companion on World Wide Web, pp. 101–102 (2016)

30. Socher, R., Chen, D., Manning, C.D., Ng, A.Y.: Reasoning with neural tensor networks for knowledge base completion. In: International Conference on Neural Information Processing Systems, pp. 926–934 (2013)

31. Song, Y., Wang, H., Wang, Z., Li, H., Chen, W.: Short text conceptualization using a probabilistic knowledgebase. In: Proceedings of the Twenty-Second International Joint Conference on Artificial Intelligence, vol. 3, pp. 2330–2336 (2011)

32. Song, Y., Wang, S., Wang, H.: Open domain short text conceptualization: a generative + descriptive modeling approach. In: Proceedings of the 24th International Conference on Artificial Intelligence (2015)

33. Teh, Y.W., Jordan, M.I., Beal, M.J., Blei, D.M.: Hierarchical dirichlet processes. Am. Stat. Assoc. **101**(476), 1566–1581 (2006)

34. Unger, C., Lehmann, J., Ngomo, A.C.N., Gerber, D., Cimiano, P.: Template-based question answering over RDF data. In: International Conference on World Wide Web, pp. 639–648 (2012)

35. Wang, Y., Huang, H., Feng, C.: Query expansion based on a feedback concept model for microblog retrieval. In: International Conference on World Wide Web, pp. 559–568 (2017)

36. Wang, Y., Huang, H., Feng, C., Zhou, Q., Gu, J., Gao, X.: CSE: conceptual sentence embeddings based on attention model. In: 54th Annual Meeting of the Association for Computational Linguistics, pp. 505–515 (2016)

37. Wang, Z., Zhang, J., Feng, J., Chen, Z.: Knowledge graph embedding by translating on hyperplanes. In: Twenty-Eighth AAAI Conference on Artificial Intelligence, pp. 1112–1119 (2014)

38. Wang, Z., Zhao, K., Wang, H., Meng, X., Wen, J.R.: Query understanding through knowledge-based conceptualization. In: International Conference on Artificial Intelligence, pp. 3264–3270 (2015)
39. Wu, W., Li, H., Wang, H., Zhu, K.Q.: Probase: a probabilistic taxonomy for text understanding. In: SIGMOD Conference (2012)
40. Xiao, H., Huang, M., Meng, L., Zhu, X.: SSP: semantic space projection for knowledge graph embedding with text descriptions. In: AAAI (2017)
41. Xiao, H., Huang, M., Zhu, X.: TransG: a generative model for knowledge graph embedding. In: Meeting of the Association for Computational Linguistics, pp. 2316–2325 (2016)
42. Xie, R., Liu, Z., Jia, J.J., Luan, H., Sun, M.: Representation learning of knowledge graphs with entity descriptions. In: AAAI (2016)
43. Xie, R., Liu, Z., Sun, M.: Representation learning of knowledge graphs with hierarchical types. In: International Joint Conference on Artificial Intelligence, pp. 2965–2971 (2016)
44. Yi, T., Luu, A.T., Hui, S.C.: Non-parametric estimation of multiple embeddings for link prediction on dynamic knowledge graphs. In: Thirty First Conference on Artificial Intelligence (2017)

Efficient Distributed Knowledge Representation Learning for Large Knowledge Graphs

Lele Chai[1], Xin Wang[1,2(✉)], Baozhu Liu[1], and Yajun Yang[1,2]

[1] College of Intelligence and Computing, Tianjin University, Tianjin, China
{lelechai,wangx,liubaozhu,yjyang}@tju.edu.cn
[2] Tianjin Key Laboratory of Cognitive Computing and Application, Tianjin, China

Abstract. Knowledge Representation Learning (KRL) has been playing an essential role in many AI applications and achieved desirable results for some downstream tasks. However, two main issues of existing KRL embedding techniques have not been well addressed yet. One is that the size of input datasets processed by these embedding models is typically not large enough to accommodate large-scale real-world knowledge graphs; the other issue is that lacking a unified framework to integrate current KRL models to facilitate the realization of embeddings for various applications. We propose DKRL, which is a distributed KRL training framework that can incorporate different KRL models in the translational category using a unified algorithm template. In DKRL, a set of primitive interface functions is defined to be implemented by various knowledge embedding models to form a unified algorithm template for distributed KRL. The effectiveness and efficiency of our framework have been verified by extensive experiments on both benchmark and real-world knowledge graphs, which show that our approach can outperform the existing ones by a large margin.

Keywords: Knowledge representation learning ·
Distributed framework · Knowledge graphs

1 Introduction

The concept **Knowledge Graph (KG)** was first officially proposed by Google in 2012, which is inherently a large-scale Semantic Web [1]. The KG approach aims to model entities and relations in the real world in form of a graph, in which vertices represent entities and edges represent relations. To this end, two entities are connected by a certain relation, i.e., the edge between two entities. Currently, with the massive growth of linked Web data, the scale of KGs has become larger and larger, even reaching billions of knowledge triples. Each triple is represented as (h, r, t), where h is the head entity, t the tail entity, and r the relation connecting head and tail. Take $(TJU, location, Tianjin)$ as an example, the statement claims a basic fact that Tianjin University (TJU) is located

© Springer Nature Switzerland AG 2019
J. Shao et al. (Eds.): APWeb-WAIM 2019, LNCS 11641, pp. 398–413, 2019.
https://doi.org/10.1007/978-3-030-26072-9_29

in *Tianjin*. Although the traditional triple representations like RDF or sparse matrix can behave well in maintaining the KG structures and expressiveness, using those forms to store KGs, on the one hand, leads to difficulties in executing the downstream tasks to some extent, such as doing link prediction; on the other hand, they also incur low computing performance and high sparsity issue.

To tackle these issues, one of the solutions is to represent the entities and relations in KGs as the low-dimensional and continuous vectors, which is called *knowledge embedding*. TransE [5] is one of the most representative models among the KRL research works. It facilitates fewer parameters to model the multi-relations between entities, and performs quite well in downstream evaluation tasks. Although some of the existing KRL models can exhibit expected results on a couple of downstream tasks over small scale benchmark KGs, two main issues to improve scalability and efficiency of KRL have not been addressed yet, one is that the existing knowledge embedding algorithms cannot be applied efficiently to large-scale KG datasets; the other issue is that lacking a unified framework to integrate current KRL models to facilitate the realization of embeddings for various applications. In fact, the differences of TransE, TransH [14], TransR [11], and TransD [9] models are insiginificant since they all belong to the category of translational knowledge representation learning. Thus, we are able to unify these models into our KRL training framework.

Currently, there have been some research works on training the KRL models in a unified framework. One category of methods is based on a single machine experimental environment [11], in which the training procedure is executed linearly, i.e., one triple is updated each time the Gradient Descent function is invoked, thus its performance cannot meet the requirement of large-scale KGs. Moreover, in [7], though GPU and parallelism mechanism can speed up the training process, it imposes high demands on the performance of the hardware, which is not suitable for most normal experimental settings; and it is a parallel training model rather than based on distributed settings. In contrast, the other category of methods trains the KRL models in a distributed environment [10], which provides wide machine learning interfaces using the Spark [15] framework. However, SANSA [10] only implements two graph embedding approaches in its beta version, and the logic and training results of the implemented codes are incorrect. In that case, we design an efficient uniform distributed framework for KRL training with relatively low requirements for hardware performance. It is worth mentioning that our framework is inspired by the Template Method design pattern [8].

To this end, in this paper, we propose a novel Distributed Knowledge Representation Learning (**DKRL**) framework to uniformly train the existing KRL models. We evaluate our proposed methods with the link prediction task and verify the running performance on public benchmark datasets, i.e., WN18 and FB15K. We have conducted extensive experiments also on the real-world large-scale KG DBpedia. Experimental results show that our approach achieves better performance.

It is worthwhile to highlight our main contributions in this paper: (1) we propose a Distributed Knowledge Representation Learning (DKRL) framework, which is regarded as an algorithm template to train the translational KRL models; (2) the training process in the DKRL framework can meet large-scale datasets and high dimensional requirement for the translational models; (3) the extensive experiments were conducted to verify the efficiency on both benchmark and real-world KGs.

The rest of this paper is organized as follows. Section 2 reviews related works. In Sect. 3, we introduce preliminaries of the distributed framework. In Sect. 4, we describe in detail the proposed algorithm for learning the embedding of the entities and relations in the knowledge graphs. Section 5 shows experimental results, and we conclude in Sect. 6.

2   Related Work

Most of the existing approaches learn the knowledge representation of the entities and relations under single machine settings, but relatively fewer works focus on KRL by utilizing the distributed approaches. Although there are a large number of research works about KRL models, to the best of our knowledge, no distributed framework has been proposed to unify these models.

The current representative KRL models can be categorized into distance models, single-layer models, energy models, and translational models. After analyzing the current research works, the efficiency of an embedded model is closely related to the complexity of a model. We classify the existing approaches into the following categories.

2.1   Traditional KRL Models

(1) <u>Distance Model.</u> The basic idea of Distance Model is to map the head entity and tail entity of a triplet into the relation-specific space, and compute the absolute value sum of the vector elements, which is used to measure the distance between two entities in the relational space. If the distance between the two is small, the relationship between the two entities holds; otherwise, the two entities do not have the relationship. The disadvantage is that the model is less collaborative. For the precise semantic relationship between two entities, the distance model cannot be accurately characterized. One of the representative models of this category is Structured Embedding [6].

(2) <u>Single-Layer Model.</u> The Neural Network Model [13] is based on the extended operator of the distance model, it adopts the nonlinear operation of the single-layer neural network. By analyzing the operators between vectors, such operations can solve the defect that the structural representation model cannot coordinate and accurately describe the semantic relationship between entities and relations. But the operators of nonlinear operations can only capture the weak semantic relationship between entities and relations. However, it incurs high computation cost, and such a model is not acceptable.

(3) Energy Model. To more accurately determine the semantics between entities and relations, more complex operations are defined in semantic matching energy models [3,4]. In these models, entities and relations are represented by low-dimensional vectors of a specified dimension. In addition, a number of projection matrices are defined in order to clearly characterize the intrinsic relationships between entities and relations. Specifically, the two models define two kinds of scoring functions for each triple. At the same time, they also define several projection matrices, which brings high computational complexity and leads to very high operation complexity.

(4) Translational Model. In TransE, if the triple (h, r, t) exists in KG, the r can be viewed as the translation from the h to the t, in that case, the equality $h + r \approx t$ holds. Compared with prior research works based on tensors or neural networks, TransE models elements in KG with fewer parameters and lower computation complexity. The experimental results show that TransE achieves better performance in the link prediction task [13].

Although the TransE model handles the 1-to-1 relations well, it is not suitable for tackling 1-to-N, N-to-1, and N-to-N relations. Then TransH is proposed, which is an embedding model that interprets the relation as a translating operation of a hyperplane, the head entity and tail entity are projected on the relation-specific hyperplane. The norm vector of the hyperplane is n_r, the target training equality are $h_p = n_r^\top \cdot h \cdot n_r$, $t_p = n_r^\top \cdot t \cdot n_r$, and $h_p + r \approx t_p$. This simple technique can overcome the flaws of TransE. In the prior two methods TransE and TranH, entities and relations are in the same projection space. Lin et al. [11] consider the heterogeneity of both entities and relations, and believe that it is not reasonable to model them in the same space. Thus they introduce the mapping matrix in TransR, which is used to project head and tail entities to the relation space. TransD converts the complex computation of TransR to relatively simple vector operations.

2.2 The Uniform Training Framework

(1) Standalone KRL framework. The work KB2E is a knowledge graph embedding toolkit but implemented in a single machine experimental environment. Thus its performance cannot meet the requirement of large-scale datasets. Another similar project OpenKE is a TensorFlow-based framework for knowledge embedding. Although one can follow the given pattern to introduce a new KRL embedding model under this framework, OpenKE depends on the high efficiency of a GPU to achieve better results. Meanwhile, OpenKE is implemented by using a multi-thread parallel mechanism instead of the shared-nothing distributed framework that can provide horizontal scalability.

(2) Distributed KRL framework. SANSA is a distributed computing stack, which provides wide interfaces for users to do various applications. The SANSA stack has implemented a distributed machine learning framework based on Spark, however, the accuracy and efficiency of the TransE implementation program cannot even produce a correct output.

Unlike the above previous works, in this paper, we focus on the distributed learning scenario, and propose a distributed algorithm template framework, called **DKRL** based on Spark for the knowledge representation learning on large-scale KGs. Furthermore, to the best of our knowledge, the **DKRL** is the first work to implement an efficient embedding approach using the distributed parallel computing framework.

3   Preliminaries

In this section, we introduce some basic background definitions about knowledge graph, RDD, and Distributed Relations which are used in our algorithms. We borrow the concept of RDD (Resilient Distributed Dataset) from the Spark framework to name the core data structure used in our method.

The knowledge graph is based on the graph model of mathematical graph theory. The graph model is $G = (V, E)$, where V is the set of vertices and E is the set of edges. The knowledge graph represents entities using vertices, and employs edges to represent relations between vertices. This general and versatile data representation is a natural way to portray the wide-ranging connections of things in the real world. Currently, there are two main types of knowledge graph data models: the RDF graph model and the property graph model.

Definition 1. (Domain) *Let DT be a finite set of data types, which are used to store the basic information of variables. Domain is a finite set of values with the same data type, denoted as $Dom = \{v_i \mid v_i \in DT(i) \wedge v_i \notin DT(j), i, j > 1 \wedge i \neq j\}$.*

Definition 2. (RDD) *The input data of the DKRL framework is loaded in the form of RDD, denoted as $RDD[DT] = \{r_1, r_2, \ldots\ldots, r_n \mid n \geq 1, r_i \in DT(i) \wedge r_i \notin DT(j), i \neq j\}$. i.e., the elements in RDD are finite sequences and can be operated on in parallel.*

Definition 3. (Distributed Relation) *Distributed Relation is logically a flat two-dimensional table stored in the distributed environment, which is defined as: (1) Distributed Relation is a relation schema, denoted as $DR(E_1, E_2, \ldots, E_n)$, (2) where $E_1 \times E_2 \times \cdots \times E_n = \{(e_1, e_2, \ldots\ldots, e_n) \mid e_i \in E_i, i \geq 1\}$, (3) and each element $(e_1, e_2, \ldots\ldots, e_n)$ is referred to as an n-tuple, (4) each value in each element e_i is a component, (5) the i-th column of DR is denoted as $Col(E_i)$, where $Col(E_i) \in Dom$.*

The DKRL is a distributed parallel framework for training KRL models on large-scale KGs. The computation in DKRL is executed on the computing cluster consisting of multiple computing sites. For each computing procedure, we define the function $\mathsf{trainRun}(\gamma, \alpha, Trans(X))$ as the center executing calculator. The detailed DKRL framework can be defined as follows:

Definition 4. (DKRL framework) *Given a triple set as the input data, each computing site $s \in Sites$ is in one of the two states, i.e., working and idle.*

The function state: *Sites* → {*working, idle*} *gets the current state of a computing site. The master site ms* ∈ *Sites. In the initial stage, all the computing sites are idle. An idle site will be invoked by receiving the signal sent to it from the master site. Within a computing procedure, the user-defined function* train-Run($\gamma, \alpha, Trans(X)$) *is executed on each computing site in parallel. The function* map() *and* reduce() *is to calculate the distance of all the sampled triples, where* map: (k_1, v_1) → [(k_2, v_2)] *and* reduce: ($k_2, [v_2]$) → [(k_3, v_3)], *and the function* map *is used to calculate the distance* v_i *of a single triple* k_i, *and the function* reduce *is to sum up all the distances of the triples. Here we use* [(k_i, v_i)] *to denote the sequence of triples. When the* trainRun($\gamma, \alpha, Trans(X)$) *is invoked on each computing site* s_i, *it (1) first generates sample triples (i.e., each* T_p, T_n ∈ S) *of the whole training dataset, (2) calculates and/or sum up values, and (3) computes the loss functions and the gradient for the update procedure. The parallel computation terminates when there is no computing site working.*

Table 1. The framework function signatures.

Function	Description
$dataPrepare(S)$	To load data, where S is the input triple set
$dataInit(DR, X)$	To initialize DR according to the model type X
$trainRun(\gamma, \alpha, X)$	To start to train model X, γ is the loss function parameter, and α is the learning rate
$sample(IS, seed)$	To generate positive samples from the dataset IS, IS denotes that its data type is $Integer$, $seed$ represents the random seed
$negative(IS, V, seed)$	To generate negative samples from the dataset IS
$D(T)$	To calculate the distance of those triples T

In the entire DKRL framework, the master site is the most crucial one, which is responsible for managing, dispatching, and scheduling resources for the worker sites. The function signatures of the proposed framework in this paper are summarized in Table 1.

4 Algorithms

In this section, we propose the distributed algorithm for training entities and relations in the knowledge graphs. First, we demonstrate the entire training procedure, then we elaborate the implementation of each concrete KRL model. Finally, we describe the overall evaluation of the algorithm.

4.1 Intuition

Our intuition is inspired by the popularity of large-scale knowledge graphs, since the scale of KGs has been growing, the existing research works cannot meet the requirements of training procedure for parallelized distributed embedding.

We first introduce the architecture of the entire framework before reporting the concrete implementation of algorithms. The design principle of the DKRL framework conforms to the Template Method design pattern, which is shown in Fig. 1 and presents the scalability of the proposed algorithms.

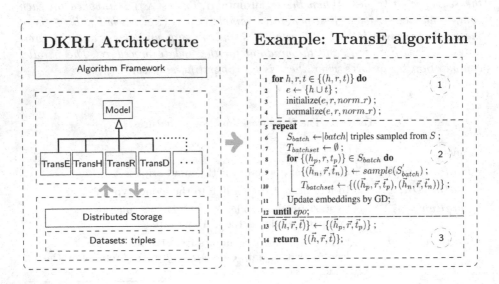

Fig. 1. The DKRL architecture.

Prior to the release of KB2E and OpenKE, the underlying implementations of the KRL models are not unified, i.e., different designs and languages are used in various research works. Although KB2E and OpenKE perform well on the training procedure to some extent, they do not have distributed implementations. In that case, it is necessary to develop a unified training framework to unify the existing works using the Template Method design pattern.

Please note that our proposed DKRL algorithm framework can be readily extended to implement all translational knowledge representation learning models other than the four exemplary models implemented in this paper. Because the translational embedding methods have similar algorithm frameworks, optimize the loss function by utilizing GD function, and only have a difference in the definition of the score function. In that case, new models can be implemented by plugging specific used-defined classes into the API provided by the DKRL framework. The training design has been shown in Fig. 2. The input dataset of DKRL framework is loaded in the distributed storage system, which provides convenience for the next in-memory-based training procedure. Due to the in-memory computation of the execution, the DKRL framework achieves better performance.

4.2 Distributed Training Model

We propose an approach to addressing the parallel training issue in the distributed computing model. The proposed training framework DKRL can incorporate different KRL models in the translational category using a unified algorithm template and show great performance in the downstream learning tasks.

Algorithm 1. DKRL-Traning /*train in each site in parallel*/

Input   : RDF triple Sets $S = \{(h, r, t)\}$, where $h, t \in V$ and $r \in E$
Parameter: max training iterations epo, embedding dimension k, learning rate
               α, margin γ
Output   : Embedding results of S: $Vec(S) = \{(\boldsymbol{h}, \boldsymbol{r}, \boldsymbol{t})\}$
1 $Vec(S) \leftarrow \emptyset$;
2 **Function** dataPrepare (S)
3      $rdd : RDD[DT] \leftarrow$ load S;
4      $DR(Ih, Ir, It) \leftarrow$ convert RDD to DR;

5 dataInit $(DR(Ih, Ir, It), Trans(X))$;   /* $X \in \{E, H, R, D\}$ */
6 **Function** trainRun $(\gamma, \alpha, Trans(X))$
7   **foreach** *computing site* s_i **do**
8   **repeat**
9              $T_p \leftarrow$ sample$(IS, rand\_seed[s_i])$;
10             $T_n \leftarrow$ negative$(IS, V, rand\_seed[s_i])$;
11             $t \leftarrow |T_p|$;
12   **while** $t > 0$ **do**
13                 $T \leftarrow T \setminus DR(Ih, Ir, It)$;   /* $T \in \{T_p, T_n\}$ */
14   map$(\emptyset, DR)$ s.t. $DR \in T$;
15   D$(T) \leftarrow$ reduce$(\emptyset, (N_1, N_2))$;   /* $N_1, N_2 \subseteq T \wedge N_1 \cap N_2 = \emptyset$ */
16   Loss $\leftarrow \gamma +$ D$(T_p) -$ D(T_n);
17   **if** *Loss* > 0 **then**
18                 $\{(\boldsymbol{h}, \boldsymbol{r}, \boldsymbol{t})\} \leftarrow$ update $\{(\boldsymbol{h}, \boldsymbol{r}, \boldsymbol{t})\}$ w.r.t. GD;
19   **else**
20                 $\{(\boldsymbol{h}, \boldsymbol{r}, \boldsymbol{t})\}$;
21   **until** epo;

22 $Vec(S) \leftarrow Vec(S) \cup \{(\boldsymbol{h}, \boldsymbol{r}, \boldsymbol{t})\}$;
23 **return** $Vec(S)$;

The overall training process of DKRL is shown in Algorithm 1, in which we take the sets of RDF triple as the input data, the dataset S is stored on a distributed storage system, the input data is loaded as an RDD, and then

converted to be the distributed relation (DR) (lines 2–4). To make preparation
for the training process, the loaded data are initialized according to a random
function, which conforms to Uniform or Bernoulli Distribution. The embeddings
of entities and relations are stored in memory since each computing site performs
updates frequently and independently. The training type Trans(X) is regarded
as a parameter, for various training model, i.e., TransE, TransR, etc., while the
initialization phases are different.

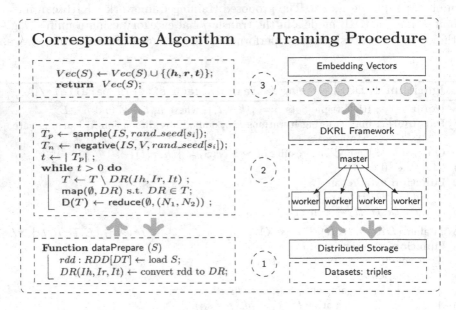

Fig. 2. The DKRL algorithm framework.

In the initial phase, all the computing sites are in the idle state. When we
submit the training job to the framework, the master site receives the task,
manages all the available resources, and dispatches the current resources to the
available computing sites (line 5). Then the computing procedures are executed
synchronously, and the training iteration is distributed in the working sites.
(1) In each iteration, the function sample($IS, rand\_seed[s_i]$) is invoked and the
random sample seeds are different among the computing sites so as to guarantee to generate a unique random sequence. For each positive triple, nega-
tive($IS, rand\_seed[s_i]$) is called to form the corresponding negative triple for the
next loss function calculation (lines 8–10). (2) Calculating the score for each
triple pair to construct the loss function $Loss$, (3) the map and reduce functions
are invoked when the calculating procedure is called (lines 12–15), and (4) the
initialized vectors will be updated when the Gradient Descent (GD) statement is
satisfied (lines 17–20). The training procedure can be terminated when reaching
the maximum number of iterations.

The computing sites finish executing and send their local computing results
to the master site, then the master site gathers all the results and returns the

Table 2. Functions of the distributed computing framework.

Trans(X) \ Function	dataInit()	negative()	D(T)
TransE [5]	t: tail vector h: head vector r: relation vector	Uniform	$-\|h + r - t\|_{L1/L2}$
TransH [14]	t: tail vector h: head vector n_r: norm vector r: relation vector	Uniform Bernoulli	$h_p = n_r{}^\top \cdot h \cdot n_r$ $t_p = n_r{}^\top \cdot t \cdot n_r$ $-\|h_p + r - t_p\|_{L1/L2}$
TransR [11]	t: tail vector h: head vector r: relation vector M_r: projection matrix	Uniform Bernoulli	$-\|M_r \cdot h + r - M_r \cdot t\|_{L1/L2}$
TransD [9]	t: tail vector h: head vector r: relation vector W_h: mapping vector W_t: mapping vector W_r: mapping vector	Uniform Bernoulli	$h_p = W_r \cdot W_h^\top \cdot h$ $t_p = W_r \cdot W_t^\top \cdot t$ $\|h_p + r - t_p\|_{L1/L2}$

final results to users. The overall distributed parallel training framework is shown in Fig. 2.

The functions incorporated in the proposed algorithms are shown in Table 2. Our DKRL framework can meet various requirements of KG embedding models, such as for the function dataInit($DR, Trans(X)$), TransE model only initializes the h, r and t, while the TransH model initializes the normal vector of the relation-specific hyperplane in addition. In TransR, an entity usually belongs to multiple properties at the same time, and different relations focus on various properties of the entity, so the projection matrix is introduced in TransR model. Although TransR outperforms the former two works TransE and TransH, it introduces high complexity during the training process, thus TransD is proposed based on the mapping vector, which improves the computing efficiency significantly.

The function negative($IS, V, rand\_seed[s_i]$) is used to sample corrupted triples from the dataset for the training process. There are two sample types used in our algorithm, the first one is to randomly generate corrupted triples, and the other method is using the Bernoulli distribution, which considers the probability of replacing head entities and tail entities. The function D(T) is developed to calculate the scores of the positive and negative triples. For different training models, the procedure for calculating distance depends on the specific parameters.

The correctness of the DKRL algorithm is guaranteed by the following theorem.

Theorem 1. *Given a knowledge graph G, Algorithm 1 gives the correct embedding vector sets $Vec(S)$ in the $O(|epo|)$ number of iterations, where epo is the total number of iterations.*

Proof. (Sketch) The algorithm correctness can be proved as follows: (i) Before executing the first training iteration of Algorithm 1, the input triples of the knowledge graph G can be initialized and normalized as $DR[IntergerTriples]$, which are transferred as the index numbers so as to better improve the training procedure; (ii) Then in each training iteration, the whole training triples in the dataset will be sampled for $|batch|$ times. The number of sample triples in each batch is represented as $|batchsize|$, i.e., $|batchsize|$ numbers of positive samples T_p are generated from the entire KG within per sample batch. Meanwhile, the function negative($IS, V, rand\_seed[s_i]$) is used to produce the corresponding corrupted triples T_n. (iii) The loss function is $L = \|\gamma + D(T_p) - D(T_n)\|_{L1/L2}$, if $L > 0$, the Gradient Descent function will be invoked to update the initialized embeddings of triples. For each time the GD is used, the distance $D(T)$ of triple (h, r, t) will be closer, which means that when the results of $h + r - t$ gradually grow, the experimental results are better. Consequently, the correct embedding vector sets $Vec(S)$ can be obtained in the number of iterations epo.   □

Theorem 2. *The complexity of the DKRL algorithm is bounded by $O(|k| \cdot |batchsize|)$, where $|k|$ is the total number of iterations, and $|batchsize|$ is the maximum number of the sampled triples.*

Proof. (Sketch) Algorithm 1 gives the correct embedding vector set $Vec(S)$ after being called in the limited iteration $|k|$. The time complexity consists of two parts: (i) Generating $|batchsize|$ number of positive and negative triples; (ii) All the sampled triples are executed in $|k|$ number of iterations. Consequently, the complexity of the DKRL algorithm is $O(|k| \cdot |batchsize|)$.   □

5   Experimental Evaluation

In this section, we evaluate the performance of our approach on the link prediction task. We have carried out extensive experiments to verify the effectiveness and efficiency of our proposed algorithms on both benchmark and real-world knowledge graphs.

5.1   Settings

In the comparison experiments, the parameters used in our approach are similar to the previous research works, and the maximum number of training iterations is at most 1000 rounds. The prototype program, which is implemented in Scala using Spark, is deployed on an 8-site cluster connected by a gigabit Ethernet. Each site has an Intel(R) Core(TM) i7-7700 CPU with 4 cores of 3.60 GHz, 16 GB memory, and 500 GB disk. We used Hadoop 2.7.4 and Spark 2.2.0. All the experiments are carried out on Linux (64-bit CentOS) operating systems.

Table 3. Datasets used in the experiments

Dataset	#Rel	#Ent	#Train	#Valid	#Test
FB15K	1,345	14,951	483,142	50,000	59,071
WN18	18	40,943	141,142	5,000	5,000
DBpedia	654	4,278,797	9,187,439	5,000	5,000

The evaluation task of link prediction is usually implemented on two popular benchmark datasets, i.e., WordNet [12] and Freebase [2]. (1) WordNet is a semantic lexical knowledge graph, which is widely used in the field of Natural Language Processing. Entities in WordNet are words, which corresponds to various semantics, and relations in WordNet indicate lexical relations among words; (2) Freebase is a famous knowledge base consisting of a large volume of general facts. In this paper, we use two subsets of the benchmark knowledge bases, i.e., WN18 and FB15K. In addition, to validate the efficiency of DKRL for large-scale KGs, we have carried out experiments on DBpedia[1], which is a real-world dataset extracted from Wikipedia. As listed in Table 3, we summarize the statistics of these datasets.

5.2 Experimental Results

The link prediction task is widely employed to predict missing entities or the entities that can be incorporated into knowledge graphs. After learning the embeddings of entities and relations in KGs, we use link prediction to evaluate the quality of the entity and relation embeddings.

Table 4. Link prediction effectiveness results on WN18 and FB15K.

Dataset	WN18		FB15K	
Metrics	Mean rank		Mean rank	
	Raw	Filt	Raw	Filt
TransE (unif)	247	254	206	117
TransE (bern)	232	302	195	102
TransH (unif)	287	295	259	127
TransH (bern)	304	382	215	94
TransR (unif)	263	245	232	106
TransR (bern)	242	238	197	89
TransD (unif)	253	241	224	84
TransD (bern)	243	218	189	96

---

[1] http://wiki.dbpedia.org/downloads-2016-10.

Exp 1. Effectiveness of the Algorithms in Accuracy

(1) *WN18 and FB15K datasets.* To validate the effectiveness of our proposed
approaches, we conducted extensive experiments on the two benchmark
datasets as mentioned above, i.e., WN18 and FB15K. The experiment
parameter settings such as learning rate, loss function margin, etc. are similar to the previous works [7,11]. As shown in Table 4, the experimental
results of our distributed training approaches can reach competitive performance to the methods in [5,9,11,14], and the experimental results are as
what we have expected. In addition, for each training model, the corrupted
triple generating strategies were implemented in two versions, one is the
uniform random distribution, the other is the Bernoulli distribution. In our
DKRL framework, we not only provide a uniform training interface but also
improve the effectiveness of the training process. The experimental results
have shown that the link prediction results of our methods have comparable
accurate scores with that of existing models.

(2) *DBpedia dataset.* The existing works have achieved good performance on
benchmark datasets, but they lack the validation on the scalability to large-
scale KGs. Our algorithms are based on the distributed environment, so it
can tackle large-scale knowledge graphs, we conducted the experiments on
large scale dataset DBpedia to verify the effectiveness. Mean Rank is to
calculate the energy score of a correct triple, and Hits@N is to calculate the
percentage of the ranks less than N, which is a validation approach usually
used to evaluate the effectiveness of experimental results. In the DBpedia
experiment, we use Hits@10, Hits@20, Hits@50, and Hits@100 metrics.

Table 5. Hits@N effectiveness results on DBpedia.

Metric	Hits@10	Hits@20	Hits@50	Hits@100
Head	4.8	5.1	5.2	5.7
Tail	21.9	25.0	32.3	44.6

From the results in Table 5, we can see, with the number N growing, the
ranking percentage grows larger, which is the same as what we expected. The
results demonstrate that even in large-scale KGs, the output of our methods is
correct and can run efficiently.

(3) *To compare various methods effectiveness on FB15K.* To further verify the
effectiveness of our methods, we conducted experiments to compare with
the state-of-the-art methods in SANSA and KB2E. The metric we use in the
experiment is Raw Mean Rank, which calculates the mean ranking of the
correct scores. However, the SANSA framework only implemented TransE
model.

Table 6. Raw Mean Rank effectiveness results on FB15K.

Metrics	Raw mean rank			
Methods	Baseline	SANSA	KB2E	DKRL
TransE	243	7438	210	**206**
TransH	**211**	—	221	215
TransR	226	—	208	**197**
TransD	194	—	**184**	189

In Table 6, it can be observed that the accuracy of SANSA is far away from the baseline, which shows that the performance of SANSA is quite low. By checking the source code, we find that the optimized procedure of SANSA is not correct, it updates all the entities in the datasets for each training iteration, which is not in accordance with the idea of the original paper [5,9,11,14]. KB2E has relatively good outputs and is slightly better than the baseline, whose reason is that, before training, KB2E preprocessed the datasets and appended unique keys to entities and relations, thus all the input data can be trained directly. The DKRL method achieves the best ranks, which demonstrates that our in-memory-based strategies can achieve the best training performance.

Exp 2. Efficiency of the Algorithms in Time

The training process executed in our DKRL framework is in-memory-based, and all the working sites are working together synchronously, which can speed up the model training process. The process will be terminated until the best training efficiency reached. For the verification of our algorithms, we conducted experiments on the FB15K dataset to compare the time efficiency with SANSA.

Table 7. Time results on FB15K.

Models	Time (s)
TransE (SANSA)	36596.940
TransE (DKRL)	**7080.633**

Because KB2E is implemented on the single machine, to be fair, we did not incorporate KB2E in the comparison experiment. As shown in Table 7, our method finished running quickly, which was about 5 times faster than TransE (SANSA).

6   Conclusion

In this paper, we propose an efficient distributed framework DKRL for KG embedding training. The DKRL is a unified training framework to incorporate various existing KRL models and can accelerate training under the strategy of distributed experimental settings. In DKRL, a set of primitive interface functions is defined to be implemented by various knowledge embedding models to form a unified algorithm template for distributed KRL. The proposed framework is verified by extensive experiments on both synthetic and real-world KGs, which shows that our approach can efficiently train the current translational KRL models.

Acknowledgments. This work is supported by the National Natural Science Foundation of China (61572353, 61402323), the National High-tech R&D Program of China (863 Program) (2013AA013204), and the Natural Science Foundation of Tianjin (17JCYBJC15400).

References

1. Berners-Lee, T., Hendler, J., Lassila, O.: The semantic web. Sci. Am. **284**(5), 34–43 (2001)
2. Bollacker, K., Evans, C., Paritosh, P., Sturge, T., Taylor, J.: Freebase: a collaboratively created graph database for structuring human knowledge. In: Proceedings of the 2008 ACM SIGMOD International Conference on Management of Data, pp. 1247–1250. ACM (2008)
3. Bordes, A., Glorot, X., Weston, J., Bengio, Y.: Joint learning of words and meaning representations for open-text semantic parsing. In: Artificial Intelligence and Statistics, pp. 127–135 (2012)
4. Bordes, A., Glorot, X., Weston, J., Bengio, Y.: A semantic matching energy function for learning with multi-relational data. Mach. Learn. **94**(2), 233–259 (2014)
5. Bordes, A., Usunier, N., Garcia-Duran, A., Weston, J., Yakhnenko, O.: Translating embeddings for modeling multi-relational data. In: Advances in Neural Information Processing Systems, pp. 2787–2795 (2013)
6. Bordes, A., Weston, J., Collobert, R., Bengio, Y., et al.: Learning structured embeddings of knowledge bases. In: AAAI, vol. 6, p. 6 (2011)
7. Han, X., et al.: OpenKE: an open toolkit for knowledge embedding. In: Proceedings of EMNLP (2018)
8. Hannemann, J., Kiczales, G.: Design pattern implementation in Java and AspectJ. ACM SIGPLAN Not. **37**(11), 161–173 (2002)
9. Ji, G., He, S., Xu, L., Liu, K., Zhao, J.: Knowledge graph embedding via dynamic mapping matrix. In: Proceedings of the 53rd Annual Meeting of the Association for Computational Linguistics and the 7th International Joint Conference on Natural Language Processing (Volume 1: Long Papers), vol. 1, pp. 687–696 (2015)
10. Lehmann, J., et al.: Distributed semantic analytics using the SANSA stack. In: d'Amato, C., et al. (eds.) ISWC 2017. LNCS, vol. 10588, pp. 147–155. Springer, Cham (2017). https://doi.org/10.1007/978-3-319-68204-4_15
11. Lin, Y., Liu, Z., Sun, M., Liu, Y., Zhu, X.: Learning entity and relation embeddings for knowledge graph completion. In: AAAI, vol. 15, pp. 2181–2187 (2015)

12. Miller, G.A.: WordNet: a lexical database for English. Commun. ACM **38**(11), 39–41 (1995)
13. Socher, R., Chen, D., Manning, C.D., Ng, A.: Reasoning with neural tensor networks for knowledge base completion. In: Advances in Neural Information Processing Systems, pp. 926–934 (2013)
14. Wang, Z., Zhang, J., Feng, J., Chen, Z.: Knowledge graph embedding by translating on hyperplanes. In: Twenty-Eighth AAAI Conference on Artificial Intelligence (2014)
15. Zaharia, M., Chowdhury, M., Franklin, M.J., Shenker, S., Stoica, I.: Spark: cluster computing with working sets. HotCloud **10**(10–10), 95 (2010)

Coherence and Salience-Based Multi-Document Relationship Mining

Yongpan Sheng and Zenglin Xu[✉]

University of Electronic Science and Technology of China, Chengdu, China
shengyp2011@gmail.com, zlxu@uestc.edu.cn

Abstract. In today's interconnected world, there is an endless 24/7 stream of new articles appearing online. Faced with these overwhelming amounts of data, it is often helpful to consider only the key entities and concepts and their relationships. This is challenging, as relevant connections may be spread across a number of disparate articles and sources. In this paper, we propose a unified framework to aid users in quickly discerning salient connections and facts from a set of related documents, and presents the resulting information in a graph-based visualization. Specifically, given a set of relevant documents as input, we firstly extract candidate facts from above sources by exploiting Open Information Extraction (Open IE) approaches. Then, we design a Two-Stage Candidate Triple Filtering (TCTF) approach based on a self-training framework to maintain only coherent facts associated with the specified document topic from the candidates and connect them in the form of an initial graph. We further construct this graph by a heuristic to ensure the final conceptual graph only consist of facts likely to represent meaningful and salient relationships, which users may explore graphically. The experiments on two real-world datasets illustrate that our extraction approach achieves 2.4% higher on the average of F-score over several OpenIE baselines. We also further present an empirical evaluation of the quality of the final generated conceptual graph towards different topics on its coverage rate of topic entities and concepts, confidence score, and the compatibility of involved facts. Experimental results show the effectiveness of our proposed approach.

Keywords: Multi-Document relationship mining ·
Graph-based visualization

1 Introduction

In today's digital and highly interconnected world, there is an endless 24/7 stream of new articles appearing online, including news reports, business transactions, digital media, etc. Faced with these overwhelming amounts of information, it is helpful to consider only the key entities and concepts and their relationships. Often, these are spread across a number of disparate articles and sources. Not only do different outlets often cover different aspects of a story.

© Springer Nature Switzerland AG 2019
J. Shao et al. (Eds.): APWeb-WAIM 2019, LNCS 11641, pp. 414–430, 2019.
https://doi.org/10.1007/978-3-030-26072-9_30

Typically, new information only becomes available over time, so new articles in a developing story need to be connected to previous ones, or to historic documents providing relevant background information. While there is ample work on news topic detection and tracking, previous work has not explored how to connect and present important facts and connections across articles.

In this paper, we propose a unified framework to extract salient entities, concepts, and their relationships, discover connections within and across them, such that the resulting information can be represented in a graph-based visualization. We rely on a series of natural language processing approaches, such as Open Information Extraction (Open IE) methods [3] readily extract large amounts of subject-predicate-object triples from the unstructured texts. However, it does not make any attempt to connect the extracted facts across sentences or even documents. Additionally, OpenIE methods tend to yield countless non-informative and redundant extractions that are not indicative of what is genuinely being expressed in a given text. E.g, we may obtain triples such as (*"they"*, *"spoke with"*, *"multiple sources"*). From this, we proposed a Two-Stage Candidate Triple Filtering (TCTF) approach to discern which of the mined candidate facts are coherent with the specified document topic, and connected them to form an initial graph. We further construct this graph by a heuristic strategy that iteratively remove the weakest concepts with relatively lower importance scores are computed by the extended TextRank algorithm, so that it ensures the final large conceptual graph only consists of facts are likely to represent meaningful and salient relationships, which users may explore graphically.

2   Related Work

GoWvis[1] is an interactive web application that generates single-document summarizations for a text provided as input, by producing a Graph-of-Words representation. Edges in such graphs, however, merely represent co-occurrences of words rather than specific relationships expressed in the text. The Networks of Names project [4] adopts a similar strategy, but restricted to named entities, i.e., any two named entities co-occurring in the same sentence are considered related. The Network of the Day project[2] builds on Networks of Names to provide a daily analysis of German news articles. The *news/s/leak* project[3] further extends this line of work by adding access to further corpora and helps journalists to analyse and discover newsworthy stories from large textual datasets. This version also attaches general document keywords as tags to relationships, but does not aim at sentence-level relation semantics as our system. Sheng et al. [15] introduce a system that can extract facts from a set of related articles. However, for the quality of extracted facts, not sufficient evaluation is performed.

OpenIE systems [3] do extract specific facts from text, but they do not aim at connecting facts across sentences or documents, and often neglect whether

---

[1] https://safetyapp.shinyapps.io/GoWvis/.

[2] http://tagesnetzwerk.de.

[3] http://www.newsleak.io/.

the extractions are meaningful on their own and indicative of what is being expressed in the input text. [12] used information extraction-based features to improve multi-document summarization. [11] investigated logical constraints to aggregate a closed set of relations mined from multiple documents. Our approach, in contrast, addresses the task of extracting and connecting salient entities and facts from multiple documents, enabling a deeper exploration of meaningful connections.

3   Problem Statement

3.1   Problem Formulation

Our task is aimed to assist users in quickly finding salient connections and facts from a collection of relevant articles, and in summary, it can be best described as a combination of three major subtasks:

- **Subtask 1: Candidate Fact Extraction.** Given a collection of documents $D = \{d_1, d_2, ..., d_M\}$ clustered around a topic T. The goal of this subtask is to extract a set of facts $F_c = \{f_1, f_2, ..., f_N\}$ from D. Each of facts is essentially (s, r, o) triple, for *subject s*, *relation r*, and *object o*. Since we need to estimate the coherence of these preferred facts for T, we refer to them as *candidate facts*.
- **Subtask 2: Topic Coherence Estimation of Candidate Facts.** Given a specified document topic T, the goal of the subtask is to find a subset of $F_c' \subseteq F_c$, and each of them should be coherent with T.
- **Subtask 3: Conceptual Graph Construction.** The goal of the subtask is to determine which of the facts from $F_c' \subseteq F_c$ generated by the previous subtask are more likely to be salient, which of their entities and concepts to merge and, when merging, which of the available labels to leverage in the final conceptual graph G.

3.2   The Framework of Our Approach

The framework of our approach can be shown in Fig. 1. It also comprise three major phases in order to address the problems associated with the subtasks as we have discussed earlier. In the candidate fact extraction phrase, given a specified document topic, we first preprocess relevant input texts in natural language, which can be more efficient for later procedures. To achieve this, we rely on a series of natural language processing methods, including document ranking, coreference resolution and sentence ranking. Then, three existing popular extractors including OLLIE [7], ClausIE [8], MinIE [9] are utilized to extract candidate facts from processed text documents, which can be expressed as subject-predicate-object triples[4].

---

[4] During the extraction, we applied a few straightforward transformations to correct two types of common errors such as wrong boundaries and uninformative extraction, which were caused by the syntactic analysis in extraction approaches.

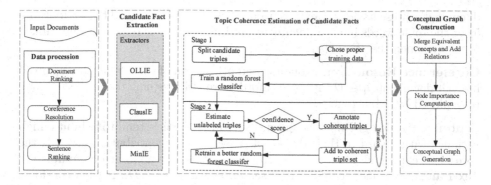

Fig. 1. The framework of our approach

The above phase can generate a set of candidate facts from the input documents. However, many of extracted relational triples are correct ones but irrelevant to the specified document topic. Therefore, we propose a Two-Stage Candidate Triple Filtering (TCTF) approach to further filter out those irrelevant triples by how coherence they are for the given document topic of interest. In the first stage, we firstly split the candidates and annotate specific smaller set of triples from them as available training data, and then used to train an random forest classifier. In the second stage, based on the trained classifier, it estimate each of unlabeled facts from the candidates by its confidence score. The triples that are annotated as coherent ones will add to corresponding coherent triple set for retaining a better random forest classifier used to annotate the unlabeled triples. The whole procedure works in an iterative manner based on a self-training framework until the convergence of the unlabeled triple set or achieving the max iteration number. Finally, we aggregate those coherent facts into an initial graph by further merging their equivalent entities and concepts, and adding synthetic relations. We further construct this graph by a heuristic strategy that iteratively removes the weakest entities and concepts with relatively lower importance scores computed by the extended TextRank algorithm, so that it ensures the final conceptual graph only consist of facts likely to represent meaningful and salient relationships where users may explore graphically.

4   The Proposed Approach

4.1   Candidate Fact Extraction

Document Ranking. As the first step, given a specified document topic T, we selected the words appearing in its relevant document collection $D = \{d_1, d_2, ..., d_M\}$ with sufficiently high frequency as topic words, and computes standard TF-IDF weights[5] for each word. The topic words are used to induce

---

[5] https://en.wikipedia.org/wiki/Tf%E2%80%93idf.

document representations. Next, we rank D by the TF-IDF weights (see footnote 5) of the topic words in each document, and by default, to obtain a final ranked top-k document list $D' = \{d_1, d_2, ..., d_k\}$.

Coreference Resolution. Pronouns and other form of coreference are resolved in each document $d_i \in D'$ using Stanford CoreNLP system [1] (i.e., *"she"* may be replaced by *"Angela Merkel"*).

Sentence Ranking. Different sentences within an article tend to exhibit a high variance with regard to their degree of relevance and contribution towards the core ideas expressed in the article. To address this, our approach computes the TextRank importance scores[6] for all sentences within $d_i \in D'$. It then considers only those sentences with sufficiently high scores, denoted by $S = \{s_1, s_2, ..., s_K\}$.

Candidate Relational Triples Extraction. As a crucial step of the high-quality fact corpus for conceptual graph construction, we therefore have to committed to the high-accuracy of information extractors. For this, we adopt an Open IE approach to extract relational triples of the form (*noun phrase, relation phrase, noun phrase*) for raw texts, rather than just focusing on named entities (e.g., *"Billionaire Donald Trump"*), as some previous approaches do, our approach supports an unbounded range of noun phrase concepts (e.g., *"the snow storm on the East Coast"*) and relationships with explicit relation labels (e.g., *"became mayor of"*). The latter are extracted from verb phrases as well as from other contributions. More specifically, We leveraged three existing excellent and popular extractors namely OLLIE [7], ClausIE [8], MinIE [9] to process all selected sentences from S after the sentence ranking, only a relational triple is obtained simultaneously by the extractors is incorporated into the collection of candidate facts[7].

Due to all of the extractors we chose are based on the dependency parsing trees, which may lead to two categories of errors: (1) Wrong boundaries, especially when triples with conjunctions in the sentence were not properly segmented, which the Open IE approach sometimes fails to do; (2) Uninformative extraction, which might occur in the *relation phrases*. i.e., the *relation phrases* are expressed by a combination product of a verb with a noun, e.g., light verb constructions. Moreover, the distance between the adjacent words in the *relation phrase* may, in fact, be distant in the original sentence. The adjacency order of the words in a triple may be different from that in the sentence, e.g., we may obtain a triple of (*"The police"*, *"are driving away"*, *"paraders"*) from such a sentence *"The police are driving paraders away from GOP front-runner Trump"*. Therefore, we applied a few straightforward transformations to correct above errors: (1) We broke down triples with conjunctions in either of the *noun phrase* into separate triple per conjunction; (2) We defined some word order constraints, i.e., the *noun phrases* in a triple (*noun phrase1, relation phrase, noun phrase2*) are ordered. All words in *noun phrase1* must appear before all words in *relation*

---

[6] https://github.com/letiantian/TextRank4ZH/blob/master/README.md.

[7] As for a triple extracted by OLLIE, ClausIE, and MinIE, only when its confidence is greater than 0.85 can it be judged that the triple is a correct extraction.

phrase. All the words contained in *noun phrase2* must appear after the *predicate phrase.* The order of words appearing in *predicate phrase* must be consistent with them appearing in the original sentence. In addition, each *relation phrase* must have appeared in the original sentence, is not the modified word or the added word.

4.2 Topic Coherence Estimation of Candidate Facts

Once above extracting process is completed, the framework will yield numerous candidate triples $F_c = \{f_1, f_2, ..., f_N\}$. However, not all the triple $f_i \in F$ is indicative of what is genuinely being expressed in a given text for the specified topic. E.g, the triple *("Trump", "will visit", "the Apple Inc")*, this is a correct one but meaningless for the news topic like *US president election.* Therefore, the task of this phrase seeks to filter out those irrelevance triples from the candidate triples for different document topic. In this work, we propose a TCTF approach based on Self-training for explaining when to annotate a triple as coherent with the specified topic. A pseudo-code of TCTFA is provided in Algorithm 1.

In the first stage, we randomly select a small fraction of triples $F_s \subseteq F_c$ under the document topic, and divide them into training set F_{str}, validation set F_{sv}, and test set F_{st} with a 8:1:1 ratio. We then train a random forest classifier M over F_s and obtain a F1-score θ (Line 3). Specifically, the triples used in the training are which both subject and object occurring as the topic words list are labeled as positive examples F_c' (i.e., initial coherent triples set), while neither nor them are labeled as negative examples. The identification of topic words rely on their sufficiently high frequency that can be computed using TF-IDF (see footnote 5), which we described earlier for the document ranking part in Sect. 4.1. To measure the topic coherence for each $f_i \in F_s$ from different aspects, we defined several features which are divided into three groups namely *topic features*, *text features* as well as *source features*, as described in Table 1. Besides, we further illustrate the computations of a series of critical features within above groups, including the following:

- *In_Titles*, the frequency of candidates in document titles is divided into three cases: greater than 13.40, less than 3.16, and between 3.16 and 13.40. Three binary features are used to represent these cases.
- *Redundance*, this is defined as the ratio of the size of candidates that have same *subject, relation* and *object* with the candidate triple f to the total number of candidates, these extracted redundant facts are from different sentences within and across the documents.
- *Similarity*, this is defined as the ratio of the number of candidates that are similar to the candidate triple f over the total number of candidates, i.e.,

$$Sim(f, l) = \frac{count_{sim}(f, l)}{count(l)} \tag{1}$$

Algorithm 1. Two-Stage Candidate Triple Filtering Approach based on Self-training

---

Input: F_c: Candidate triple set;
    F_s: A small fraction of training triple set;
    F_c': Coherent triple set;
    F_{sv}: Validation set;
    kF: flag for the first update for the model;
Output: F_l^*: Combined coherent triple set;
1: **Initialization:** $F_l^* \leftarrow F_c', \theta \leftarrow 0, \theta_{new} \leftarrow 0, kF \leftarrow true$;
2: **The first stage** → **Train a Random Forest Classifier**
3: Learn a random forest classifier M from F_s and obtain F1-score θ;
4: **The second stage** → **Extend Coherent Triple set**
5: **repeat**
6:    **if** kF is $false$ **then**
7:       $\theta \leftarrow \theta_{new}$;
8:       $M \leftarrow M*$;
9:    **end if**
10:    **for** each unlabeled triple $f \in F_c$ **do**
11:       Use M to label f and obtain f';
12:       **if** $confidence(f') > \alpha$ **then**
13:          $F_l^* \leftarrow F_l^* + f'$;
14:          $F_c \leftarrow F_c - f$;
15:       **end if**
16:    **end for**
17:    Retrain model $M*$ on F_l^*;
18:    Test model $M*$ on F_{sv} and obtain a new F1-score θ_{new};
19:    $kF \leftarrow false$;
20: **until** $\theta_{new} - \theta < \epsilon$
21: **return** F_l^*

---

where $count_{sim}(f, l)$ denotes the number of candidates that are similar to f in the whole candidates[8], $count(l)$ is the total number of candidates.

- *Relation_Context*, the ratio of the size of context of the relation r in the candidate triple f over the total number of candidates, i.e.,

$$RelCxt(r_f, l) = \frac{count_{context(r_f)}}{count(l)} \tag{2}$$

where $count_{context(r_f)}$ denotes the size of *relation context* of r in f, which is consist of the candidates that have same *relation type* with f, and $count(l)$ is the total number of candidates.

---

[8] The similarity scores between two candidate fact f_i, f_j is computed as $sim(f_i, f_j) = \gamma \dot{s}_k + (1 - \gamma)\dot{l}_k$, where s_k, l_k denote the semantic similarity and literal similarity scores between the facts, respectively. We compute s_k using the *Align, Disambiguate and Walk* algorithm [2], while l_k are computed using the Jaccard index. $\gamma = 0.8$ denotes the relative degree to which the semantic similarity contributes to the overall similarity score, as opposed to the literal similarity.

Table 1. The features for candidate triples classification

#	Advanced features	Comment	Value range
Topic features			
1	Is_Topic_Word	Whether both subject and object in a candidate fact occurring as the topic words list	0 or 1
2	Is_Subject_tw	Whether subject in a candidate fact occurring as the topic words list	0 or 0.5
3	Is_Object_tw	Whether object in a candidate fact occurring as the topic words list	0 or 0.5
4	In_Titles	Some binary features based on the frequency of occurrence of a candidate triple in document title in the relevant documents	0 or 1
5	Redundance	The ratio of redundant candidates with the candidate fact	[0, 1]
6	Similarity	The ratio of candidates are similar to the candidate fact	[0, 1]
7	Relation_Context	The ratio of candidates involving the same type of relation with the candidate fact	[0, 1]
8	Compatibility	The compatibility between the relation context of the candidate triple and the semantic information itself	[0, 1]
Text features			
9	Is_In_Title	Whether a candidate triple appears in the document title	0 or 1
10	Is_In_Abstract	Whether a candidate triple appears in an automatic summarization of the document	0 or 1
11	Is_In_MaxSent	Whether a candidate triple appears in the sentence with maximum TextRank importance score in the document	0 or 1
12	Sum_tfidf	Sum of TF-IDF of subject and object in the candidate triple in the relevant documents	[0, 1]
13	Avg_tfidf	Average of TF-IDF of subject and object in the candidate triple in the relevant documents	[0, 1]
Source features			
14	Source_Num	The number of sources where the candidate triple is extracted	1 or 2
15	Sentence_Num	The number of sentences where the candidate triple is extracted	1, 2, ... 50
16	Relevant_Docs	The ratio of documents which contain the candidate triple	[0, 1]

- *Compatibility,* the compatibility between the *relation context* of the relation r in the candidate triple f and the semantic information of f itself, i.e.,

$$Cmp(r_f, f_{ht}) = RelCxt(r_f, l) \cdot (1 - \epsilon + \epsilon \cdot Sem(f_{ht}, context(r_f))) \quad (3)$$

Here, the first term $RelCxt(r_f, l)$ denotes the *relation context* of r in f as computed in Eq. 2; the second term denotes the ratio of the number of candidates that have same *subject* or *object* with f from $RelCxt(r_f, l)$, which is calculated by $Sem(r_f, f_{ht}) = \frac{count_{f_{ht}}}{count_{context(r_f)}}$. Parameter ϵ is used for smoothing as well as to control the influence of the *relation context*, and is fixed to 0.5 in our implementation.

In the second stage, based on the trained classifier M, it calculates the confidence score for every unlabeled triple $f \in F_c$ using classifier's confidence and regard the triples that are assigned with the score above a fixed threshold α as the coherent triples. They will make up the next iteration's labeled triple set F_l^* for retraining a better model $M*$ (Line 10–17). Correspondingly, the parameters of $M*$ are tuned according to the precision metric on F_{sv}, and obtain a new F1-score θ_{new} (Line 18). We perform $\theta_{new} - \theta$ and compare it to a threshold, ϵ. If the result to be smaller than ϵ, it will enter the next iteration of the learning process based on a self-training framework. After several numbers of iterations, the final extended coherent triple set F_l^* will be generated and used to form a conceptual graph in the next phrase of our approach.

4.3   Conceptual Graph Construction

Merge Equivalent Concepts and Add Relations. In order to establish a single connected graph that is more consistent, we further merge potential entities and concepts in F_l^* stemming from former process which works in two steps: (1) We made use of Stanford CoreNLP entity linker which is based on [5] for identifying entity or concept mentions and link them to Ontological KBs such as Wikipedia, Freebase entity linking. Roughly, in about 30% cases, we get this information for the entities. If two entities and concepts are linked to the same Wikipedia entity, we assume them to be equivalent as per this information. e.g., *US* and *America* can get linked to the same Wikipedia entity *United_States*; (2) It was found that approximately 53% pairs of entities or concepts, whose labels present slightly similarity in terms of their literal meaning, to suggest possible connection. When we can't obtain sufficient context information, it's difficult to decide to whether should merge to form a label that is appropriate for both of them, e.g., *all the Democratic candidates* and *US Democratic Parties*. The identification of them requires more human-crafted knowledge. For this, we decided to use three expert annotators with NLP background for such subtask: they could connect these entities and concepts according to their background knowledge step by step, and observed an agreement of 84% ($\kappa = 0.66$[9]). To support the annotators, once again the *Align, Disambiguate and Walk tool* [2] is used for semantically similarity computation between concepts for coreference.

After that, on average, there remains not more than 5 subgraphs that can further be connected for different document topics. Hence, for each topic, annotators were allowed to add up to three synthetic relations with freely defined labels to connect these subgraphs into a fully connected graph G', observing 87% ($\kappa = 0.71$) agreement.

Node Importance Computation. A relational triple is more likely to be salient if it involves important entities and concepts of the sentence. Motivated in part by the considerations given by [14], we illustrate the node importance computation, seeking to retain only the most salient facts to include in the final concept graph for different document topics. Formally, let $G' = (\mathcal{V}, \xi)$ denotes a weighted directed graph generated by former step, where $\mathcal{V} = \{v_1, v_2, ..., v_R\}$

---

[9] Kappa implementation: https://gist.github.com/ShinNoNoir/9687179.

represent a set of preferred nodes which correspond to entities and concepts in G', and ξ is a directed edge set, associated with each directed edge $v_i \rightarrow v_j$ representing a dependency relation originating from v_i to v_j. We assign a weight $w_{ij} = 1$ to $v_i \rightarrow v_j$ and its reverse edge $v_j \rightarrow v_i$ with $w_{ji} = 0.5$. By adding lower weighted reverse edges, we can analyze the relation between two nodes which are not connected by directed dependency links while maintaining our preferences toward the original directions.

TextRank [10] is a ranking algorithm can be used to compute the importance of each node within G' based on graph random walks. Similarity, suppose a random walker keeps visiting adjacent nodes in G' at random. The expected percentage of walkers visiting each node converges to the TextRank score. We assign higher preferences toward these nodes when computing the importance scores since entities and concepts are more informative for G'. We extend TextRank by introducing a new measure called "back probability" $d \in [0, 1]$ to determine how often walkers jump back to the nodes in $\mathcal{V}$ so that the converged score can be used to estimate the relative probability of visiting these preferred nodes. We defined a preference vector $\mathbf{p}_R = \{p_1, p_2, ..., p_{|\mathcal{V}|}\}$ such that the probabilities sum to 1, and p_k denotes the relative importance attached to v_k. p_k is set to $1/|\mathcal{V}|$ for $v_k \in \mathcal{V}$, otherwise 0. Let I be the $1 \times |\mathcal{V}|$ importance vector to be computed over all nodes in G' as follows.

$$I_i = (1 - d) \sum_{j \in \mathcal{N}(i)} \frac{w_{ji}}{\Sigma_{k \in \mathcal{N}(j)} w_{jk}} I(j) + d \cdot p_i, \tag{4}$$

where $\mathcal{N}(i)$ stands for the set of the node v_i's neighbors.

Conceptual Graph Generation. The recommended [3] maximum size of a concept graph is 25 concepts, which we use as a constraint. We rely on a heuristic to find a full graph that is connected and satisfies the size limit of 25 concepts: We iteratively remove the weakest concepts with relatively lower importance score is computed using Eq. 4 until only one connected component of 25 entities and concepts or less remains, which is used as the final conceptual graph G. This approach guarantees that the graph is connected with salient concepts, but might not find the subset of concepts that has the highest total importance score.

5 Experiment

5.1 Dataset

Our dataset include 5 categories, and for each category we have 2 popular events and each of which represents a document topic. Every topic cluster comprises approximately 30 documents with on average 1,316 tokens, which leads to an average topic cluster size of 2,632 tokens. It is 3 times larger than typical DUC[10]

---

[10] https://duc.nist.gov/.

clusters of 10 documents. With these properties, our dataset presents an interesting challenge towards real-world application scenarios, in which users typically have to deal with much more than 10 documents. The articles in our dataset stem from a larger news document collection[11] released by Signal Media as well as crawled from Web Blogs by ourselves, we rely on event keywords to filter them so as to retain related ones for different topics. The overall statistics of the resulting dataset are shown in Table 2.

Table 2. Dataset description

Category	Topic ID	Document topic	Time period	Docs	Doc.Size	Source
Armed conflicts and attacks	1	Syria refugee crisis	2015-09-01–2015-09-30	30	2179 ± 506	News, Blog
	2	North Korea nuclear test	2017-08-09–2017-11-20	30	1713 ± 122	News
Business and economy	3	Chinese cooperation with Sudan	2015-09-01–2015-09-30	30	768 ± 132	News, Blog
	4	Trump TPP	2016-12-23-2017-02-23	30	879 ± 306	News
Politics and elections	5	US presidential election	2016-06-14–2016-08-14	30	1175 ± 207	News, Blog
	6	US-China trade war	2018-03-23–2018-06-15	30	2412 ± 542	News, Blog
Arts and culture	7	Muslim culture	2013-02-01–2013-05-01	30	972 ± 161	News, Blog
	8	Turing Award winner	2019-03-15–2019-04-01	30	1563 ± 464	News, Blog
Information technology and application software	9	Next-generation search engine	2016-11-07–2017-01-03	30	729 ± 280	News, Blog
	10	Program repair for Android system	2018-02-01–2018-05-10	30	772 ± 453	Blog

5.2 Experimental Setting

We conduct two types of experiments on above datasets for validating the effectiveness of our proposed approach: (1) The first experiment focuses on sentence-level extractions. Therefore, we first randomly sample 10 documents from every document topics (100 documents in total) and perform coreference resolution. Then, once again a random sample of 10 sentences from every extracted document (1,000 sentences in total) for further analysis. Each sentence is examined by three expert annotators with NLP background independently to annotate all of correct triples[12]; (2) We further conduct an empirical study to investigate the quality of the final generated conceptual graph towards different document topics on its coverage rate of topic entities and concepts, confidence score, and the compatibility of involved facts.

---

[11] http://research.signalmedia.co/newsir16/signal-dataset.html.

[12] A triple is annotated as correct if the following conditions are met: (i) it is entailed by its corresponding clause; (ii) it is reasonable or meaningful without any context and (iii) when these three annotators mark it correct simultaneously (The inter-annotator agreement was 82% ($\kappa = 0.60$)).

OpenIE Baseline Methods. We compare our extraction method with the baselines of using just the OpenIE systems: OLLIE [7], ClausIE [8], MinIE [9], and OpenIE-4.x [6]. Apart from this, we also evaluate two variants of the approach to study the effect of coreference resolution and several transformations used to correcting errors caused by OpenIE methods.

Evaluation Metrics. The metrics used in our experimental analysis are Precision (P), Recall (R), F-score (F1), coverage rate of topic entities and concepts, confidence score, compatibility of involved facts in the graph, defined as follows:

- Three standard metrics are Precision (P), Recall (R), F-score (F1), respectively

$$P = \frac{\# \; correct}{\# \; extractions}, R = \frac{\# \; correct}{\# \; relations}, F1 = \frac{2PR}{P+R} \qquad (5)$$

where "# correct" denotes the number of extractions deemed as correct, "# extractions" denotes the total number of extractions, and "# relations" denotes the number of triples are annotated as correct extractions (see footnote 12).

- The coverage rate of topic entities and concepts, i.e.,

$$TopicCon\_Coverage = \frac{\# \; topic\_concepts}{\# \; concepts} \qquad (6)$$

where "# topic_concepts" denotes the number of entities and concepts for which annotated as topic concepts[13], and "# concepts" denotes the total number of all entities and concepts in the conceptual graph.

- Confidence score, i.e.,

$$Avg\_Confidence(f_i, n) = \frac{\sum_{i=1}^{n} conf(f_i)}{n}, \qquad (7)$$

where $conf(f_i)$ denotes the confidence score[14] of each fact f_i, n is the number of facts which involved in the final conceptual graph.

- Compatibility of involved facts in the graph, i.e.,

$$Avg\_Compatibility(f_i, f_j, n) = \frac{\sum_{i=1}^{n} \sum_{j>i} cmp(f_i, f_j)}{c_n^2}, \qquad (8)$$

where f_i and f_j are any facts are in the final conceptual graph, which contains n facts. $cmp(f_i, f_j)$ denotes the compatibility between f_i and f_j, similar to Eq. 3, it can be calculated using $cmp(f_i, f_j) = (RelCxt(r_{f_i}, n) + RelCxt(r_{f_j}, n)) \cdot (1 - \epsilon + \epsilon \cdot sim(f_i, f_j))$, where $sim(f_i, f_j)$ denotes the similarity scores (see footnote 8) between fact f_i and fact f_j, parameter ϵ is used for smoothing as well as to control the influence of the relation context, and is fixed to 0.5 in our implementation.

---

[13] An entity or concept is regarded as topic concept when it occurs in the topic words list.

[14] For popular OpenIE systems such as OLLIE, ClausIE, and MinIE, we use the confidence value computed by each system itself as the confidence score of each of facts.

Table 3. Evaluation of precision, recall, and F-score on five independent document topics (including topic 1 to topic 5) from two datasets

OpenIE methods	#Topic 1			#Topic 2			#Topic 3			#Topic 4			#Topic 5		
	P	R	F	P	R	F	P	R	F	P	R	F	P	R	F
OLLIE [7]	0.72	0.34	0.67	0.28	0.37	0.32	0.84	0.35	0.49	0.75	0.43	0.55	0.62	0.29	0.40
ClausIE [8]	0.63	0.62	0.66	0.58	0.55	0.56	0.80	0.53	0.64	0.59	0.55	0.57	0.79	0.52	0.63
MinIE [9]	0.70	0.69	0.72	0.61	0.64	**0.62**	0.86	0.58	0.69	0.70	0.62	0.66	0.81	0.66	**0.73**
OpenIE-4.x [6]	0.74	0.46	0.54	0.35	0.80	0.49	0.79	0.41	0.54	0.76	0.40	0.52	0.69	0.34	0.46
Our approach (without coref)	0.43	0.29	0.56	0.44	0.27	0.33	0.65	0.24	0.35	0.47	0.33	0.39	0.45	0.30	0.36
Our approach (without trans)	0.79	0.70	**0.82**	0.61	0.55	0.58	0.92	0.68	**0.78**	0.82	0.71	**0.76**	0.81	0.67	**0.73**
Our approach	0.86	0.85	**0.85**	0.78	0.74	**0.76**	0.95	0.92	**0.93**	0.95	0.82	**0.88**	0.92	0.78	**0.84**

5.3 Evaluation and Results Analysis

Performance Analysis of Extraction Approaches. We selected and presented the evaluation results of our method and OpenIE baselines on ten document topics in Tables 3 and 4[15]. We can draw from that our method has more superiorities compared with the baseline methods across all metrics as illustrated in Equation. In particular, our approach enhanced F-score with an average improvement of 2.4% compared with the baseline methods. The reason is mainly that: (1) Our method takes advantage of the results (see footnote 7) of three extractors including OLLIE [7], ClausIE [8], MinIE [9], which may achieve better performance compared with using single extractor; (2) Using a few straightforward transformations, we observed our approach is better able to identify the boundary of triples for long sentence with conjunction structure, but generally other methods including OLLIE [7] and Open IE-4.x [6] cannot. Moreover, we observed the number of uninformative extractions especially appearing in the relation phrases can significantly reduce by using a relaxed constrain for the word order, i.e., decreasing the frequency of this type of error from 36.8% to 17.1%, this illustrated the effectiveness of the above operation. It also indicated that two types of extraction errors as above in our approach caused by depending on intermediate structures such as dependency parses could also be well figured out. Additionally, it can be seen that our approach (without coref) performs the worst among all the methods, which is due to many cases in which the selected sentences from different topics are not readable without the context when we do not perform coreference resolution, so that it is difficult to correctly identify the extractions relaying on the output of our approach.

Feature Selection and Parameter Tuning. In the process of estimating the topic coherence of candidate facts, although there are fifteen designed features for measuring the topic coherence for each fact from different aspects. To our best knowledge, simply combining all of them will not lead to the best triple classification performance. Therefore, we study the effectiveness of each feature

---

[15] We mark top-2 performance results in F-score in bold face.

Table 4. Evaluation of precision, recall, and F-score on five independent document topics (including topic 6 to topic 10) from two datasets

OpenIE methods	#Topic 6			#Topic 7			#Topic 8			#Topic 9			#Topic 10		
	P	R	F	P	R	F	P	R	F	P	R	F	P	R	F
OLLIE [7]	0.62	0.27	3.38	0.69	0.44	0.54	0.68	0.35	0.46	0.81	0.24	0.37	0.59	0.21	0.31
ClausIE [8]	0.70	0.53	0.60	0.73	0.55	0.63	0.59	0.43	0.50	0.66	0.64	0.65	0.61	0.49	0.54
MinIE [9]	0.82	0.55	**0.66**	0.77	0.64	**0.70**	0.71	0.62	0.66	0.81	0.60	0.69	0.78	0.55	0.65
Open IE-4.x [6]	0.73	0.51	0.60	0.64	0.30	0.41	0.66	0.58	0.62	0.74	0.59	0.66	0.71	0.65	0.68
Our approach (without coref)	0.43	0.29	0.35	0.44	0.32	0.37	0.47	0.30	0.37	0.55	0.42	0.48	0.40	0.29	0.34
Our approach (without trans)	0.73	0.57	0.64	0.71	0.62	0.66	0.82	0.71	**0.76**	0.81	0.70	**0.75**	0.76	0.71	**0.73**
Our approach	0.90	0.73	**0.81**	0.78	0.69	**0.73**	0.95	0.78	**0.86**	0.88	0.73	**0.80**	0.78	0.74	**0.76**

for the trained random forest classifier. We selected χ^2 and information gain [13] as the classification criteria, and ranked features by χ^2 are shown in Table 5. The results show that only Is_Topic_Word is the top-1 feature ranked by both two measures. Moreover, we further evaluate the contribution of each feature for the classifier when top-k features (sorted by χ^2) are used, all performance results including accuracy, recall and F-score on average are reported in Fig. 2(a). We can observe that the top-7 features dominate the performance of the classifier, i.e., all of measures will converge to an upper limit after the top-7 features are leveraged. The results suggest that the classification accuracy is better when a few effective features are included.

Subsequently, we compared three different thresholds (i.e., $\alpha = 0.67$, $\alpha = 0.93$, $\alpha = 0.96$) used for control noise and select proper triples at the second stage of the proposed TCTF approach. The results are shown in Fig. 2(b). We can observe that, based on the self-training framework, improper thresholds not only could not improve the model (i.e., the random forest classifier), but also render the model fail to annotate improper triples as coherent ones for the specified document topic. The reasons for such phenomena probably are as follows: (1) If the threshold is set to be too small (e.g., $\alpha = 0.67$), the training of the model might be easily affected by the noise data (i.e., added triples), leading to less improvement of triple classification task; (2) If the threshold is set to be too large, on the other hand, the training speed of the model could decrease on the certain of level. E.g., we set α as 0.96, the model converges for a total of 35 epochs approximately. Hence, through considering both training speed and the performance of self-training process in the TCTF approach, we set confidence α as 0.93, which could achieve better performance at the expense of few training epochs.

Quality Analysis of Conceptual Graph. The result of the empirical evaluation of quality of the final generated conceptual graph is shown in Fig. 3. Our approach achieved 100% coverage rate of topic entities and concepts ($TopicCon\_Coverage$), 87% confidence score ($Avg\_Confidence$), and 58% fact compatibility ($Avg\_Compatibility$) over ten document topics. The results

Table 5. Effectiveness of features (sorted by χ^2)

#	Feature	χ^2	IG%	#	Feature	χ^2	IG%
1	Is_Topic_Word	53.94	2.90	11	Redundance	16.93	0.68
2	Is_Subject_tw	41.32	1.72	12	Is_In_Title	0.82	0.51
3	Is_Object_tw	40.02	1.73	13	Sentence_Num	0.74	0.06
4	Relation_Context	38.20	1.32	14	Relevant_Docs	0.45	1.62
5	Similarity	37.07	2.03	15	Source_Num	0.44	0.07
6	Compatibility	23.09	2.24				
7	Is_In_Abstract	24.41	1.48				
8	Is_In_MaxSent	23.20	1.70				
9	Sum_tfidf	20.07	0.52				
10	Avg_tfidf	19.58	0.48				

indicates that: (1) The proposed TCTF approach is capable to retain only coherent triples from the candidates towards different document topics[16] whereas the threshold selection which is significant tough; (2) The extracted facts have higher confidence, which demonstrate the importance of node importance computation in conceptual graph construction; (3) Obviously, our approach may not guarantee that the extracted facts have better compatibility, which needs to be further explored.

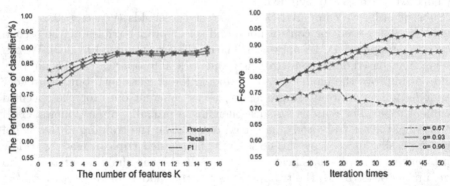

(a) The performance of top-k features for triple classification task

(b) Learning curve of self-training process in TCTF approach at three different confidence thresholds α on the validation set

Fig. 2. The performance evaluation of the model and parameter tuning

---

[16] The random forest classifier has an average absolute 87% higher on the F-score metric for different topics when the model has converged.

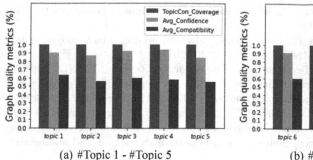
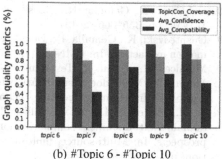

| (a) #Topic 1 - #Topic 5 | (b) #Topic 6 - #Topic 10 |

Fig. 3. The quality analysis of the final generated conceptual graphs for ten document topics from three aspects (i.e., $TopicCon\_Coverage$, $Avg\_Confidence$, and $Avg\_Compatibility$).

6 Conclusion

In this paper, we presented a novel framework that aids users in quickly discerning coherence and salience-based connections in a collection of documents, via graph-based visualizations of relationships between concepts even across documents. Experiments on two real-world data sets demonstrate the effectiveness of our proposed approach. In the future, we will give greater exploration to fact fusion problem before fully automated conceptual graph construction for specified domain is possible.

Acknowledgments. This paper was partially supported by National Natural Science Foundation of China (Nos.61572111 and 61876034), and a Fundamental Research Fund for the Central Universities of China (No.ZYGX2016Z003).

References

1. Manning, C., Surdeanu, M., Bauer, J., Finkel, J., Bethard, S., McClosky, D.: The Stanford CoreNLP natural language processing toolkit. In: ACL, pp. 55–60 (2014)
2. Pilehvar, M.T., Jurgens, D., Navigli, R.: Align, disambiguate and walk: A unified approach for measuring semantic similarity. In: ACL (Volume 1: Long Papers), pp. 1341–1351 (2013)
3. Banko, M., Cafarella, M.J., Soderland, S., Broadhead, M., Etzioni, O.: Open information extraction from the web. In: IJCAI, pp. 2670–2676 (2007)
4. Kochtchi, A., Landesberger, T.V., Biemann, C.: Networks of names: visual exploration and semi-automatic tagging of social networks from newspaper articles. In: Computer Graphics Forum, pp. 211–220 (2014)
5. Spitkovsky, V.I., Chang, A.X.: A cross-lingual dictionary for English Wikipedia concepts. In: LREC, pp. 3168–3175 (2012)
6. Mausam, M.: Open information extraction systems and downstream applications. In: IJCAI, pp. 4074–4077 (2016)
7. Schmitz, M., Bart, R., Soderland, S., Etzioni, O.: Open language learning for information extraction. In: EMNLP-CoNLL, pp. 523–534 (2012)

8. Del Corro, L., Gemulla, R.: Clausie: clause-based open information extraction. In: WWW, pp. 355–366 (2013)
9. Gashteovski, K., Gemulla, R., Del Corro, L.: Minie: minimizing facts in open information extraction. In: EMNLP, pp. 2630–2640 (2017)
10. Mihalcea, R., Tarau, P.: TextRank: bringing order into text. In: EMNLP (2004)
11. Mann, G.: Multi-document relationship fusion via constraints on probabilistic databases. In: Human Language Technologies 2007: NAACL, pp. 332–339 (2007)
12. Ji, H., Favre, B., Lin, W.P., Gillick, D., Hakkani-Tur, D., Grishman, R.: Open-domain multi-document summarization via information extraction: challenges and prospects. In: Multi-source, multilingual information extraction and summarization, pp. 177–201 (2013)
13. Fuchs, C.A., Peres, A.: Quantum-state disturbance versus information. Uncertainty Relat. Quantum Inf. Phys. Rev. A 53(4), 20–38 (1996)
14. Yu, D., Huang, L., Ji, H.: Open relation extraction and grounding. In: IJCNLP (Volume 1: Long Papers), pp. 854–864 (2017)
15. Sheng, Y., Xu, Z., Wang, Y., Zhang, X., Jia, J., You, Z., de Melo, G.: Visualizing multi-document semantics via open domain information extraction. In: ECML-PKDD, pp. 695–699 (2018)

Correction to: A Framework for Image Dark Data Assessment

Yu Liu, Yangtao Wang, Ke Zhou, Yujuan Yang,
Yifei Liu, Jingkuan Song, and Zhili Xiao

Correction to:
Chapter "A Framework for Image Dark Data Assessment"
in: J. Shao et al. (Eds.): *Web and Big Data*, LNCS 11641,
https://doi.org/10.1007/978-3-030-26072-9_1

The original version of the chapter "A Framework for Image Dark Data Assessment", starting on p. 3 was not correct. The abstract section and the keywords have been exchanged. This have been now corrected.

The correct abstract and keywords as follows:

Abstract. Blindly applying data mining techniques on image dark data whose content and value are not clear, is highly likely to bring undesired result. Therefore, we propose an assessment framework which includes offline and online stages for image dark data. In offline stage, we first transform images into hash codes by Deep Self-taught Hashing (DSTH) algorithm, then construct a semantic graph, and finally use our designed Semantic Hash Ranking (SHR) algorithm to calculate the importance score. During online stage, we first translate the user's query into hash codes, then match the suitable data contained in the dark data, and finally return the weighted average value of these matched data to help the user cognize the dark data. The results on real-world dataset show our framework can apply to large-scale datasets, help the user conduct subsequent data mining work.

Keywords: Image dark data · Deep self-taught hashing ·
Semantic hash ranking · Assessment

The updated version of this chapter can be found at
https://doi.org/10.1007/978-3-030-26072-9_1

© Springer Nature Switzerland AG 2019
J. Shao et al. (Eds.): APWeb-WAIM 2019, LNCS 11641, p. C1, 2019.
https://doi.org/10.1007/978-3-030-26072-9_31

Author Index

Printed in the United States
By Bookmasters